Introduction to Remote Sensing of the Environment

A National Council for Geographic Education Pacesetter Book

Series Editor, Joseph M. Cirrincione

1

Remote Sensing: An Overview

Benjamin F. Richason, Jr.
Carroll College, Wisconsin

From Saddles to Satellites[1]

Emerging Concepts of the Environment

The first assessment of the environment of the United States began in a limited manner with the Land Ordinance of May 20, 1785. This law provided for the survey of lands north and west of the Ohio River into townships, ranges, and sections for the purpose of inaugurating a rational basis for disposing of the Public Domain. The early surveyors who were engaged for this task were required to traverse each mile by foot or on horseback, to map and describe the suitability of the land for agriculture, and to locate the corners of townships and sections with markers. These deputy surveyors, for the most part, carried out their work on the frontier of settlement, and the first inventory of the environment of the Northwest Territory came from their maps and field notes. In the nineteenth century, the survey of lands was continued by the United States Geological Survey. Features of the landscape, in many cases, were mapped from the saddle by use of crude instruments and "eyeball" sketching.

What did people learn about the environment from these early surveys? It is suggested that two concepts emerged which have had lasting implications concerning our perception of the earth and its resources. In the first place, the earth appeared to almost every person at that time as immensely large. Inland from the Atlantic seacoast of the United States were the forests, and beyond them were the prairies. The prairies gave way to western mountain ranges and alpine pastures. The coast of the Pacific Ocean was months away from St. Joseph, Missouri, the outfitting station for most western migrations; and many additional months from the "Forks of the Ohio." Large oceans separated North America from the European hearthlands, and eastern Asia was too remote even to be considered.

Secondly, because of the earth's apparent immensity, natural resources were equated with unlimited resources, the attitude of the time being that everything which comes from nature must be abundant and free. Forests, which stood in the way of advancing settlement were cut down to clear a million acres for agricultural land, and logs were sent down the rivers or hauled overland to build hundreds of cities and thousands

1

of miles of railroad tracks. Natural gas wells in Ohio and Indiana were ignited for sport to see which one would flame the highest. The railroads brought settlers to the prairies where the sod was "busted," lands were overgrazed, and minerals were exploited for immediate gratification. These limited examples illustrate how the mental and physical horizons of that time were restricted by views from the saddle. To our forefathers it was a large world filled with resources which would last forever.

The twentieth century, however, witnessed a sophistication in surveying and mapping procedures which resulted in refinements of knowledge concerning the world and its resources. Governments inaugurated extensive aerial photography programs in order to rapidly and accurately inventory croplands, wetlands, forests, and mineral resources. Within the past decade a new dimension in assessing earth features has emerged from the cornucopia of science as both manned and unmanned space vehicles have carried cameras and other sensors hundreds, and even thousands, of miles above the earth, as well as to other planets.

From saddles to satellites we have moved into a position in the last quarter of the twentieth century where the whole earth and its resources can be imaged, mapped, and analyzed with multistage, multiscale, multiplatform, multidate, multispectral, and multidisciplinary capabilities. The people of the earth are being given a new mental and ethical perspective that is causing them to view the earth and its resources as a home to be husbanded. Data from modern-day satellites equipped with sophisticated sensors show us that we are living on a small, finite, fragile earth.

Remote sensor data are providing almost immediate information about the location of ores and mineral fuels, about sea-ice floes and ocean shipping routes, about regions of potential earthquake and volcanic hazards, and about water and forest resources, urban expansion, crop and rangeland distributions, plant diseases, and insect infestations. In the matter of monitoring crop acreage, for example, the 1976 world wheat acreage and yields were predicted with more than 90 percent accuracy by use of satellite data.

What are scientists discovering about the earth when they study it in light of data collected from satellites rather than from saddles? This question was partly resolved by Astronaut William A. Anders when, peering at the earth from deep space, he mused:

> The Earth looked so tiny in the heavens that there were times during the Apollo 8 mission when I had trouble finding it. If you can imagine yourself in a darkened room with only one clearly visible object, a small blue-green sphere about the size of a Christmas-tree ornament, then you can begin to grasp what the Earth looks like from space. I think that all of us subconsciously think that the Earth is flat or at least almost infinite. Let me assure you that, rather than a massive giant, it should be thought of as a fragile Christmas-tree ball which should be handled with considerable care.[2]

If remote sensing from space has taught us nothing else, it has changed our concepts of the earth and the conditions of its atmosphere, hydrosphere, and lithosphere. People have become aware of the finite nature of the world and the fragility of its environments. We have come to realize that the four billion people who live on the earth are close neighbors, and that the condition of the landscapes of the earth reflects either the artistry or the carelessness of those who have lived or are living within those landscapes.

The Problem

The need for an intensive inventory of earth resources never has been greater than in the latter part of the twentieth century. Devastating wars have left individual countries with depleted resource bases. The affluence of some societies in the post-World War II period has created additional demands for mineral fuels and other nonrenewable resources. Unprecedented rates of population growth on a world-wide scale characterized the mid-twentieth century, and placed exceptional burdens on arable lands to produce more food. Marginal lands were brought into production in order to partially satisfy the appetites of millions of people. Environmental sciences were being called upon to solve the world's ills which were associated with famine, pestilence, disease, and dwindling resources.

The failures of conventional scientific assaults on the environment are self-evident. By the last quarter of the twentieth century, more than two-thirds of the people of the world were going to bed hungry every night. Food products remained unevenly distributed between the Haves and Have-Nots. Industrialized societies found that prices of durable goods were increasing rapidly, not only because of rising labor costs, but because of premiums which had to be paid for the specter of shortages of fuels and other resources, combined with rising expectations of comfortable living. Extensive and detailed mapping projects were demanded to explore for new ore fields, water resources, and agricultural lands; yet many areas of the world remained inadequately mapped because of remoteness or political inaccessibility. Despite the surveillance of planning agencies, cities were expanding into agricultural areas in a haphazard, uncontrolled, and accelerated manner. Pollution of the earth's atmosphere, hydrosphere, and lithosphere has resulted from careless disposal of increasing amounts of industrial, agricultural, and domestic wastes. In short, too many people and too few, often poorly distributed resources demanded the type of attention and surveillance which conventional scientific methods and exploration could not provide.

The New Science

A new science was demanded; that is, a method by which large areas of the earth could be viewed and studied on a repetitive schedule. Aerial photography has been used for surveillance and mapping purposes since the early part of the century. Because of aircraft ceiling limitations, however, aerial photographs cannot provide the small scale needed for viewing large areas or small images. Furthermore, aerial photography coverage of critical areas is dependent upon clear weather and bases close to target areas. Aerial photography missions are expensive, and many planning agencies, for example, are unable financially to up-date the aerial photography coverage of their areas at frequent intervals.

Following World War II, science began to adapt wartime techniques to peacetime needs. Infrared and color aerial photographs had been used for purposes of camouflage detection, troop deployment surveillance, bomb-spotting, and coastal- zone mapping. In the past two decades cameras have been mounted in high altitude aircraft and orbiting space vehicles for peaceful surveys. Radar, which was almost entirely devoted to detecting military targets, was later employed to image earth terrain scenes through clouds and at night. Multispectral scan-

ners were mounted in manned and unmanned orbiting satellites to simultaneously view and image earth phenomena. The new science of remote sensing was born.

What Is Remote Sensing?

Imaging Objects at a Distance

In its broadest definition, remote sensing refers to any technique of imaging objects without the sensor being in direct contact with the object or scene itself. The eye, for example, is a remote sensor, but it does not have the ability to permanently record events or scenes except as these images are stored in the brain to be recalled by memory. The most commonly used remote sensing device is the camera. In an effort to bring the real world into the classroom for study, teachers have utilized photographs and slides as part of their audiovisual instructional strategy for decades. There is nothing pedagogically new about the use of slides, ground photographs, and aerial photographs as instructional instruments in classrooms or public lectures. By employing these visual materials, not only teachers and lecturers, but businessmen and planners, too, are, in fact, making use of a remote sensing technique.

Aerial photographs, imaged by lenses, films, and filters in concert with conventional cameras, are one of the most commonly used products of remote sensing (discussed in chapter 3). The altitude of aircraft or manned satellite flights provides camera platforms from which scaled-down images of earth features can be obtained. Multiple camera stations on airplanes or satellites, by varying the films and filters in several cameras which are programmed to image the same scene at the same moment, can provide multispectral imagery for study (discussed in chapter 4).

More precisely, Luney and Dill clarified the parameters of remote sensing by stating that, "Broadly defined . . . remote sensing denotes the joint effects of employing modern sensors, data processing equipment, information theory and processing methodology, communications theory and practice for the purposes of carrying out aerial and space surveys of the earth's surface."[3] The *Manual of Remote Sensing* defines remote sensing as, "the measurement or acquisition of information of some property of an object or phenomenon, by a recording device that is not in physical or intimate contact with the object or phenomenon under study . . . (and) the practice of data collection in wavelengths from ultraviolet to radio regions."[4] It is in these two definitions that restrictions are placed on the modern concept of remote sensing of the environment.

Cameras versus Other Remote Sensors

Although cameras with varying combinations of lenses, films, and filters remain the remote sensors with the greatest imaging capabilities in terms of versatility, economy of operation, and high resolution of detail, they possess certain liabilities in terms of modern remote sensing needs. Photographic systems image targets in a narrow band of the spectrum, essentially in those wavelengths which the human eye can see. Furthermore, camera systems must operate in daylight or with flash accessories, and for optimum imagery production they must operate in clear weather and on days with minimal atmospheric haze so that images will not be degraded by atmospheric scattering (discussed in chapter 2).

Perhaps the major disadvantage of the conventional camera as a remote sensor, in the modern use of the term, is that photography is not a "real time" system. Photographic images

are not immediately available for study. The film and paper on which photographic latent images are registered must be processed in a laboratory before they are in a form suitable for observation and analysis. This lag between the photographic exposure and the production of a photochemical print may be several days, or even longer for photographs made by cameras in space vehicles.

On the other hand, modern remote sensing detectors, such as scanners, microwave detectors, radar units, and other sophisticated remote sensing equipment are "real time" systems; that is, they have the capabilities of telemetering images of objects to earth receiving stations where the signals can be stored on magnetic tapes which, in turn, can be processed and analyzed by computer systems, or generated into photographlike prints.

Scarcely less impressive is the ability of modern remote sensing systems to detect phenomena and events which exist beyond the range of human vision. The electromagnetic spectrum, discussed in chapter 2, is considered to be a continuum of energy-developed wavelengths which extend from the ultraviolet regions at one end to the infrared and radio bands on the other end, but the human eye and conventional film in conventional cameras can detect only a very small, narrow band of the total spectrum. One scientist stated that if the total span of the electromagnetic spectrum, as we know it, is conceived to be analogous to the circumference of the earth at the equator, the human eye and conventional film would be able to see only that portion of it which is equal to the width of a pencil. This analogy serves to make clear that there are many phenomena and events in the earth's environment which lie beyond the perception of the human eye and conventional film in cameras. The new science of remote sensing allows people to receive information about the earth quickly and in areas of the spectrum they have never observed before.

Remote Sensing—the Multiscience

In order to convey the concept of remotely and immediately imaged phenomena and events in portions of the spectrum beyond the range of human vision, the term "remote sensing" was coined; however, remote sensing refers to a science that has many more capabilities. Remote sensing is a powerful new tool which is employed by geographers, geologists, oceanographers, hydrologists, agronomists, engineers, cartographers, foresters, urban planners, regional analysts, and many other scientists. Almost since the term "remote sensing" appeared as part of the jargon of new science, investigators working with photographic and nonphotographic imagery began to refer to it as the "multiscience" because of its multifaceted potential and capabilities.[5]

In the modern sense, remote sensing techniques are multidate, multispectral, multiplatform, multiscale, multienhancement, multiregional, and multidisciplinary. Many remote sensors, particularly those which operate from orbiting satellites, image the same area of the earth on a regular, repetitive schedule. Changes in events can be observed at frequent intervals. Remote sensing detectors have capabilities of imaging the same scene in different bands of the spectrum, thereby providing more data and more accurately identified information than when only a narrow, or single, wavelength is employed. This results from the fact that some targets reflect or radiate energy more effectively in certain bands of the spectrum than in others.

Remote sensing instruments are used at ground level, in cranes or cherry pickers, from conventional aircraft and high-altitude airplanes, or from space satellites. The altitude of the remote sensing detector provides images at different scales, ensuring greater accuracy in the inventory of earth resources than when a single scale is available for study. The images obtained from remote sensing equipment may be processed by computers or by optical, photochemical enhancements in order to slice out of the imaged scene those features which are to be identified, measured, and analyzed. Remote sensors which operate from high-altitude aircraft, such as the U-2 or RB-57F (discussed in chapter 10), or from space satellites (discussed in chapters 7, 8, and 9), image large sections of the surface of the earth; therefore, several discrete regions may appear on a single frame.

Because of the multidate, multiplatform, multiscale, multispectral, multienhancement, and multiregional capabilities of remote sensing, earth and life scientists have turned to this new tool for much of their research and studies, along with planners and others who have need for information concerning the earth and the ways in which people use their environments and resources. Remote sensing has become multidisciplinary in its approach to inventorying and analyzing earth phenomena and the physical and cultural environments of the world. In all of its multifaceted aspects, remote sensing techniques involve an analyzation of phenomena by use of different wavelengths of energy in the electromagnetic spectrum.

The Electromagnetic Spectrum

The Need to Understand the Electromagnetic Spectrum

One of the principal advantages of remote sensor imagery is that it may be obtained simultaneously from several portions, or wavelengths, of the spectrum which are not normally apparent to the human eye. Because users are presented with this kind of imagery, it becomes necessary for them to understand the principles of electromagnetic energy and the ways by which various wavelengths and frequencies of energy in the electromagnetic spectrum are sampled and recorded.

Just as it is possible to drive a car without being an automotive mechanic, so a person can be trained to gain some useful information from remote sensor images without knowing the physics of the interactions of energy between a sensor and a target. On the other hand, motorists often find it useful to understand the rudimentary aspects of an engine's operation in order to drive more effectively; and certainly, a driver should know how to read and understand the indications of a speedometer, odometer, or other guages on a dashboard, as well as to be able to change a tire.

A person may be able to take acceptable photographs with a camera without understanding the physics of optics, the mechanics of a shutter, or the photochemical processes involved in producing a latent image on film or paper. On the other hand, creative and scientific photographers must know something about lighting, reflectance, and image production in order to obtain the desired results of imaging scenes in the landscape. The good photographer must know what he is doing and why he is doing it.

Basic to an ability to interpret a remotely sensed image from a sophisticated detector is a knowledge of the physics of energy and the nature of energy in the electromagnetic spectrum (discussed in chapter 2). Without this type of working background, the remote sensing interpreter is like the scientist who did not feel it was necessary to study mathematics, claiming that if he ever had a mathematical problem he could get a mathematician to solve it for him. The point is, of course, such a person will never have a mathematical problem because he will never know the power of the mathematical tool. If users are to be able to accurately interpret remote sensor imagery, or if they are to be able to plan projects for the use of remote sensor surveillance of prescribed targets or events, they must become familiar with the energy relationships of the electromagnetic spectrum, the capabilities of different sensors, the characteristics of different targets, and the interactions between sensors and targets to electromagnetic energy.

Detection of Electromagnetic Energy

By definition, the electromagnetic spectrum is a continuum of electric and magnetic wavelengths which extend from the cosmic short waves of high frequency at one end of the spectrum to long radio waves of low frequency at the other end. Scientists do not know the limits of the electromagnetic spectrum, nor are they able to discretely categorize bands of wavelengths within the spectrum. The ends of the electromagnetic spectrum may lie at infinity, and the wavelengths of energy within the spectrum merge imperceptibly into each other. Wavelengths in the cosmic-ultraviolet areas of the known portion of the electromagnetic spectrum are measured in millionths (10^{-6}) or billionths (10^{-9}) of a meter. In the long wavelength regions of the electromagnetic spectrum, wavelengths are measured in centimeters, meters, or even in kilometers.

As previously stated, only a small portion of the wavelengths of the electromagnetic spectrum are discernable by our human senses; however, life on earth is constantly bombarded by energy of which we are not aware. People may not be conscious of ultraviolet energy, but extensive exposure to the sun and a resultant painful sunburn should be enough to alert them to the presence of this energy. People are not aware of radio wavelengths, but when a television set is turned on the invisible electromagnetic energy which is transmitted by a TV station miles away is converted into wavelengths of energy which are audible and visible.

Remote sensor images are produced by detectors which are sensitive to certain wavelengths and frequencies of energy which are emitted or radiated from an object. In the case of conventional film in conventional cameras, the wavelengths and frequencies of energy which are recorded are the same as those perceived by human eyes. In other words, the pictures are records of the scenes we see. Other types of remote sensors, however, are designed to image objects and events which reflect or radiate energy in those bands of the electromagnetic spectrum which are different from those which we are capable of seeing. The problem which is presented to users of remote sensing is one of searching through the electromagnetic spectrum to find those wavelengths which are most suitable for imaging and identifying a target or event. When a remote sensor image is studied, the person must be familiar with the radiation of the electromagnetic energy which contributed the image of the scene, as well as with the capabilities of the sensor which generated the image, and the varying transparency of the atmosphere as it transmits different wavelengths of energy.

Passive and Active Remote Sensors

Most energy which is imaged by remote sensing devices is reflected sunlight. A passive sensor is one which images radiation reflected by an object. This principle is familiar to anyone who has taken photographs with a camera. Wavelengths of light which are reflected from an object or scene are incident upon the lens of the camera. The lens collects these wavelengths of light and transmits them to a plane where they will be in sharp focus on the film. On the other hand, if the camera is carried above the earth in an airplane, for example, the energy which is reflected from targets is scattered by the atmosphere because the wavelengths pass through large areas of the air and are scattered by the different types and sizes of particles in the ocean of air. This scattering results in degraded images on the film. Remote sensors (and certain types of film/filter combinations in cameras) which operate in those areas of the spectrum which are least sensitive to atmospheric scattering of wavelengths of energy have been used effectively in the surveillance of earth resources. As previously stated, using several sensors which operate individually in different bands of the electromagnetic spectrum may serve useful purposes in producing multispectral images. These taken collectively may provide accurate identification of and information about objects or events within a scene.

In order to be able to isolate a target or event at the earth's surface for identification and study, objects must be differentiated from their background. Passive sensors may be selected and programed to be responsive to the wavelengths of a target, but at the same time to be relatively insensitive to the radiated energy of the materials surrounding the object. For example, some passive sensors are responsive to radiation from healthy vegetation and less responsive to vegetation which is suffering the effects of drought, insect infestation, disease, or salinity (discussed in chapters 10, 12, and 13). Other passive sensors measure heat radiation from targets (discussed in chapters 4 and 5). In each case the energy which arrives at the remote sensor is either reflected or emitted by the object being sensed.

Active remote sensors provide their own illumination of a target. Radar is this type of sensor (discussed in chapter 6). The radar sensor generates and emits its own energy pulse which is directed at a target or scene. The target is illuminated by the energy pulse and backscatters some of the transmitted energy to the receiving antenna of the radar sensor. The antenna relays the energy to a display unit where the image may be made permanent for study. Active sensors, like radar, operate with complete disregard of daylight or weather conditions. In addition, the transmitted and returned pulses of energy may be polarized to enhance the images. A classic example of the use of radar in imaging earth scenes was a mapping project in the Darien Province of Panama. Here, in an area of tropical rainforest which was almost continuously cloud-covered, making aerial photography nearly impossible, an airborne radar remote sensor produced high fidelity images of the topography through the cloud deck.

Remote Sensing from Space

Space Photographs from Gemini and Apollo Missions

Photography from manned space vehicles was initiated with the mission of Gemini III in 1965 (discussed in chapter 7). Space photography was continued in the Apollo missions which began in 1968 and ended in 1972. Many of the photographic assignments of the Gemini and Apollo astronauts were concerned with general views of weather and landforms, along with certain cultural features on the surface of the earth. These space photographs can be used effectively in studies of regional geography, and in weather, climate, vegetation, and landform analysis.

Weather and terrain photographs have been assembled in five books published by the National Aeronautics and Space Administration (NASA), and distributed by the Superintendent of Documents, Washington, D.C. 20402. These books, listed below, provide an inexpensive source of selected and annotated space photographs which can be used by students of remote sensing to become familiar with space photograph interpretation.

Earth Photographs from Gemini III, IV, and V. NASA SP-129, 1967.
Exploring Space with a Camera. NASA SP-168, 1968.
Earth Photographs from Gemini VI through XII. NASA SP-171, 1968.
Ecological Surveys from Space. NASA SP-230, 1970.
This Island Earth. NASA SP-250, 1970.

Several commercially prepared books contain excellent space photographs accompanied by interpretations and thematic maps of the areas imaged. These books can be used in courses on regional geography, or for interpretation of landforms, hydrology, transportation patterns, urban configurations, and rural land uses. Among the books are:

Paul D. Lowman, Jr. *Space Panorama.* Zurich, Switzerland: Weltflugbild, 1968. This book contains space photographs categorized into regions of North America, South America, Africa, Asia, Australia, and the Oceans.

J. Bodechtel and H. G. Gierloff-Emden. *The Earth from Space.* New York: Arco Publishing Company, 1974. This book contains units of space photographs on the atmosphere, oceans, and weather forecasting, as well as space illustrations of movements of the earth's crust, landscape types, and the environment in relation to its surroundings.

Several catalogs have been prepared which list individual photographic frames imaged from Gemini and Apollo spacecrafts. These catalogs provide information on the mission, date, and area imaged by each photograph, along with information on the altitude at which the photograph was made, percentage of cloud cover, and percentage of overlap. Among the catalogs are:

Gemini Synoptic Photography Catalog. Technology Application Center, University of New Mexico, Albuquerque, New Mexico 87131.

Apollo Synoptic Photography Catalog. Technology Application Center, University of New Mexico, Albuquerque, New Mexico 87131.

Individual Gemini and Apollo photography prints and slides can be ordered from the same source as the catalogs; or from Pilot Rock, Inc., Box 470, Arcata, California 95521.

Users must be aware of problems which arise in the interpretation of earth features from space photographs (discussed in chapters 7 and 8). Carter and Stone point out that certain precautions must be taken in the use of space photographs for interpretation and explanation of earth features.[6] These include the absence of a perfect vertical view, resulting from the tumbling path of the space vehicle, and the very small scale of the space photograph. Also, sun glitter may interfere with interpretations. Space photographs are imaged above the atmospheric zone of weather; therefore, clouds may mask certain landscapes, and cloud shadows may be mistaken for lakes or other surface features. Carter and Stone cite tone, color, texture, pattern, size, and shapes of objects as requiring more careful consideration in space photograph interpretations than when these parameters are used for interpretation of suborbital, aerial photographs (discussed in chapters 3 and 9).

Space Photographs from Skylab

The Gemini and Apollo missions were preludes to the development and implementation of the Skylab Orbiting Space Station in 1973 and early 1974 (discussed in chapter 8). Skylab was operational for eight months, during which time three different three-man crews lived and worked in space on missions which lasted for one month, six weeks, and two months, respectively.

Skylab experiments were carried out in the physical sciences, biomedical sciences, and earth sciences. Information on teaching modules and suggestions concerning the relevance of these experiments to school curricula have been published in the following manuals which may be purchased from the U.S. Government Printing Office, Washington, D.C. 20402.

NASA. *Skylab Experiments.* Physical Science, Solar Astronomy, vol. 1. May 1973.
NASA. *Skylab Experiments.* Remote Sensing of Earth Resources, vol. 2. May 1973.

Skylab experiments and information of particular interest to students include techniques of remote sensing from space such aspects of the environment as agriculture, forestry, geology, geography, oceanography, air and water pollution, land use, and meteorology. Students can obtain other ideas concerning the use of space imagery from:

The Journal of Aerospace Education. National Aeronautic Association, Suite 610, Shoreham Bldg., 806 15th Street, N.W., Washington, D.C. 20005.
NASA Report to Educators. Educational Programs Division, Office of Public Affairs, Code FE, National Aeronautics and Space Administration, Washington, D.C. 20546. This report is ''published four times per year for the community of educators, especially at the elementary and secondary school levels.''

Another publication which may assist potential users understand Skylab and other remote sensing methods and interpretation is:

Earth Trak. IDEL, 2008 Westport Avenue, Sioux Falls, South Dakota 57107. Six issues of *Earth Trak* are published annually. These issues address themselves to the promotion of education in the use of remote sensing imagery as a tool for the earth scientist.

Images from Nonphotographic Remote Sensors

Nonphotographic Images

Nonphotographic remote sensing systems extend the vision of observers into the long-wave and short-wave regions of the electromagnetic spectrum. In the short wavelengths, nonphotographic sensors permit observation in the ultraviolet regions of the spectrum. The long wavelengths of the spectrum are imaged in the near, medium, and far infrared regions by means of solid state detectors in scanners and radiometers, and in the radio wavelengths regions by radar.

A problem concerning classroom or other learning use of images from nonphotographic sensors is that the photographlike prints are difficult to obtain from private corporations and researchers and from government agencies. In addition, interpretations of images from nonphotographic sensors should be carried out in connection with radiometric ground data (discussed in chapter 5). In many cases this data is not provided with the imagery.

Nonphotographic remote sensing imagery is used to study such phenomena as groundwater discharges and salt water incursions from infrared scanners; oil spills by use of ultraviolet imagery; landforms, hydrology, and urban areas from multispectral infrared scanners; as well as landform, drainage, and vegetation studies by use of radar imagery. Satellite images of weather systems may be obtained from:

U.S. Department of Commerce, National Oceanic and Atmospheric Administration (NOAA), National Environmental Satellite Service, Washington, D.C. 20233.

Landsat Imagery

In an effort to collect information about the earth and its resources on a regular, repetitive basis, the U.S. Department of the Interior established the Earth Resources Observations Systems (EROS) in 1964 (discussed in chapter 9). On July 23, 1972, ERTS-1 (now referred to as Landsat 1) was launched by NASA. Landsat 2 was launched on January 22, 1975, into a similar orbit. Both vehicles were in nearly polar, sun synchronous orbits around the earth. Because of progressive mechanical failures on Landsat 1, its payload operations were terminated on January 6, 1978. Landsat 2 continued to operate successfully, and on March 5, 1978, Landsat 3 was launched from Vandenburg Air Force Base in California.

A Landsat platform circles the earth in about 103 minutes, or 14 times each day. After 252 orbits, or every 18 days, it returns to the same position above the earth. Because Landsat 1 and 2 followed each other 180° apart, every place on earth came under the surveillance of their sensors every 9 days. Landsat 3 follows Landsat 2 in a similar path so that, as long as they continue to operate, all places on earth will continue to be imaged every 9 days, atmospheric conditions permitting.

The sensors on board the Landsat platforms include Return Beam Vidicon (RBV) television cameras and multispectral line-scanning systems (MSS). (Illustrated in chapter 9 by figures 9.4 and 9.5.) The MSS systems image scenes simultaneously in the green, red, and two infrared bands of the electromagnetic spectrum.

Imagery from the Landsat sensors is easy to obtain. Photographlike prints from the four bands, and false-color composites prepared from three of the bands (plate 1), can be obtained for individual domestic and foreign scenes from:

Chief, User Services, EROS Data Center, Sioux Falls, South Dakota 57189.

Photographlike paper prints, negatives, and film positives may be obtained for each Landsat spectral band at scales of 1:3,369,000 (the original scale of the imagery); 1:1,000,000; 1:500,000; or 1:250,000. A map illustrating the dates when each Landsat vehicle crosses areas in the United States can be obtained from:

Missions Utilization Office, Landsat Newsletter, Code 902, NASA-GSFC, Greenbelt, Maryland 20771.

Landsat tracks over Canada can be obtained from:

Integrated Satellite Information Services Ltd. P.O. Box 1630, Prince Albert, Sasketchewan S6V 5T2.

A photomap of conterminous United States was prepared from mosaics of Landsat by the U.S. Soil Conservation Service (fig. 1.1). The photomap is a controlled mosaic which was constructed from about 600 black and white Band 5 (discussed in chapter 9) Landsat images obtained on cloud- and snow-free days between July and October, 1972. Also, a winter season photomap was prepared from imagery obtained between December, 1972, and March, 1973. Because of the general absence of a vegetation cover on the winter photomap, topographic features are more pronounced. In addition to the complete United States maps, smaller sections of the country are portrayed on separate, larger-scale photomaps. Information about these mosaic images can be obtained from:

Aerial Photography Field Office, U.S. Department of Agriculture, P.O. Box 30010 (2222 West 2300 South), Salt Lake City, Utah 84125.

Individual cloud-free frames from Landsat 1 images may be obtained for all areas of the United States along each satellite track. Information about these products can be obtained from the "Single ERTS Coverage Map" which is distributed by the EROS Data Center.

Dr. James C. Fletcher, former Administrator of NASA, stated, "If I had to pick one spacecraft, one Space Age development, to save the world, I would pick Landsat and the satellites which I believe will be evolved from it."[7] Typical applications of Landsat imagery support Dr. Fletcher's statement. Landsat imagery is being used for the identification of regional topographic forms and rock structures as an aid in exploring for mineral deposits; for the identification of snow packs in mountain areas as an aid in determining potential water supply; for monitoring strip mining operations; for surveilling forest and rangeland conditions; for studying changes in coastline morphology as an aid in monitoring seashore parks and wetland resources; for evaluating changes in land use for city and regional planning purposes; and for identifying ground cover and the vigor of vegetation, especially crops.[8]

Prospects for the Future

The multitude of environmental and population problems which afflict the world are awesome, but the arsenal of remote sensing weapons to help solve these problems in an orderly, progressive, rational, and humane way is equally spectacular. For example, progress is being made in the refinement and use of thermal infrared scanners, radar, and multispectral scanners. Landsat 3, which was launched in 1978, carries improved sensors. The Space Shuttle laboratories will be concerned, in part, with monitoring environmental problems (discussed in chapter 8). Some of the dreams of scientists will be found in chapter 20.

Admittedly, remote sensing from space platforms was inaugurated as part of the challenge presented by Sputnik 1. The successful missions of Mercury, Gemini, Apollo, Skylab, and the joint United States-Soviet Union Apollo-Soyuz, however, demonstrated that people can live and work in space. The Space Shuttle, of the Space Transportation System (STS), is an outgrowth of those previous successes. Dr. Alan Lovelace asserted that, "A significant contribution to the (Space) Shuttle payload will be Spacelab, the first European contribution to this new adventure of establishing a Space Transportation System. . . . The Shuttle/Spacelab combination is going to launch and return weather communications satellites . . . and carry out surveys of Earth resources from its vantage point of viewing our planet in its entirety."[9]

Those who study this book today well may be the remote sensing specialists of the next generation. Certainly, they will spend their lives in an age when more and more information about the earth will be garnered from space platforms and from space sensors. As this is being written, the science of remote sensing is about a decade old. In light of the progress anticipated for science in the next several decades, the present reader may look back on what the authors of this book now consider to be the "cutting edge" of environmental science and see that it is merely a continuation of studying the earth from the saddle, with the saddle having turned into a satellite.

NOTES

1. The subtitle was suggested by a subtitle on the cover of: R. H. Rogers, L. E. Reed, and N. F. Schmidt, *Landsat 1: Automated Land-Use Mapping in Lake and River Watersheds* (Ann Arbor, Mich.: Bendix Aerospace Systems Divisions, 1975).
2. Orin W. Nicks, ed., *This Island Earth,* NASA SP-250 (Washington, D.C.: National Aeronautics and Space Administration, 1970).
3. Percy R. Luney and Henry W. Dill, Jr., "Uses, Potentialities, and Needs in Agriculture and Forestry," chapter 1 in *Remote Sensing with Special Reference to Agriculture and Forestry* (Washington, D.C.: National Academy of Science, 1970), p. 1.
4. Robert G. Reeves, ed.-in-chief, *Manual of Remote Sensing* (Falls Church, Va.: American Society of Photogrammetry, 1975), p. 2102.
5. Ibid., pp. 5-11.
6. Louis D. Carter and Richard O. Stone, "Interpretation of Orbital Photographs," *Photogrammetric Engineering* 40 (1974): 193-97.
7. James C. Fletcher, "Landsat Wins New Believers," *NASA Report to Educators* 5, no. 2 (May 1977), p. 6.
8. U.S. Geological Survey, *EROS Reprint,* no. 167, 1973.
9. Alan Lovelace, "The STS-Toward Human Occupation of Space," *NASA Report to Educators* 5, no. 2 (May 1977), p. 3.

2

The Electromagnetic Spectrum and Its Use in Remote Sensing

John E. Estes
University of California-Santa Barbara

Introduction

The detection, recording, and analysis of electromagnetic energy is the foundation of remote sensing. Detection and recording of electromagnetic energy are made possible because of the series of complex energy-matter-environment interactions which combine to produce recorded contrasts by remote sensor systems between a target and its background. Students seeking to derive useful information from remotely sensed data should acquire a basic understanding of these interactions which are involved in image formation. It is important to understand the basic physics of the processes, techniques, and methodologies associated with the formation of remote sensor images.

Energy may be conveyed by means of electromagnetic waves which occur in a continuum of energy frequencies or by quanta. Quanta are minimum energy units of electromagnetic radiation, known as photons. The frequency range associated with such energy progresses from long, low-frequency radio waves to the very short, high-frequency gamma and cosmic waves. In common practice, this continuum is subdivided into various wavelength bands or spectral regions, such as X-ray, ultraviolet, visible, infrared, and microwave (fig. 2.1). In reality, none of these regions has distinct boundaries, and the limits of the electromagnetic spectrum on both the short and long wavelength areas are not known precisely.

The level of energy reflected or emitted from objects normally varies with frequency or wavelength throughout the electromagnetic spectrum. The *signature* of the imaged object is basically governed by the different amounts of energy incident upon and reflected from the object, along with the wavelength sensitivity of the sensor at the time the image is acquired. A unique signature of an object, therefore, can often be identified if this energy is subdivided into carefully chosen wavelength bands. While systems with broad-band response may tend to diminish object-to-background differentiation, selective recording of energy within particular wavelength bands may improve the discrimination of objects. It is these concepts which form the basis of multiband multispectral remote sensing.

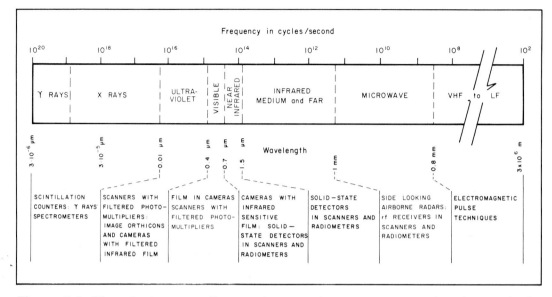

Figure 2.1. The electromagnetic spectrum and some sensors that image in it. (Adapted from Estes, 1974)

In order to understand the variations in signatures on various types of remote sensor images, the principles of electromagnetic energy radiation must be understood. The physical formulas which have been included in these discussions aid in the quantitative interpretation of electromagnetic radiation, but the student may gain considerable understanding about the processes involved in the use of the electromagnetic spectrum in remote sensing from the descriptive material in the chapter. Remote sensors, including the human eye and conventional cameras, detect objects because of the way in which electromagnetic radiation in various wavelengths is transmitted.

Generation and Transmission of Electromagnetic Radiation[1]

Basically, electromagnetic radiation is generated by changing the size or direction of an electric or magnetic field with time. This is accomplished by changes in the source of the electric and magnetic fields. In the *Manual of Remote Sensing* electromagnetic radiation is defined as energy propogated through space, or through material media, in the form of an advancing interaction between electric and magnetic fields. Also, the term radiation is commonly used when referring to this type of energy, although it actually has a broader meaning.

Electromagnetic radiation is emitted whenever an electric charge is accelerated. The wavelength of electromagnetic radiation which is generated depends on the length of time over which acceleration occurs along with the number of accelerations per second (i.e., frequency). The electrically charged particles involved in generating electromagnetic radiations are atoms, electrons, ions, and molecules. Electrons are very small, negatively charged particles which revolve around the positively charged nucleus of an atom. Atoms of different substances are made up of a varying number of electrons arranged in a variety of ways.

As will be seen later in this chapter, electrons, like light, sometimes behave as particles, and at other times they behave as waves. Interaction between a positively charged nucleus and a negatively charged electron keeps the electron in orbit around the nucleus. While the orbit of an electron is not fixed, its motion is restricted to a definite range from the nucleus. The orbital paths of electrons about an atom can be thought of as determined by related energy levels. In order for one electron to attain a higher level some work must be done on the atom; however, not just any amount of work will do.

Unless sufficient energy is available to move the electron up at least one energy level, it will accept no work. If enough energy is received and the electron jumps to a new level, the atom is said to be excited. Once the electron is in a higher orbit, it possesses additional potential energy. After about one hundred millionth (10^{-7}) of a second, the electron falls back to the lowest empty energy level (or orbit) of the atom, at which time it gives off radiation. The wavelength of radiation given off is a function of the amount of work done on the atom, or the quantum of energy it absorbed to cause the electron to be moved to this higher orbit. Occasionally the atom will absorb sufficient energy to cause the electron to escape completely from the orbit. When this occurs, the atom has a positive charge equal in magnitude to the negatively charged electron which escaped. The electron becomes a free electron, and the atom is called an ion.

In the ultraviolet and visible portions of the electromagnetic spectrum, radiation is produced by changes in the energy levels of the outer electrons. The wavelengths of energy produced in these interactions are a function of the orbital energy levels of the electrons involved in the excitation process. If atoms absorb sufficient energy to become ionized, and if a free electron drops in to fill the vacant energy level, the radiation given off is a continuous spectrum rather than a band, or a series of bands, of electromagnetic radiation.

In molecules, orbital electron motion changes produce short wavelength radiation; vibrational changes produce short and intermediate range infrared radiation; and rotational changes produce long wavelength infrared radiation extending into the short wavelength microwave range. When metals are heated, they tend to radiate a continuous spectrum. As a metal heats, it will first glow a dull, then a bright red (red hot). If heating continues, metal will glow orange, then yellow, and finally white (white hot) as a total spectrum of light is given off. Matter then can be heated to such high temperatures that electrons, which typically move in captive orbits, are broken free.

The hot surface of the sun produces radiation of all wavelengths, that is, it produces a continuous spectrum. The radiant power peak within this continuous spectrum produced by the sun approximates a 6000 °K (Kelvin) blackbody. As a result, peak solar radiance, or the highest level of radiation striking a surface per unit area of earth (fig. 2.2), is about $0.5\,\mu$m (in the green band of the visible portion of the electromagnetic spectrum, fig. 2.1). The sun, then, is an excellent energy source for measuring the reflectance of objects or scenes in the visible ($0.4\,\mu$m to $0.7\,\mu$m) portion of the electromagnetic spectrum. When emission phenomena are being measured for objects that are near the average ambient temperature of the earth (300 °K), the radiant power peak of the earth is at a wavelength of about $9.6\,\mu$m (fig. 2.3). The earth itself, therefore, is an excellent energy source of passive energy for remote sensing of the earth's surface in the thermal infrared region.

All objects above absolute zero which absorb the radiant energy of the sun tend to increase in temperature and, for the most part, reemit energy in the infrared, or long wavelength, portion of the spectrum. This radiation can be defined as energy per second per centimeter

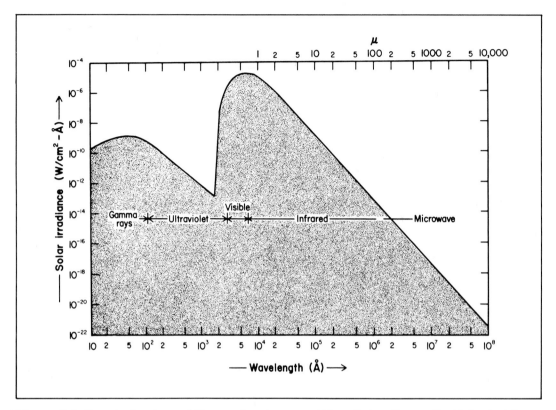

Figure 2.2. Spectral nature of the sun's energy as it irradiates the earth. (Adapted from Estes, 1974)

squared. According to the Stefan-Boltzmann law for a *blackbody,* the total emissive power, or radiant energy per unit time, is proportional to the fourth power of its absolute temperature. That is:

$$W = \sigma T^4 \tag{1}$$

where: W = radiant exitance (watts cm^{-2})
             σ = Stefan-Boltzmann Constant (5.67×10^{-12} watts cm^{-2}°K^{-4})

The relationship described in this formula applies to all wavelengths of the electromagnetic spectrum shorter than microwave. In the microwave region of the spectrum the relation is:

$$W = F(T) \tag{2}$$

That is, microwave radiant energy varies directly as a function (F) of the temperature of an object in degrees Kelvin.

No actual substance behaves as a true blackbody although some soots closely approximate this ideal. In accordance with Kirchoff's law, a blackbody absorbs all wavelengths and emits at all wavelengths with maximum intensity for any given temperature. Since there is no such thing

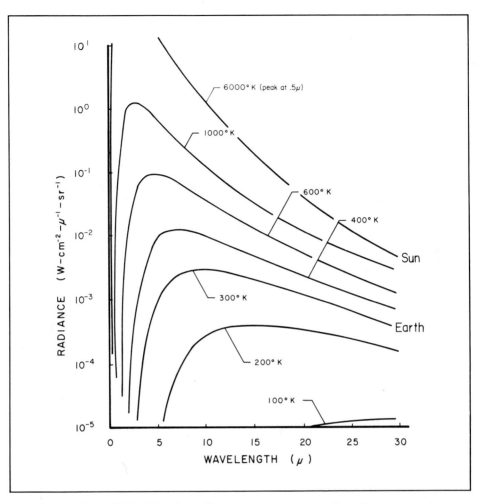

Figure 2.3. Plot of blackbody radiance and wavelength with temperature as a variable. (Adapted from Estes, 1974)

as a true blackbody, emissivity (ϵ) is defined as the ratio w/w_b, where w is the exitance of a real body and w_b is the exitance of a black body at exactly the same temperature, or $\epsilon = w/w_b$.

Once energy in the form of electromagnetic radiation is given off by an atom, molecule, or the sun, it will move off through space. Electromagnetic radiation is the only form of energy transfer which can take place through free space. To fully understand the method by which this transfer occurs, radiation in both particle and wave forms must be considered.

Nature of Electromagnetic Radiation—the Particle Model

Early in the twentieth century, Max Planck recognized the discrete nature of radiant energy exchanges and proposed the quantum theory of electromagnetic radiation. His experiments showed that energy is transferred in short wave trains or bursts (fig. 2.4, C) in which each burst carries a radiant energy, Q, proportional to the frequency, of the wave, so that:

$$Q = h\nu \tag{3}$$

where: h = the universal, or Planck's, constant, with a value of 6.625×10^{-34} joule second. [A joule, in physics, is a unit of work equal to 10 million ergs.]

The radiant energy carried by such wave bursts is not delivered uniformly to a receiver but is delivered to a location on a probablistic basis. The receipt of radiant energy at each location along the wave, therefore, is proportional to wave amplitude at that location. If large numbers of such wave bursts are incident on a receiving plate, the time average rate of energy delivered will approximate that which Maxwell's formulation predicts for the average wave amplitude (see below). From instant to instant, however, the rate of energy flow to the plate will be continuously fluctuating because of the quantum and statistical nature of electromagnetic radiation. The nature of quantum radiation is illustrated by figure 2.4, A and C.

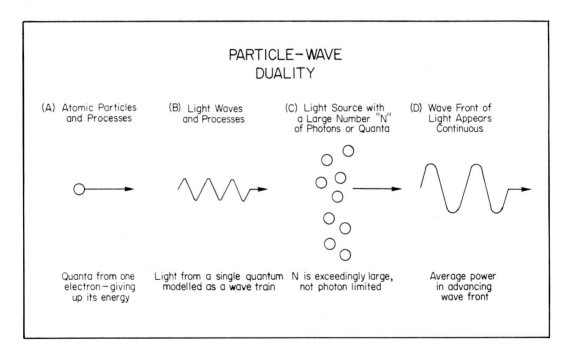

Figure 2.4. Electromagnetic radiation (EMR) possesses characteristics associated with both wave and particle theory. This particle wave duality means that under given conditions EMR may either be modeled as individual quanta as shown in (A) above, or as a wave train shown in (B). A light source which emits a large number "N" of photons or quanta can be either modeled as separate particles as in (C), or as a continuous wave format as seen in (D). The radiant energy which is contained in each quanta or wave train depicted in (A), (B), (C), and (D) can be determined by the formula: $Q = h\nu$ where radiant energy Q is proportional to the velocity of the wave ν times Planck's constant h.

The term "photon" is used to emphasize the quantized and statistical properties of radiation, while the term "wave" is used to emphasize the overall, average effects of radiation. According to the quantum theory of radiation, the elementary quantity, or quantum, of radiant energy is the photon. It is regarded as a discrete quantity have a momentum of P, equal to:

$$P = \frac{h\nu}{c} \tag{4}$$

where: h = Planck's Constant
ν = frequency of the radiation
c = the speed of light in a vacuum

The photon is never at rest, it has no electric charge or magnetic moment, but it does have a spin moment. When a measurement of radiation lacks precision primarily because of the inherent statistical nature of energy delivery by radiation, the measurement is said to be photon-noise limited.

Although Planck's mathematical formulation of electromagnetic radiation as discrete quanta accounts for many phenomena associated with energy/matter interactions, it fails to account for a number of significant phenomena which become more pronounced at lower wavelengths. These are generally accounted for by use of the wave model of electromagnetic radiation proposed by Maxwell in the 1800s.

Nature of Electromagnetic Radiation—the Wave Model

There are three basic measurements which define the wave character of electromagnetic radiation: wavelength (distance from one crest of a wave to the next); wave velocity (speed and direction at which the wave crests advance); and wave frequency (number of wave crests which pass a given point in a specified period of time. According to the wave model, the velocity of energy is constant in a vacuum, or free space. Consequently, a reciprocal relationship exists between wave frequency and wavelength. This relationship is shown by:

$$\lambda = \frac{c}{\nu} \tag{5}$$

where: λ = wavelength
ν = frequency
c = universal constant speed of light

From the formula it can be seen that frequency is inversely proportional to wavelength and directly proportional to wave velocity. When electromagnetic radiation passes from one substance to another of different density (as when passing through the atmosphere), the velocity and wavelength change, but the frequency does not change.

Electromagnetic radiation emitted by a source such as the sun consists of two fluctuating fields: one electric and the other magnetic. Each of these fields is at right angles to the other and perpendicular to the direction of propagation. Waves of electromagnetic radiation, therefore, can be represented by two transverse wave forms which spread from a source at the speed of light (fig. 2.5).

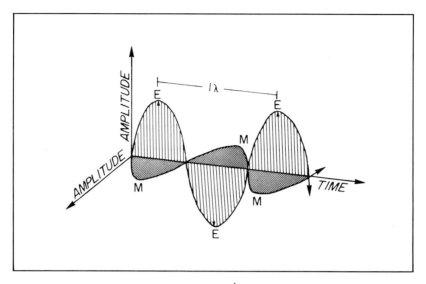

Figure 2.5. Electric (E) and magnetic (M) vectors of an electromagnetic wave. (After Nunnally, 1969)

By employing the principle of wave superposition, it is possible to predict the kind of waves which will result when an incident wave reflects from or penetrates into matter. If the sinusoidal component waves which comprise the incident field and how each component will reflect or penetrate are known, the composite or net reflected or penetrating wave will be the sum (the superposition) of the reflecting or penetrating component waves. The material properties which specify the response of the material (to these sinusoidal component waves) as a function of frequency are called material spectral properties.

The principle of superposition of waves states that: the composite wave, produced in a region alternatively described by two separate waves, has an amplitude which is the sum of the amplitudes of the two separate waves. The view, therefore, that complicated wave forms are actually composed of superimposed sinusoid curves is justifiable regardless of the manner in which the wave form was generated. Those sinusoids which are used to characterize a complicated wave-time relation are called the spectral components of the wave. Figure 2.6 illustrates the composition of a complex wave form with only two spectral components.

In remote sensing applications, the mathematical analysis of wave forms is rarely performed by calculations from graphs. Instead, direct measurements are made with spectrographic devices. The purpose of spectrographic instruments is to measure the amount of radiant flux in each of the sinusoidal components of any complex wave form. The term "flux" as applied to radiation refers to the time rate with which radiant energy passes a point in space. Remote sensing techniques use radiant flux measurements to infer "how much energy is where in the spectrum." The inherent limit to the precision of radiant flux measurements frequently forces a compromise between the rapidity and the precision of such measurements. Because of the essentially quantized and statistical properties of electromagnetic radiation, there is a limit to the precision of the measurement of radiant-energy flow rates. The precision diminishes as

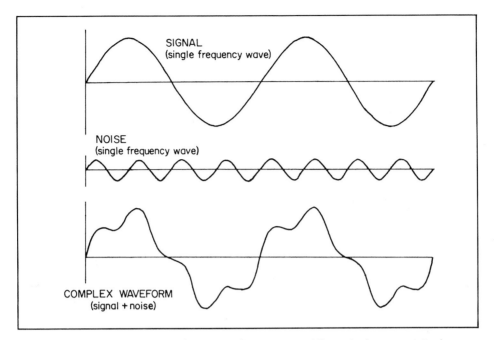

Figure 2.6. Composition of a complex wave with only two spectral components. The complex waveform seen below is formed by the superposition of the signal seen above and noise seen in the center.

the number of incident wave trains arriving in the measurement time interval decreases. At this point it is important then that we go further into the concepts involved in the interactions of radiant energy.

Radiant Energy

In order to fully understand both active and passive remote sensing techniques (see chapter 1) and the inherent problems involved in the optimization of such techniques, a study of radiant energy is necessary. Table 2.1 contains terms and definitions associated with the radiant energy.

Energy is a measure of the capacity to do physical work. Work is accomplished by moving an object by force, or by heating an object, or by causing a change in the state of matter. Radiant energy is the energy carried by electromagnetic radiation. Radiant energy causes a sensor-detecting element (e.g., a photographic emulsion or solid state detector) to physically change, with the change being taken as the evidence of the cause.

Radiant Flux

The units of radiant flux are either joule/second or most commonly the watt. Flux, in remote sensing, can also describe the time rate of flow of the quantized energy (photons) delivered to a detector. This is called "photon flux." The rate of flow of visible radiant energy is termed "luminous flux." When the context makes clear what energy flow is being used, it is

TABLE 2.1

Summary of Radiometric Concepts

Name	Symbol	Units	Concept
Radiant energy	Q_e	Joules, J	Capacity of radiation within a specified spectral band to do work
Radiant flux	ϕ_e	Watts, W	Time rate of flow of energy on to, off of, or through a surface
Radiant flux density at surface			
Irradiance	E_e	Watts per square meter, Wm^{-2}	Radiant flux incident upon a surface per unit area of that surface
Radiant exitance	M_e	Watts per square meter, Wm^{-2}	Radiant flux leaving a surface per unit area of that surface
Radiant intensity	I_e	Watts per steradian, Wsr^{-1}	Radiant flux leaving a small source per unit solid angle in a specified direction
Radiance	L_e	Watts per steradian per square meter, $Wsr^{-1}m^{-2}$	Radiant intensity per unit of projected source area in a specified direction
Hemispherical reflectance	ρ_e	Dimensionless	ϕ_e reflected / ϕ_e incident for any surface
Hemispherical transmittance	τ_e	Dimensionless	ϕ_e transmitted / ϕ_e incident for any surface
Hemispherical absorption	α_e	Dimensionless	ϕ_e absorbed / ϕ_e incident for any surface

Note: The subscript, e, is not used when context indicates that the symbols represent radiometric quantities. The word "Spectral" precedes the name of every radiometric quantity to obtain the names of the spectral quantities. Subscripts λ or υ are required on spectral quantity symbols to denote the differential properties and to emphasize the dimensional differences except in the case of spectral reflectance, transmittance, and absorptance, which remain dimensionless. (From *Manual of Remote Sensing*, vol. 1, chap. 3. Reprinted with permission.)

not necessary to repeat the modifying terms: radiant, photon, or luminous. When the radiant flux, ϕ, is constant, the total radiant energy, Q, which passes in time, t, is simply:

$$Q = \phi t \qquad\qquad (6)$$

When changes from time to time occur, the total energy is then:

$$Q = \int_{t_1}^{t_2} \phi(t)\, dt \qquad\qquad (7)$$

Energy delivered to a sensor system by radiant flux causes the detector to operate and record an image of the scene from which the flux emanated. A certain minimum quantity of energy, however, is needed for the detector to provide evidence that any radiant flux from a given scene was intercepted.

Radiant Flux Density, E and M

When a radiant flux is intercepted by a plane surface, there is often need to know the flux intercepted per unit of area of surface. If the surface is at right angles to the radiant flux intercepted, the flux divided by the area of the plane equals the average angles to the radiant flux intercepted, the radiant flux density at the plane. Each segment of the plane may be visualized as intercepting a small portion of the flux. When the segments are very small, this ratio can be thought of as the radiant flux density at each segmented point on the surface.

Radiant flux density, or flux incident upon a plane surface, is called *irradiance*. The direction of flux is not specified as long as it arrives at the surface from any or all directions within a hemisphere over the surface (fig. 2.7). If the irradiance is constant from point to point, then the intercepted flux is simply:

$$\phi = E\,A \tag{8}$$

where:　A ＝area of the surface
　　　　E ＝irradiance

However, irradiance will typically vary from point to point with the variation summed in incremental fashion. The flux density or radiant flux leaving the planar surface is called exitance (M). The directions taken by the flux are not specified so long as it leaves the point on the surface in any or all directions within a hemisphere over the surface.

Figure 2.7. The concept of radiant flux density (irradiance). Energy striking surface area (A) arrives from all parts of the hemisphere. The amount of this radiant energy striking the surface in a given unit of time is radiant flux; while the flux per unit area is the flux density. (Copyright 1975 by the American Society of Photogrammetry. Reproduced with permission.)

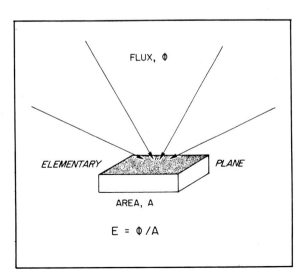

Hemispherical Reflectance

Hemispherical reflectance (ρ) is defined as the dimensionless ratio of the exitance from a plane of material to the irradiance on that plane:

$$\rho = M_{reflected}/E \qquad (9)$$

The concept of reflectance is depicted in figure 2.8. In the figure, the radiant flux is the sum of both direct solar radiation and the atmospherically scattered radiation. This incident energy may then be reflected, transmitted, or absorbed by the surface.

Reflection refers to the process whereby radiation "bounces off" a surface. Reflection usually involves the reradiation of photons by atoms or molecules in a surface layer approximately one-half wavelength deep from a given object. Reflection exhibits a number of basic characteristics which are of fundamental importance in remote sensing:

- The direction of incident radiation, the reflected radiation, and a vertical to the reflecting surface all lie in the same plane.
- The angle of incidence and the angle of reflection as measured from this vertical are always equal.

Reflectivity of a surface is dependent upon three factors: the angle of incident energy, the refractive index, and extinction coefficient of the reflecting material. A substance's index of refraction is a measure of the amount of refraction (a property of a nonconducting substance). It is the ratio of the wavelength or phase velocity of electromagnetic radiation in a vacuum to that in the substance. The ability of an object to refract electromagnetic radiation is a function of wavelength, temperature, and pressure. The extinction coefficient of a substance is that small portion of incident light that interacts with a particle of that substance. Reflectivity at any

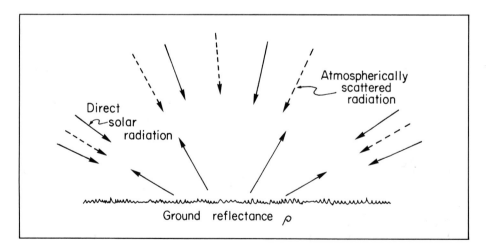

Figure 2.8. Reflection received at the surface is ϕ; where ϕ is the sum of direct and atmospherically scattered solar radiation incident on the surface. (Copyright 1975 by the American Society of Photogrammetry. Reproduced with permission.)

given angle is directly proportional to the refractive index. When light strikes a transparent medium at an angle of incidence of 0°, the amount of reflection is almost solely a function of the refractive index. Furthermore, since the extinction coefficient is proportional to the absorption coefficient, the greater the absorption, the lower the reflectivity. Metals which tend to be low absorbers typically exhibit high reflectivity.

Specular versus Diffuse Reflection

The effects discussed above are most easily observed for spectral or "mirror-like" reflection. In specular reflection the surface from which the radiation is reflected is essentially smooth. If the surface is rough, reflected rays may leave it in many directions, depending on the individual orientations of the many small reflecting surfaces. This process, called diffuse reflection, does not yield a mirror image but instead gives rise to a spreading out or scattering of radiation. White paper, certain types of snow, white powders, and other similar materials reflect light in this manner.

Many natural terrain features roughly approximate diffuse reflectors; however, water surfaces and man-made objects are not diffuse reflectors and often cause a distribution of reflected radiant flux to lie between the two extremes depicted in figure 2.9. Such nondiffuse reflectors typically exhibit a peak in their reflectance distributions (the so-called specular reflectance component) where the angles of incidence and reflectance are equal.

Hemispherical Transmittance

Hemispherical transmittance (τ) is the dimensionless ratio of the transmitted exitance leaving the opposite side of a plane exposed to the irradiant energy. Thus:

$$\tau = M_{transmitted}/E \qquad (10)$$

where: M = radiant exitance
E = irradiance

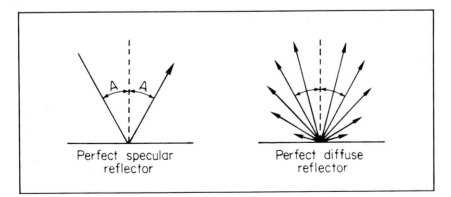

Figure 2.9. Distribution of reflected radiant flux from different types of reflecting surfaces. (Copyright 1975 by the American Society of Photogrammetry. Reproduced with permission.)

Hemispherical Absorption

Hemispherical absorption (α) is seen by the dimensionless relation:

$$\alpha = 1 - \tau - \rho \qquad (11)$$

where: τ = hemispherical transmittance
ρ = hemispherical reflectance

These definitions imply that radiant energy must be conserved. Incident flux is either returned back by reflection, transmitted through, or is transformed into some other form of energy inside the substance. The net effect of absorption of radiation by most substances is that some of the energy is converted into heat, causing a subsequent rise in its temperature, and the radiation of longer wavelength radiation.

Concept of Solid Angle

The concept of the solid angle of a cone is fundamental to two quantitative measures of radiant flux: radiant intensity (I) and radiance (L). The cone angle subtended by a part of a spherical surface of area, A, is equal to the area divided by the square of the radius of the sphere. The unit of cone angle or solid angle is the steradian. Since the surface area of a sphere is $4\pi r^2$, there are 4π steradians of solid angle in a sphere. The solid angle concept is illustrated by figure 2.10.

In remote sensing, there is frequent need to know the solid angle subtended by a flat plate as viewed from a long distance. From a knowledge of the solid angle both the directional measures of radiant flux, radiant intensity and radiance can be obtained. This concept can be used to illustrate the law that the intensity of spherical waves varies inversely as the square of the distance from the source. This inverse square law is in reality a consequence of the law of

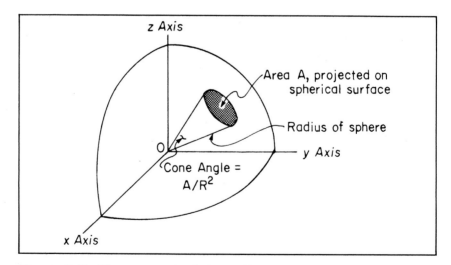

Figure 2.10. The concept of the solid angle in angular measure. (Copyright 1975 by the American Society of Photogrammetry. Reproduced with permission.)

conservation of energy. The concept of the spherical wave is that the energy flow per unit spherical area, per unit time, is proportional to $1/r^2$. An example of the effect of the law of inverse squares can be illustrated by moving to and from a fire. At some distance from the fire one is comfortable. At one-half the distance the radiation density is four times greater. At twice the distance the radiation density is one-fourth that encountered at the original comfortable distance.

Several examples of the practical importance of this inverse square law in remote sensing can be illustrated with respect to thermal sensing of forest fires and the imaging of terrain by a real-aperture side-looking airborne radar. With respect to the thermal sensing of forest fires, table 2.2 illustrates the reduction in energy incident on an infrared detector element from a 1-m^2 force at 800 °C. The effect on imagery taken from an aircraft at progressively higher altitudes is due to the inverse square law acting to reduce the energy incident on the detector. With respect to real-aperture, side-looking, airborne radar viewing a strip of terrain, it can be noted that the energy leaving the radar transmitter diminishes as $1/r^2$ to the target. Subsequently, a portion of the energy incident upon the target is returned to the antenna. This radiation, in turn, also diminishes at $1/r^2$ to the transmitter. The radiant flux leaving the antenna incident on the target, which is reradiated back to the antenna, is reduced as $1/r^4 \times \sigma$, where σ is the backscattering coefficient of the target directed at the antenna.

TABLE 2.2

Variation of Energy Incident Upon an Infrared Detector with Altitude

Height (km)	Energy Flow (W/cm^2)	Energy Ratio	Area Ratio	Product
0.5	1×10^{-5}	1	1	1
1.0	2.5×10^{-6}	4	4	16
2.0	6.25×10^{-7}	16	16	256

Reproduced from *Remote Sensing of Environment,* edited by Joseph Lintz, Jr., and David S. Simonett, with permission of publishers, Addison-Wesley/W.A. Benjamin Inc. Advanced Book Program, Reading, Mass., USA.

Radiant Intensity

Radiant intensity of a point source in a given direction can be defined as the radiant flux per unit solid angle leaving the source in that direction. If a source is isotropic (i.e., it radiates equally in all directions), then the radiant intensity (I) is:

$$I = \phi/4\pi \qquad (12)$$

Radiance

Radiance is the radiant flux per unit solid angle leaving a source in a given direction per unit projected source area in that direction (fig. 2.11). The concept of radiance corresponds to the concept of brightness. Much of remote sensing is concerned with the amount of radiant flux leaving areas (surfaces, objects) in a direction toward the sensor. Consequently, the concept of radiance is often used.

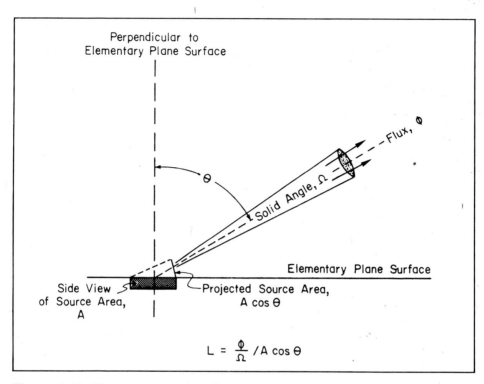

Figure 2.11. The concept of radiance. (Adapted from *Manual of Remote Sensing,* vol. 1, chap. 3. Reproduced with permission.)

As seen in figure 2.11 the projected area in a direction which makes a polar angle, θ, with the perpendicular to an elementary plane segment of area A, is A cos θ. If the plane segment is small and is considered to be a point source, then the radiant intensity of the plane segment in a direction is:

$$I = LA \cos \theta \qquad (13)$$

The plane source for which the radiance, L, does not change value as a function of angle of view is called Lambertian source (after Lambert's laws of illumination). A piece of white matte paper, when illuminated by diffuse skylight, is a good approximation to a Lambertian source, since the perceived visual brightness of such a piece of paper does not change with angle of view. Cameras and other remote sensing devices record the radiance (L) (i.e., the radiant flux per solid angle per square meter) reflected or exited from a given scene.

Energy transmittance through, and reflectance and exited energy from ground targets varies considerably for different types of materials. It is this differential characteristic of terrain materials which permits us to identify, classify, and differentiate terrestrial objects, features, or phenomena. Their reflected or exited energy is expressed as a gray tone level or color on remote sensing imagery. The tonal or color variations between objects and their backgrounds are typically referred to as contrasts. The relative magnitude of these tonal differences are termed contrast ratios. The reflectance from a number of ground objects has been measured: black asphalt typically reflects only 2 percent of the solar radiation, timberland 3 percent, open grassland 6 percent, concrete 36 percent, and snow 80 percent. When contrast ratios are quite high—for example, a cleared asphalt highway passing through an area covered with snow—an image interpreter can easily differentiate between the two. But if the highway were light concrete, the interpreter might not be able to make this differentiation.

The sun angle is also a very important consideration because it affects not only the amount of illumination being either emitted or reflected to the sensing system, but also the spectral quality. As the angles and intensity of illumination change, object-to-background contrast ratios vary. In addition, shadows are emphasized at lower sun angles enabling the interpreter to see lineations. The intensity and duration of illumination for a given locality varies with latitude, season of the year, time of day, and local topography. For example, latitude 40°N receives about 11,000 footcandles in June during high sun but only 5,000 footcandles when the sun is low in December. At these latitudes, similar drops in illumination occur approximately four hours before or after noon. For this reason, aerial photographic missions are generally flown within a period two hours before or after noon local sun time.

Thermal contrasts (as opposed to tonal contrasts based on reflective energy differences) between land and water may be masked diurnally by differential heating caused by a change in illumination. Just as changing angles of illumination may decrease object-to-background contrast ratios, making interpretation more difficult when passive sensing systems are used, it can also affect the "signature" of objects recorded by active sensing systems. It has been shown that variations in slope of basically similar terrain can produce significant variations in the amount of illumination received by an area. These variations cause a definite spectral response on radar imagery.

Because of the variability caused by changes in the angle of illumination, care should be exercised in the study of different remote sensor images which were acquired at different times. Some interpreters have gone so far as to suggest that since levels of solar illumination are subject to variation, not only as a function of time but from area to area, the best method of obtaining precise and consistent information from photographic density measurements is to use sensors which create their own energy or a calibrated light source for illumination. While artificial illumination might alleviate part of the problem of variation in the level of solar illumination, there is still a need for further study and a better understanding of such sophisticated sensors.

Atmospheric Effects

A major consideration in the interpretation of data acquired by means of remote sensing is the effect of the earth's atmosphere on electromagnetic radiation. As the altitude of the sensor platform (aircraft, spacecraft) above the earth increases, the amount of atmosphere through which the energy from terrestrial materials must pass increases. As much as 50 percent of the earth's atmosphere lies within 17,500 feet (5,334 meters) of the earth's surface, 75 percent within 35,000 feet (10,668 meters), and 99 percent is believed to lie within 25 miles (40.2 kilometers). The atmosphere is an extremely complex medium through which electromagnetic energy must pass in order to be remotely sensed. The changes in the character of solar radiation as it passes through the atmosphere affects the energy levels which can be measured on the surface, as well as those recorded by a sensor system. A summary of factors affecting the utility of remote sensing in various parts of the spectrum is presented in table 2.3.

TABLE 2.3
General Characteristics of Remote Sensor Systems

Spectral Region and Sensor Systems	Approximate Wavelength Interval (micrometers)	Approximate Spatial Resolution Attainable (milliradians)	Atmospheric Penetration Capability*	Day-Night Capability	Real-Time Capability+	Geometric Rectification++
Ultraviolet (Optical-mechanical scanners, image orthicons, and cameras w/IR film)	0.01-0.4	0.01-0.1		Day only	Yes	Good
Visible (Optical-mechanical scanners, conventional cameras with film, and vidicons)	0.4-0.7	0.01-0.001	H	Day only**	Generally no[x]	Potential metric quality
Reflectance IR (Conventional cameras w/IR sensitive film, solid-state detectors in scanners and radiometers)	0.7-3.5	0.01-0.1	H, Sg	Day only	Generally no[x]	Potential metric quality
Thermal IR (Solid-state detectors in scanners and radiometers, quantum detectors)	$3.5\text{-}10^{-3}$	1.0	H, S	Day or night	Yes	Good
Microwave (Scanners and radiometers, antennas and circuits)	$10^3\text{-}10^6$	10	H, S, F	Day or night	Yes	Poor/fair
Radar# (Scanners and scatterometers, antennas and circuits)	8.3×10^3 1.3×10^6	10	H, S, F, Rϕ	Day or night	Potential exists	Fair

TABLE 2.3 (Continued)

*   Denotes the atmospheric conditions which can be penetrated by energy in this portion of the electromagnetic spectrum where:

    H = haze, S = smoke, Sg = smog, F = fog or clouds, R = rain.

\+   This refers to the ability to evaluate a sensor system's output as the original information is acquired.

\#   While radar operates within the microwave region, its utility is significantly different than that of radiometers.

ϕ Penetration capability increases with increasing wavelength.

x   The potential for real-time viewing exists in scanner systems, and panchromatic film could be viewed in near-real-time utilizing a Bimat type of process.

++   Denotes the potential for plannimetric mapping.

**   Discounting the use of active optical systems such as the Edgerton flash units or laser line tracers or light amplification systems.

Resolution in the short wavelengths is limited primarily by atmospheric scattering

$$S = k_1 \cdot \frac{1}{\lambda^4} \qquad (14)$$

where S = scattering
       K = constant

Resolution in the long wavelengths is limited primarily by aperture of the sensors

$$R = k_2 \cdot \frac{\lambda}{D} \qquad (15)$$

where R = resolution
       D = diameter of the "collection optics"

As electromagnetic radiation passes through the atmosphere, a complex process of scattering, reflection, and absorption alters the amount of solar energy striking the earth. Also, alteration occurs as energy reflected from or emitted by a feature on the earth's surface travels back through the atmosphere and is recorded by a sensing device, such as a camera or radiometer. In the visible and near-visible portions of the spectrum, scattering is the chief cause of the reduction of energy. In the infrared, absorption is the chief cause. At wavelengths longer than 18 mm (in the microwave portion of the electromagnetic spectrum), there is no appreciable atmospheric attenuation (fig. 2.12).

Scattering

One very serious effect of the atmosphere is the scattering of radiation. Scatter differs from reflection in that the direction associated with scatter is unpredictable, whereas the direction of reflection is predictable. From the standpoint of remote sensing, there are essentially three types of scatter: Rayleigh scatter, Mie scatter, and nonselective scatter.

Rayleigh scatter is the result of the presence in the atmosphere of molecules, and other very small particles, many times smaller than the wavelength of radiation under consideration. Rayleigh scatter involves reradiation by atoms. Basically, this is true of all scatter; that is, it is accomplished through absorption and reemission of radiation by atoms or molecules. It is impossible to predict the direction in which a specific atom or molecule will emit a photon. Rayleigh scattering is inversely proportional to the fourth power of the wavelength. As an example, ultraviolet light (which is about one-fourth of the wavelength of red light) is scattered 16 times as much as red light and about 4 times more than blue light. Rayleigh scatter is most ob-

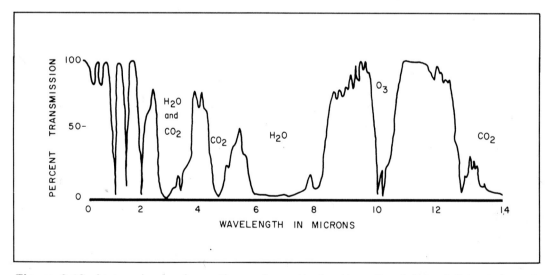

Figure 2.12. Atmospheric absorption schematic for the ultraviolet, visible, infrared, and that portion of the thermal infrared recorded by remote sensors ($3.5\,\mu m - 14\,\mu m$). (Adapted from Estes, 1974)

vious on clear days when there are few water vapor and dust particles. Since the scattering of blue light is greatest, it appears to be coming at an observer from all parts of the hemisphere, giving the sky its blue appearance.

Mie scatter occurs when there are essentially spherical particles present in the atmosphere with diameters approximately the wavelength of visible radiation. Thus, water vapor, dust, smoke, and other particles ranging from a few tenths of a micron to several microns in diameter are the main scattering agents. When the atmosphere contains large amounts of such particles, the effects of Mie scatter may exceed those of Rayleigh scatter. As the wavelengths scattered by Mie particles are longer, the sky will take on a reddish appearance. Mie scattering is restricted generally to the lower atmosphere—below 15,000 feet (4,572 meters); Rayleigh scattering occurs to an altitude of some 30,000 feet (9,144 meters). Above that altitude, there is little atmospheric scattering.

Nonselective scatter, with respect to wavelength, occurs when particles in the atmosphere are several times the diameter of the radiation being transmitted. Water droplets, which make up clouds, scatter all wavelengths of visible light equally. This causes clouds and fog to appear white (a mixture of all colors of light in approximately equal quantities produces white).

Atmospheric scatter must be considered when planning any remote sensing mission or in the analysis of any remotely sensed data. An understanding of the types of scatter is necessary before a student can choose film or filter systems to accomplish special effects or to prevent degradation of the image under certain conditions. By knowing the film/filter configuration from which images are acquired an analyst can evaluate the potential signature degradation which might be caused by atmospheric scattering.

Absorption

Absorption is the process by which radiant energy that is neither transmitted nor reflected is converted into other forms of energy. A substance which absorbs energy may also be a medium of refraction, diffraction, or scattering; these processes, however, involve no energy retention or transformation, and are to be clearly differentiated from absorption. Certain wavelengths of radiation are affected far more by absorption than by scatter. This is particularly true of infrared wavelengths shorter than visible light. Figure 2.13 illustrates the absorption-transmittance effects of the atmosphere as well as the principal substances responsible for the absorption bands. These absorption bands are ranges of wavelengths (or frequencies) in the electromagnetic spectrum within which radiant energy is significantly absorbed by gases such as CO_2, H_2, O_2, etc.

Absorption occurs principally when incident energy is of the same frequency as the resonant frequency of an atom or molecule and produces an excited state. If a photon of the same wavelength is not excited, the energy is transformed into heat motion and is eventually reemitted at a longer wavelength.

Refraction

Refraction refers to the bending of "light" when it passes from one medium to another. Refraction occurs when electromagnetic radiation passes through media which are of differing densities, modifying the speed of propagation. The index of refraction of a given substance is a

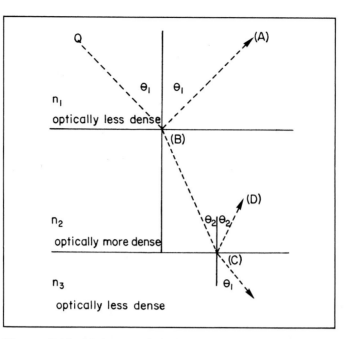

Figure 2.13. Light passing from n_1 to optically more dense n_2 may be either reflected as in (A) or refracted as in (B). Similarly, refraction will also occur as the light ray passes again into optically less dense n_3 again at (C), with reflection seen again at (D). (Adapted from Nunnally, 1969)

measure of optical density. It is the ratio of the speed of light (c) in a vacuum to the speed of light in the substance (c_n). Thus:

$$n = c/c_n \qquad (14)$$

where: n = the index of refraction

Refraction as described by Snell's Law states that for a given frequency of light, the product of the index of refraction and the sine of the angle between the ray and a line normal to the interface is a constant:

$$n_1 \sin\theta_1 = n_2 \sin\theta_2 \qquad (15)$$

From figure 2.13 the important role refraction plays in the atmosphere is obvious. A non-turbulent atmosphere can essentially be thought of as a series of layers of gases of differing densities. Anytime energy at an angle of incidence other than 0° is passed through such a medium, refraction occurs.

The specific amount of refraction that will occur is a function of the angle made with the vertical, the distance involved (typically in the atmosphere, the greater the distance, the more changes in density occur), and the densities of the air involved. Serious errors of location due to the refraction can occur in images formed from energy detected at high altitudes or at acute angles. These locational errors, however, are generally predictable and can be significantly removed during processing of the images. A much more difficult problem is refraction in a turbulent atmosphere. Turbulent motions are random; therefore, under these conditions the bending of the wave front is unpredictable, and accurate location of ground points is difficult.

Finally, illustrating the complex energy interactions which occur as energy passes through the atmosphere, figure 2.14 presents two generalized total atmospheric energy flow diagrams: a simplified clear sky example, and a more complex (more typical) normal sky example. Let us examine the two basic hypothetical energy flow examples presented in figure 2.14. In both we will assume that the total radiance of the sun directed toward and arriving at the outer reaches of our atmosphere to be 100 units.

Clear Sky

On a clear day 80 of the solar radiance units will penetrate to the earth's surface; 6 units will be lost to scattering and diffuse reflection returning to free space as short wavelength radiation; 14 units will be absorbed by atmospheric molecules and dust.

Normal Sky

Under normal atmospheric conditions, of the 100 solar radiance units arriving at the outer limits of the atmosphere, 50 units are transmitted through to reach the ground; 32 units are reflected: 5 by atmospheric scatter (Rayleigh and Mie), 21 by clouds (nonselective scatter), and 6 by the surface. Of our original 100 solar radiance units, 18 units are directly absorbed in the atmosphere: 15 by atmospheric molecules and dust particles, and 3 by absorption from clouds. These are added to 20 units from the 50 transmitted to the ground produced by latent heat of transfer from the surface to the atmosphere, and 9 units of mechanically transferred energy from the ground to the air. From the original 50 units striking the surface, a total of 47 energy units are absorbed directly into the atmosphere. It is important to note that our atmosphere also acts as a collector and recirculator of the energy within it. At any given time more than the original 100 solar radiance units may be present in the atmosphere. Typically the figure is on the order of 137 units.

At this point there are 21 energy units of the original 50 transmitted to the ground left. To these 21 units are added 77 units of counter, or indirect, radiation from the atmosphere to the earth. This produces a total of 98 radiance units absorbed by the surface at this point. These 98 radiance units are radiated as longer wavelengths of radiation to the atmosphere, and 90 of these units enter into the collection/recirculation process of the atmosphere (this, along with the 47 units of absorbed energy from our original 100 solar radiance units, gives us the atmospheric collection/recirculation total of 137 radiance units). Finally, 8 of the 98 exited units escape directly into space; 60 of the 137 atmospheric units (balancing the 77 returned to earth) also escape into space. This total of 68 long wavelength radiation units which are lost to space combined with the 32 short wavelength radiation units produce a total of 100, balancing the total energy flow depicted in figure 2.14.

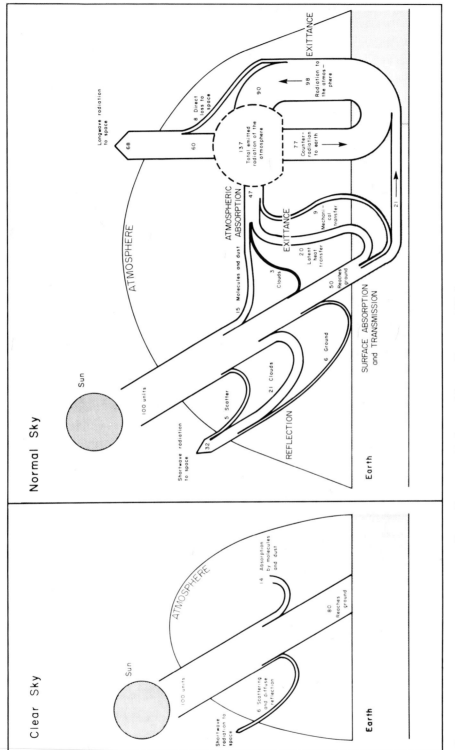

Figure 2.14. Atmospheric energy flow diagram. (Adapted from Estes, 1974)

Conclusion

This chapter has presented the energy concepts and parameters important for a basic understanding of imaging with photographic and nonphotographic sensor systems. Multiband/multispectral scanning may enhance the detection and identification of objects since improved contrast ratios may exist in discrete bands of the electromagnetic spectrum. Differential absorption, reflectance, and transmission may also facilitate or impede information extraction and ease of interpretation. Remote sensing is an important environmental analysis technique/tool and users should be aware of both the potentials and pitfalls associated with its use. An understanding of the concepts presented will move the user significantly closer to such and understanding.

Readers are encouraged to refer back to specific concepts in this chapter when studying the sensor systems and methods of remote sensing interpretation which are presented in the following chapters. By doing this, they will begin to make the relationship between the characteristics of the electromagnetic spectrum and the images which it produces at a remote detector.

Notes

1. In this and later sections of this chapter a number of sources have been drawn on heavily. Acknowledgment is made here of materials from: J. E. Estes, "Imaging with Photographic and Nonphotographic Sensor Systems," in *Remote Sensing: Techniques for Environmental Analysis,* eds. J. E. Estes and L. W. Senger (Santa Barbara, Calif.: Hamilton Publishing Co., 1974), pp. 15-50; G. Feinberg, "Light," *Scientific American* 219 (1969):50-59; R. K. Holz, ed., *The Surveillant Science, Remote Sensing of Environment* (Boston: Houghton Mifflin Co., 1973); N. Jensen, *Optical and Photographic Reconnaissance Systems* (New York: John Wiley and Sons, 1968); N. R. Nunally, *Introduction to Remote Sensing: The Physics of Electromagnetic Radiation* (Johnson City, Tenn.: East Tennessee University, under U.S. Geological Survey contract to the Association of American Geographers, Commission on Geographic Applications of Remote Sensing, 1969); R. J. Pogorzelski and K. A. Shapiro, "Introduction to the Physics of Remote Sensing," in J. Lintz, Jr. and D. Simonett, *Remote Sensing of Environment* (Reading, Mass.: Addison-Wesley Publishing Co., 1976); and in R. G. Reeves, ed.-in-chief, *Manual of Remote Sensing* (Falls Church, Va.: American Society of Photogrammetry, 1975) the following chapters: chap. 3, G. H. Suits, "The Nature of Electromagnetic Radiation;" chap. 4, F. J. Janza, "Interaction Mechanisms"; chap. 5, R. S. Fraser, "Interaction Mechanism within the Atmosphere"; chap. 6, P. N. Slater, "Photographic Systems for Remote Sensing."

3

Aerial Photograph Interpretation as Remote Sensing

Gary Whiteford

University of New Brunswick, Canada

Introduction to Aerial Photographs

The use and analysis of aerial photographs was meaningfully developed during World War II as a direct result of the sophisticated war photographic intelligence systems. During the past twenty-five years, however, the uses of aerial photography have become more commonplace in the nonmilitary professions. With this increase in civilian use of aerial photographs, there has also been important technological progress in types of aircraft, aerial cameras, and viewing mechanisms. All of these contributions have made the science of aerial photograph interpretation an integral part of many educational and research programs, with the anticipated goal of providing scientists with a spatially balanced mental image with which to view the real life environment.

The aerial photograph is an important vehicle by which observational and perceptual capabilities can be heightened. When used in conjunction with satellite images, aerial photographs provide practical and realistic methods to give observers the opportunity to view, interpret, and understand the landscape features of the earth. The ability to effectively understand an aerial photograph becomes a common denominator for such diverse fields as transportation planning, recreational management, anthropological studies, and many more. Table 3.1 indicates some areas of study in which aerial photographs are employed.

It is evident that aerial photography has far-reaching applications. With it, the various subject fields are readily bridged out and a more ecological understanding of the earth's systems is gained.[1] In no other way can the interplay of man/land relationships be fully realized and appreciated. The prerequisite for all this is normal vision and an aerial photograph, or stereoscopic pairs of aerial photographs. Barry Commoner, a noted biologist/environmentalist stated, ". . . the natural tendency to think of only one thing at a time is a chief reason why we have failed to understand the environment and have blundered into destroying it."[2]

The use of the aerial photograph allows large areas of the surface of the earth to be imaged on a single frame, which at a scale of 1:20,000, the common format, would

TABLE 3.1

Some Uses of the Aerial Photograph

Beach studies
Disease and insect identification
Fire monitoring
Flood control
Forestry studies
Geologic studies
Glacial studies
Irrigation analysis
Land resource potential
Land use analysis
Property boundaries and valuations
River movements
Soil surveys
Tidal studies
Traffic studies
Transportation studies
Urban studies
Wildlife protection and management

represent a ground area of about 23 square kilometers (9 square miles). High altitude photography can yield scales of 1:60,000, registering a ground area of about 189 square kilometers (73 square miles). By overlapping pairs of aerial photographs, a three-dimensional view of the physical and cultural features of the earth can be obtained. The aerial photograph provides a permanent record of an earth scene, and it can be stored without deterioration. It allows opportunities for time-sequential studies of the same area to be made over an extended period, and, of course, different seasons of the year can be compared using time-sequential photographs.

Aerial photographs not only permit a sizeable and significant portion of the surface of the earth to be viewed and studied, but at the same time they permit the astute observer to identify and analyze relationships among features on the ground that cannot be observed from a ground level perspective. The aerial photograph images the ground scene as a continuous mosaic, with all of the cultural and physical features interrelated. Those features which occur together on the ground are seen together on the aerial photograph.

Aerial photographs are easily obtained both in Canada and the United States. More detailed information about how such photographs can be ordered appears in Appendix A of this book. It is best to order materials through local state, provincial, or federal governmental or private agencies, thereby insuring a professionally consistent quality of the imagery. Although it is possible for individuals to take their own aerial photographs,[3] for most persons a government agency becomes the quickest, cheapest, and most efficient source of aerial photographs.

Vertical and Oblique Aerial Photographs

Aerial photographic interpretation from low to medium altitude images probably is the best method by which to introduce people to remote sensing. Aerial photographs may be either vertical or oblique in orientation. Vertical aerial photographs are those imaged with the camera axis perpendicular, or nearly perpendicular, to the surface of the earth. Figures 3.1 and 3.2 are vertical images at different scales of the same area. Vertical aerial photographs can be used effectively to study earth features because they are easily obtained at relatively small cost and they are available at relatively large scales. Furthermore, surface features on vertical aerial photographs are relatively easy to identify with some experience and practice, and measurements, including distances, can be made on them.

Successive overlapping vertical aerial photographs along a flight line provide stereoscopic views of physical and cultural features. Large areas can be studied by preparing uncontrolled mosaics from vertical aerial photographs. Figure 3.3 represents an uncontrolled mosaic of part of a county in Wisconsin. Regional detail can be interpreted when vertical aerial photographs are assembled into mosaics. The mosaic in figure 3.3 is an index sheet from which individual aerial photograph frames can be ordered from the appropriate government agency.

Uses of vertical aerial photographs include:

1. Studies of landform features and patterns
2. Studies of cultural features and patterns
3. Studies of vegetation types and areas
4. Studies of soil patterns and associated land uses
5. Studies and measurements of drainage areas and patterns
6. Measurements of heights of objects or elevation of topography
7. Studies of correlations between map symbols and the reality of phenomena imaged on the photographs

Oblique aerial photographs are made when the camera axis is positioned at an oblique angle to the surface of the earth (fig. 3.4 and plate 2). Low oblique aerial photographs are those on which the horizon does not appear (fig. 3.4), and high obliques are those where the horizon is included on the photograph (fig. 3.5). Oblique aerial photographs can be made from any high-wing aircraft and with the use of any type of hand-held camera.

Uses of oblique aerial photographs include:

1. Topographic and drainage basin studies in areas of low relief
2. Limnologic studies
3. Rural and urban landscape studies
4. Crop and vegetation identification
5. Landscape changes from season to season and from year to year

How Aerial Photographs Are Made

Figure 3.6 illustrates the position of the aerial camera for the purpose of obtaining the various types of aerial photographs. In Canada, most low altitude imagery is obtained from such aircraft as the Beechcraft and Apache, whereas high altitude imagery is obtained using the

Figure 3.1. Vertical photo of St. John, New Brunswick, and mouth of the St. John River. (Photo courtesy of Maritime Resource Management Service, Council of Maritime Premiers, Amherst, Nova Scotia)

Figure 3.2. Vertical photo of St. John, New Brunswick, and mouth of the St. John River. (Photo courtesy of Maritime Resource Management Service, Council of Maritime Premiers, Amherst, Nova Scotia)

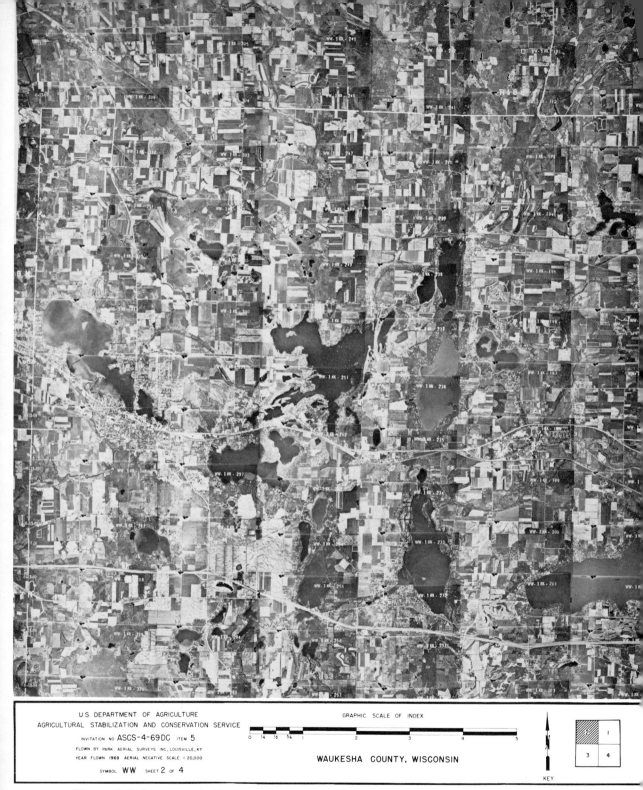

Figure 3.3. An uncontrolled mosaic prepared from vertical aerial photographs of part of Waukesha County, Wisconsin. (U.S. Department of Agriculture, Agricultural Stabilization and Conservation Service)

Figure 3.4. Low oblique of Fredericton, New Brunswick, and part of the St. John River. (Photo courtesy of Maritime Resource Management Service, Council of Maritime Premiers, Amherst, Nova Scotia)

Figure 3.5. High oblique aerial photograph of the Milwaukee harbor area, April, 1975. The horizon is included in high obliques. (Photo by Aerial Photo Interpretation class, Carroll College, Wisconsin)

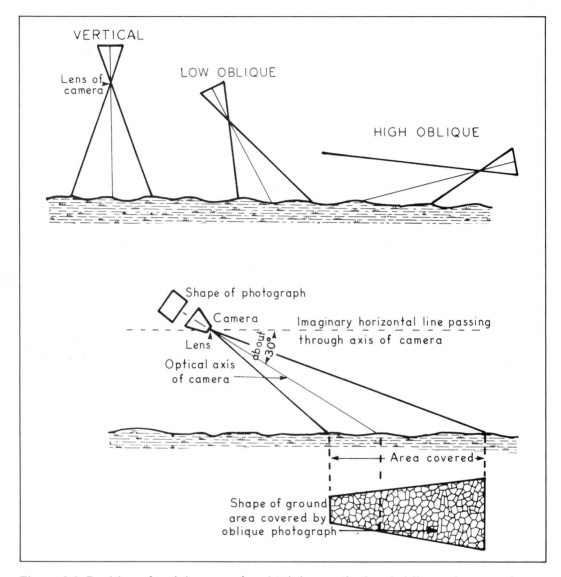

Figure 3.6. Position of aerial camera for obtaining vertical and oblique photographs.

Falcon or Convair 580. In the United States, high altitude imagery is made from such aircraft as the U-2 and RB-57F.

Figure 3.7 shows an aerial photograph taken in 1883 of the Halifax Citadel area in Halifax, Nova Scotia. Even at this early date the clarity is remarkable. The feat of obtaining an aerial photograph such as this at that date with unsophisticated airborne platforms and cameras was an accomplishment in itself.

The advantage of vertical aerial photographs is that they can be used easily with road maps, topographic maps, military maps, and thematic maps. Vertical aerial photographs pro-

Figure 3.7. First known aerial photograph taken in Canada from a captive balloon by Captain H. Elsdale of the Royal Engineers in August, 1883. Camera was suspended from the balloon and the picture shows the Ellerslie Barracks of the Halifax Citadel, Halifax, Nova Scotia. (Photograph courtesy of National Air Photo Library, Ottawa, Ontario)

vide easy yet meaningful transposition of information concerning features on the earth. On the other hand, the vertical (or orthogonal) view of the earth is not the natural way in which people view objects. For this reason, images of objects on vertical aerial photographs may not be recognized easily.

High oblique aerial photographs give a view similar to looking from an aircraft window out toward the horizon. Both the distant horizon and the immediate terrain are in evidence, whereas with the low-oblique, only the terrain is viewed. Scale differences, however, are very noticeable in both, so that distant objects appear smaller than objects in the foreground.

Camera Lenses

Most domestic aerial photographic coverage seems best accomplished by use of long camera focal-length lenses of 210 millimeters (8.25 inches) or longer and flying at altitudes giving scales of 1:20,000 or larger. The focal length is the distance measured along the lens axis between the camera lens and the focal plane where the film is located. Short focal-length lenses of 152 millimeters (6 inches) or shorter and flying at appropriate heights give photographic scales generally of 1:20,000 or smaller.

The flight path for taking aerial photographs consist of a series of parallel flight lines so arranged as to obtain overlapping photographs that can be used for stereoscopic viewing. Generally aerial photographs are taken sequentially to provide a 60 percent forward overlap. Adjacent flight lines are spaced to give a sidelap of 25 to 30 percent. This sidelapping of flight assures that some continuity of photographic coverage will be made across a given area.[4] Figure 3.8 illustrates the pattern of an idealized flight plan.

Aerial Photographic Film

Conventional aerial photographs are imaged on panchromatic film which is sensitive to all wavelengths of the visible portion of the electromagnetic spectrum—that is, it "views" approximately the same range of visible light as the human eye.

All objects on the surface of the earth have characteristic reflectances, only a part of which are visible to the human eye—the colors violet, blue, green, yellow, orange, and red. People can only identify the visible light region of this electromagnetic energy. Figure 2.1 diagrams this electromagnetic spectrum and the limited field of "vision" a person is able to comprehend. This visible region consists of short wavelengths and long wavelengths of energy as illustrated in figure 3.9.

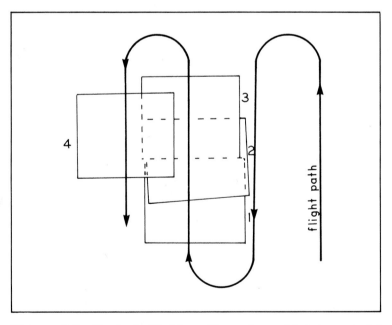

Figure 3.8. Typical flight pattern for obtaining aerial photographs for stereoscopic viewing.

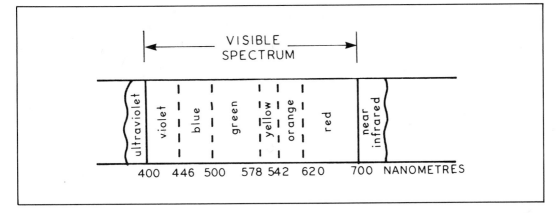

Figure 3.9. Diagram illustrating visible part of spectrum.

In aerial photography it is necessary to block out certain light rays of the visible spectrum because they prevent some features from registering on the film negative. Because blue light rays, which are short wavelength, tend to give a haze to the film, obscuring ground details, a minus-blue (or red) filter is placed over the lens. Such a filter allows long wavelengths, such as green and red, to pass through to the film. Images on panchromatic black and white photographs are shown in varying shades of gray, with each tone comparable to the reflectance from an object in the visible spectrum. Panchromatic film is excellent for identifying features of different colors, but is insensitive to green light rays.[5]

Infrared film is also used in aerial photography. It makes possible the sensing of images in the infrared portion of the electromagnetic spectrum.

Filters for color films are different from those used with black-and-white film. As Avery asserts, "Scattering of the short, invisible wavelengths of ultra violet light increases the haze effect on color film; thus a desirable filter should absorb all ultraviolet and as much blue light as is required for a correctly 'balanced' color transparency."[6]

Format of Aerial Photographs

The conventional vertical aerial photograph is made on a film format which measures 22.86 centimeters (9 inches) on a side. Variations in dimensions of paper positive prints result in slight cropping in reproduction. People who use aerial photographs should become familiar with the symbols on the format which provide information about the frame.

Figure 3.10 illustrates one frame of a vertical aerial photograph taken along a mission flight line. Observe the black registration marks at the geometric center on each side of the image. These are called *fiducial marks*. The intersection of lines drawn from opposite fiducial marks locates the *principal point* (PP) of the photograph (fig. 3.11). On figure 3.1, the fiducial marks are V-shaped indentations in the edges of the photograph. On some vertical aerial photographs, Xes are placed in the corners for locating the principal point. The PP of a photograph can be located by connecting opposite fiducial marks or opposite registration marks.

Figure 3.10. Vertical aerial photograph of coal strip mining near Coal City, Illinois. (U.S. Department of Agriculture, Agricultural Stabilization and Conservation Service)

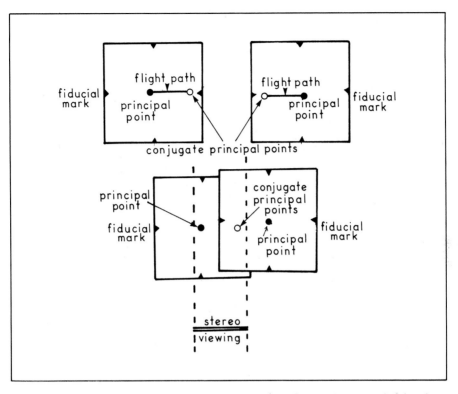

Figure 3.11. Orientation of aerial photographs for stereoscopic viewing.

The principal point of a vertical aerial photograph is the geometric center of the frame. If the airplane in which the aerial camera was mounted was in level flight at the time of exposure, the principal point of the photograph will correspond to the *nadir*. The nadir is that point on the ground which is directly below the lens of the aerial camera. An image at the principal point of one aerial photograph will also appear on a successive photograph taken along the flight track because of the interval at which exposures are made to produce overlap between photographs. The point on a vertical aerial photograph which corresponds to the principal point on a preceding frame along a line of flight is called the *conjugate principal point* (CPP) (fig. 3.11) A line connecting the PP and CPP on a vertical aerial photograph is parallel to the line of flight. The distance between the PP and CPP on a single frame is the *photograph base length,* or *absolute stereoscopic parallax.*

The numerals and letters which appear along the top edge of a vertical aerial photograph provide information about the frame. From left to right, the following data is given in order: date of photography, time of exposure, agency for which the photography was done, the nominal scale of the image, project code, magazine number (or flight line), and frame number. This information appears only on frames at the beginning and ending of a flight line. For example, the photograph in figure 3.10 was made on June 2, 1967, at 11:16 A.M. The agency for which the photograph was made was the Agricultural Stabilization and Conservation Service of the U.S. Department of Agriculture. The nominal scale of the photograph is 1:20,000. The project P is the code for an area near Coal City, Illinois, while the magazine number was 3HH, and the frame number (exposure along the flight line) is 139.

On all other frames along the flight line only the date, project code, magazine number, and frame number are given. The uncontrolled mosaic index sheet illustrated in figure 3.3 shows the coding of frames along a flight line. The beginning and ending of flight lines were beyond the limits of this index sheet. When aerial photographs are ordered from a government agency, an index sheet, such as the one illustrated in figure 3.3, would be consulted. From the index sheet individual frames or overlapping frames can be ordered by their codes. In order to determine how much area a single frame includes, take a divider and lay off the width of one frame, such as frame WW-LKK-317 in the lower left of the index sheet. That dimension will be equal to the dimensions of all sides of the photograph. It can be seen that some images on frame 317 also will be present on frames 318 and 319.

Parallax

An important property of vertical aerial photographs is *parallax*. The only part of the vertical aerial photograph that is orthogonal is at the principal point. If the nadir and the PP are the same, an object viewed at the point will be seen directly from above. For example, a smokestack at the principal point would appear as a circle. All other images on vertical aerial photographs, however, show linear displacements, similar to images on oblique aerial photographs. In other words, all images on vertical aerial photographs, except those at the principal point, will "lean" away from the center of the photograph, and the top and bottom of objects, such as buildings, smokestacks, trees, and cliffs, can be seen. The displacement of images on vertical aerial photographs is parallax.

Also, parallax is produced on successive vertical aerial photographs because of the change in position along a flight line by the camera-carrying aircraft. The displacement, or parallax, is caused by the change in observation point. Conjugate principal points which represent the principal point on the previous photograph are caused by aircraft progression.

Because of parallax, a vertical aerial photograph cannot be thought of as a map. On maps all points appear orthogonal, that is, directly below the observer. Parallax of images on vertical aerial photographs is not a liability, however. It makes possible stereoscopic vision by which successive overlapping vertical photographs along a flight line can be seen in three dimensions by use of a stereoscope.

Two measures of parallax can be obtained in determining heights of objects on stereoscopic pairs of photographs. Absolute stereoscopic parallax is the average of the distances between the principal points of corresponding images and conjugate principal points on successive, overlapping vertical photographs. Differential parallax is the difference in the displacement on the photograph of the top and bottom of a feature being measured. Using parallax measurements, the height of an object (Ho) can be obtained using the formula:

$$Ho = \frac{H \times dP}{P + dP}$$

where, H = height of aircraft above ground (in meters or feet)
   P = absolute stereoscopic parallax (in millimeters or inches)
  dP = differential parallax (in millimeters or inches)

For example, if the flying height of a camera-carrying aircraft was 7,000 meters (22,966 feet) at the time of exposure, and the absolute stereoscopic parallax, or photo base length, is found to be 90 millimeters, and the linear measurement on the photograph between the top and bottom of a cliff is 10 millimeters, the height of the cliff is 700 meters, or 2,297 feet.

Scale

The scale of a vertical aerial photograph is found by dividing the focal length of the camera lens by the height the aircraft was flying. Both measurements should be in the same unit. Because scale is expressed as a fraction, both sides of the fraction must be divided by the numerator. For example, if the focal length is 152.4 mm and the flying height is 3,048 meters, the scale would be:

$$S = \frac{F}{H} = \frac{152.4}{3,048} = \frac{\dfrac{0.1524}{0.1524}}{\dfrac{3,048}{0.1524}} = \frac{1}{20,000}$$

Figure 3.12 shows the relationship between focal length and height above ground of the aircraft. All relationships in the geometry of vertical aerial photographs can be referred to

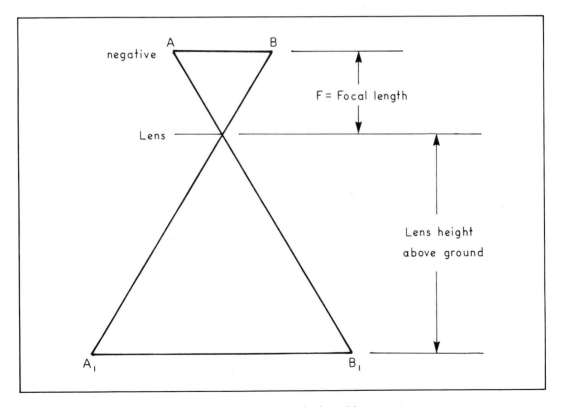

Figure 3.12. Focal length and ground height relationship.

similar triangles. For example, if the focal length or flight altitude are unknown, then the scale of the aerial photograph can be calculated as follows:

$$\text{Representative Fraction (RF)} = \frac{\text{photographic distance between two images}}{\text{ground distance between corresponding objects}} \text{ or } \frac{AB}{A_1B_1}$$

All measurements must be in the same unit.

The selection of the two points should be made such that they are aligned in a straight line, using highways, railways, or section lines. For example, if the image distance is 10.0 millimeters between two roads, and the two roads are 200 meters apart on the ground, then the scale would be:

$$S = \frac{AB}{A_1B_1} = \frac{10.0}{200} = \frac{10.0}{200,000} = \frac{1}{20,000}$$

The scale of a vertical photograph is not the same for all parts of the photograph, primarily because of differences of relief. The scale on a vertical photograph will not be the same at the top of a hill as at the location of an adjacent valley or gorge. Because of variations in elevation of topography from one area to another on the same vertical aerial photograph, the nominal scale is usually computed. The nominal scale is determined by measuring distances between identifiable images in different parts of the photograph. The average of the computed scales gives the nominal scale.

Stereoscopy

A very important characteristic of aerial photographs is that by overlapping successive pairs of aerial photographs and viewing the pairs through a stereoscope, a three-dimensional view of the surface of the earth is obtained. By performing a simple "floating finger test," it is possible for potential viewers to determine in advance if they can see stereoscopically. Pointing the index fingers at each other, place the tips together and hold them about one-third of a meter (about one foot) from your eyes. Focus the eyes not on the fingers, but on some distant object. Viewers who can see stereoscopically will see a "third finger" develop between the ends of the index fingers. By pulling the index fingers slightly apart, a "floating finger" should appear.

The preparation of photographs for stereo-viewing and the use of the stereoscope are familiar to many high school and college students. The reader is referred to Avery's brief explanation if review of the method is necessary.[7]

Implementation of Aerial Photographic Flights

Aerial photographic missions comprise detailed and sometimes very elaborate procedures. Especially important is the time of year when the photographs are taken. Summer flights allow for maximum coverage of various field crops and vegetation patterns, late fall gives an indication of terrain features normally hidden by forest and brush cover, and winter photographs allow the observer an opportunity to chart ice flows and snow cover extent. Most planning offices, however, prefer spring imagery because the ground is not obscured by foliage.

The normal flight plan consists of parallel runs so organized as to obtain an overlap of the photographs so that stereo-viewing can be achieved (fig. 3.8). Beforehand, the study area and flight lines are marked on a flight map, with the flight lines parallel to the study area boundaries. A straight and level flight is desired, with the elevation height kept as constant as possible. Cross winds and drift do occur; therefore, a careful check must be made of ground observation points either directly or by use of such visual aids as drift meters and view finders. The use of filters over the camera lens allows the operator to select the light wavelengths for stated tone contrast necessary for the selection of certain cultural features.[8]

There seems little doubt that aerial photography is important for a variety of reasons. At best, it becomes a most expedient means to update, confirm, and reaffirm material contained on conventional and special purpose maps. It allows observers to use their perceptual abilities and skills to draw their own conclusions about selected areas of the surface of the earth. The aerial photograph is a valuable aid in ground truthing for high altitude and space imagery. Used with the conventional topographic map, the user can become a "walking interpreter" of the landscape. Discovery becomes all the more meaningful because the "doing" process is an involvement process.[9]

Interpretation of the Aerial Photograph

Once the photo interpreter has mastered the techniques of observation, the photographic qualities of the man/land features can be investigated. Viewed by the unaided eye, with a magnifying glass, or under the lens stereoscope, the observer can begin to appreciate and understand landscape arrangements. Objects viewed on most aerial photographs occur at small scales, and because of the factors of scale and the vertical view, some objects assume greater or less importance than in the ground level view. A number of special characteristics of aerial photographs can be discerned which will lead the interpreter to understand what is being viewed.

Size

Size of an object is a useful clue to help in identifying objects. By using shadow and parallel measurements, some careful considerations can be made concerning identification. In general, cultural features have some degree of uniformity as compared to natural features; for example, consider telephone poles and clumps of trees. Size of objects on aerial photographs is related to scale. Before relative sizes can be considered, the scale of the photograph will have to be computed. In this way houses, barns, or warehouses might be distinguished by their size.

Shape

The top view of an object seen vertically can be quite different from the ground level or oblique view. Yet it is the vertical view which can provide the experienced interpreter with more information than that gained from the ground level perspective. Railroads and roads reveal a definite form and function when viewed from above. Railroads appear as narrow bands with smooth gradual curves, whereas roads are wider with sweeping curves. A volcanic cone from above reveals spectacular tectonic and erosional forces at work.

Shape is important because it allows for frequent and conclusive identification. In stereo-viewing, many topographic features (hills, valleys, cliffs), vegetational features (shrubs, trees), and cultural features (buildings, roads) are readily identified because of a characteristic three-dimensional shape.

Tone

In black-and-white aerial photography objects are observed in tones of gray. The tones in any given photograph are influenced by many factors which are directly related to the reflective capacity of the object viewed at any given time. For example, a body of water can range from a white to black tone depending on the angle of the sun and wave reflectivity. Under wet conditions an area may appear dark, while in dry periods the same area may appear white. Observe the different tones of water resulting from sun angle between the spoil banks in the strip mines on figure 3.10.

In general, objects reflecting considerable sunlight back to the camera will appear light in tone on positive prints, as compared to objects which appear dark in tone on positive prints because they absorb light. Smooth surfaces such as lawn, open fields, and roads appear as light shades of gray. Rough surfaces such as forests, crops near harvest, and pastures appear as dark tones. It is apparent that tone by itself is not a consistent characteristic, and the same area at different times may register a wide variety of tones in part because of the influence of the sun.[10] For color aerial photography, however, tone is a very valuable guide to interpretation because the spectral reflectivity plays a more prominent role than on black and white images.

Shadow

Many features can be recognized from their shadows, such as water towers, monuments, trees, bridges, and oil storage tanks. If objects are small and their tones are unclear, then the investigation of shadow profiles may be the necessary clue for identification. Shadow identification becomes important for urban areas containing significant open space, such as parking lots, where the shadow detail is not lost. Of course, the use of the stereoscope is very helpful in recognizing what the shadow represents. Observe the railroad bridge on figure 3.2. Much more can be determined about the bridge from its shadow than from the image of the bridge itself. The shadow shows that the bridge is supported by nine pilings, and that the largest over-water span is across the surging tidal rapids of the St. John River. This type of information could not be obtained if the shadow of the bridge was not present.

Texture

Texture may be described as smooth, fine, rough, coarse, lumpy, stippled, mottled, rippled, and so on (fig. 3.13). Like tone, it may vary from part of an area to another depending on the relative location of the sun and the camera. Texture, for example, can be very important in the study of fields because mottled textures indicate variations in soil moisture content. A lined texture results if plowing, planting, and harvesting are in parallel rows. Further, texture can be used to distinguish age of forest stands, with a young stand being finely textured and a mature growth more coarsely textured. Observe illustrations of these types of texture on figures 3.14 and 3.15.

Pattern

Spatial arrangements of objects on the surface of the earth are important in any analysis of photograph interpretation. For example, settlement patterns, rock outcrop patterns, and forest patterns give appropriate clues concerning possible origins and functions (fig. 3.16). Patterns which might be indistinguishable at ground level are apparent on the aerial photography. Pattern identification emphasizes the interrelationships of the cultural and natural features, which

Figure 3.13. Vertical aerial photograph Grand Falls area of New Brunswick. The new forested plantation areas show as light texture, the dotted texture indicates cut-over area, and the darker texture, covering most of the image, indicates nature forest. (Province of New Brunswick, Department of Natural Resources)

Figure 3.14. This oblique aerial photograph illustrates different textures and patterns of land in southeastern Wisconsin. The photograph was taken on April 24, 1974. Mottled soil patterns indicate differences in soil drainage. The land in the lower right is a marsh. The fields in the lower left and upper left have been planted in corn. The field in the center is pasture. Most of the deciduous trees are not in leaf, but the willows show leaf development. (Photo by B. F. Richason, Jr.)

Figure 3.15. Texture and pattern, as well as association, aid in the interpretation of this low oblique aerial photograph. The field in the lower left is alfalfa-grass hay. The farmer is cutting hay at the top of the photograph. The coarse texture of the field to the right indicates pasture. Cattle paths from the barn help in this identification, along with the cattle in the field. Understanding the scale of the photograph leads the interpreter to classify the animals as cows rather than sheep. The photograph was taken June 12, 1974. (Photo by B. F. Richason, Jr.)

Figure 3.16. Vertical air photo of the area northeast of Hartland, New Brunswick. This is a potato area of New Brunswick showing potatoes harvested in white and the darker field crop areas not yet harvested. (Province of New Brunswick, Department of Natural Resources)

can lead to a reliable assessment of the encroachment of people on the environment. For example, motor traffic patterns at rush hours over a small area may be better understood if such patterns are viewed using aerial photographs. The pattern of insect infestation in forests is more discernable from an aerial photograph, and certain areas can be marked as potential contamination zones. Agricultural lands exhibit distinct patterns and textures which, combined with association of features, makes possible the identification of agricultural activities and uses of land (figs. 3.17 and 3.18).

Figure 3.17. Texture, pattern, and association are well-illustrated in this vertical aerial photograph of dairy farms in southeastern Wisconsin. Located along the main roads are large, complex farmsteads, each with several large barns and silos. Mottled soil patterns result from alternating poorly and well-drained soils in an area of rolling ground moraine. The coarse-textured pattern in the lower left results from fields from which corn was picked the previous fall. Corn stalks are cut from these fields in the spring for cattle bedding. The uneven tracks in fields to the right of center were produced by a manure spreader. Farmers clean cow stalls and haul the manure from the barns to fields, sometimes depositing the manure on top of a snow cover. The finer textured, light colored fields have been plowed and will be planted in oats. The fine textured, dark colored fields will be planted in corn. The dark colored, coarse textured areas are hay fields. (Southeastern Wisconsin Regional Planning Commission photo)

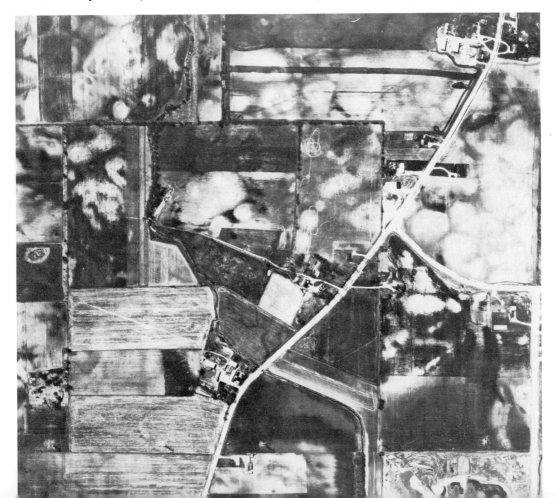

Figure 3.18. Plowing, planting, cultivation, and harvesting patterns give distinctive textures to agricultural fields. Line patterns are evident where hay has been cut from fields. Small grains show a finer grained, lined pattern. If plowing and planting have been done at right angles to each other, a screen pattern is developed. Long established fields are bounded by hedge or tree rows. Coarse textured, random patterns indicate pasture if cattle trails can be seen, otherwise the land is probably idle. (Photograph source unknown)

There is little doubt that the ability to recognize patterns from aerial photographs is a necessary requirement for explaining the spatial arrangement of features on the surface of the earth.

Association

Association is the ability of the interpreter to relate certain objects to other objects and certain land areas and, thereby, to identify a feature by its association with other features which are close to it. The presence of wide sweeping highway interchanges is perhaps indicative of a fairly large urban area; the cluster of farm buildings gives indications that silos, sheds, cow paths, water tanks, and farm machinery may be identifiable in the aerial photograph; and certain geologic formations are associated with certain drainage and vegetation patterns. This ability to associate objects to other objects is best facilitated if interpreters have at least familiarized themselves with ground level observations.

The skill of interpreting aerial photographs is a function of the ability of the observer to master selected skills, with pattern recognition and association being perhaps the two most important keys to developing competent interpretation of aerial photographs. Such abilities are the building blocks necessary for the preparation of a geographic perspective for a selected area.

Values of Aerial Photographs as Remote Sensed Images

In some cases, the use of the aerial photograph is simply a function of what is available through local supply agencies. Interpreters, however, must be aware that certain sacrifices are made if they accept such imagery. Imagery can never be considered as the "real thing." It is merely the representation of objects produced by a camera and processed by chemical means. The image is a function of many different variables operating differently at any given time period. Such variables as focal length, flight altitude, camera attitude, film type, filter type, season and sun angle all determine the image characteristics—scale, parallax, distortion, tone, shadow—present in any given aerial photograph.

In spite of the above restrictions, the aerial photograph is an effective means to detect, recognize, interpret, and analyze earth features by using a distant recorder. In essence, these features are remotely sensed either by aerial cameras on planes or by special cameras aboard orbiting satellites. For users, the aerial photograph is a very practical and economical way to become more familiar with the local environment, inasmuch as the aerial photograph readily lends itself to ground level observation, supplemented by special thematic maps.

Any investigation of the aerial photograph is really a function of resolution, that is, the ability of the entire photographic system to produce a sharply defined image. Of all remote sensors, the aerial photograph produces the best resolution of targets. This, then, is a decided advantage for photo interpreters. Depending on the size of an object to be identified, interpreters can be provided with superior ground resolution of objects. There are certain values of ground resolution at which only certain objects can be recognized. If concern is with selected aspects in topical geography, resolutions of particular objects may be critical. In regional geography, any increase in resolution increases the ability to extract information. Thus, depending on the specific purpose for the specific interpreter, it is highly likely that a given resolution of the aerial photograph may provide for a wide range of objects, extending from a large water body (RF 1/70,000) to a small dam (RF 1/10,000) for the same area.[11]

NOTES

1. Evon Z. Vogt, ed., *Aerial Photography in Anthropological Field Research* (Cambridge, Mass.: Harvard University Press, 1974). A very interesting and exciting use of aerial photographs can be seen in this book.
2. Barry Commoner, *The Closing Circle* (New York: Bantam Books, 1972), p. 23.
3. American Society of Photogrammetry, *Manual of Photographic Interpretation* (Washington, D.C., 1960), pp. 27-51. The reader can refer to these pages for an explanation of how to fly for aerial photographs.
4. Lawrence H. Lattman and Richard G. Roy, *Aerial Photographs in Field Geology* (New York: Holt, Rinehart, and Winston, 1965), pp. 15-16.

5. T. Eugene Avery, *Interpretation of Aerial Photographs,* 3d ed. (Minneapolis: Burgess Publishing Co., 1977), pp. 88-89. Chapter 6 is excellent for detailed discussion and explanation concerning types of films and filters.
6. Ibid., p. 88.
7. Ibid., pp. 25-29.
8. American Society of Photogrammetry, *Manual of Photographic Interpretation,* pp. 28-33. A more detailed explanation on executing a typical photo mission is contained on these pages.
9. Gary Manson and Ann Granacki, "Geography Is Doing," *Instructor,* October 1976, pp. 51-57.
10. Robert N. Colwell, "A Systematic Analysis of Some Factors Affecting Photographic Interpretation," *Photogrammetric Engineering* 20 (1954):442-48.
11. American Society of Photogrammetry, *Manual of Photographic Interpretation,* pp. 772-75. Kirk H. Stone has provided an elaborate outline of geographic features which can be detected using aerial photographs at various scale levels.

4

Multispectral Remote Sensing

John B. Rehder
University of Tennessee

History of Multispectral Remote Sensing

The history of multispectral photographic techniques covers the past 115 years of photographic history. As early as 1861, photographic experiments were being carried out by James C. Maxwell and Thomas Sutton in which a scene was photographed on three photographic plates through red, green, and blue filters. By superimposing the three finished plates in registration, it was possible to produce crude color photography. By 1900, a multispectral three-lens camera had been developed which focused three separate images onto a single photographic plate. In Germany during the 1930s, twin-lens aerial cameras with red and green filters were used to produce "color" images. A binocular viewing device held each finished photograph through which green light, red light, and blue daylight were projected to form the color image.

During the 1940s, the British Army Air Photo Research Unit experimented with multiband aerial photography to determine the depth of coastal waters. Red and green filters were built into a single lens aerial camera. Each filter was superimposed over the other to provide a 50 percent overlap. After film processing, comparisons were made of film densities—dark tones vs. light tones. Using a ratio of density differences between the red, green, and combined filtered scene and known water depths within a test area, the British interpreters were able to determine water depths for many shallow coastal zones.[1]

By the 1960s, multiple cameras in groups of two to four and as many as twenty-six were employed with a variety of film and filter combinations for both experimentation and practical multispectral aerial surveys. Furthermore, cameras with multiple lenses were developed. Since the 1960s, multispectral techniques and hardware have been rapidly evolving and are constantly improved. Many of the principles for acquiring the imagery remain almost unchanged, such as the use of multiple films and filters, while other developments are unique, such as the use of optical scanners.

The Concept and Rationale of Multispectral Remote Sensing

Multispectral remote sensing involves the acquisition and interpretation of remotely sensed imagery taken in different spectral bands from the electromagnetic spectrum (see chapter 2). Although the acquisition of different sets of imagery conventionally comes from photographic techniques using different film/filter combinations in several cameras, it can also come from optical mechanical scanners using a variety of detectors. The end result is a set of imagery of the same target area in a variety of spectral bands from the electromagnetic spectrum.

Every object on the earth, including its atmosphere, emits, reflects, or transmits electromagnetic energy. Because electromagnetic energy varies among different objects, at different wavelengths, this energy variable can be used to aid in the identification of objects. In short, each object on earth can be identified by the distinct electromagnetic energy it emits, reflects, or otherwise transmits. This is called the spectral signature of the object. Just as personal signatures are characteristic of each individual, the spectral signature can be proof of its identification. For example, within the visable portion of the electromagnetic spectrum, from $.35\mu$ to $.70\mu$ = microns, a green painted car may reflect green light energy in the wavelength vicinity of $.50\mu$ to $.57\mu$, while a red car may reflect red light energy in the wavelengths of $.61\mu$ to $.70\mu$. These spectral wavelength differences, therefore, tell us simply that green cars have different spectral signatures from red cars and that they can be distinguished from each other.

The interpretation of remotely sensed imagery, however, is not based exclusively upon the color or spectral signature of objects. Other signatures, such as the tone (lightness or darkness), texture (surface roughness or smoothness), shape, and size of objects are often as important as color for accurate object identification.

The electromagnetic spectrum can be compared to a horizontal pegboard on which are hung a variety of sensors capable of sensing in the ultraviolet, visible, thermal infrared, and microwave regions (fig. 4.1). From the pegboard a single sensing device can be selected and its image studied, or a number of sensors, with the resulting multitude of image products, can be chosen and multispectral information obtained. As positions along this horizontal pegboard are changed, different kinds of information form different spectral bands for the same target.

Rationale

Among researchers and students alike, questions concerning the rationale of multispectral remote sensing often arise, such as: (1) Why is it done?; (2) What kinds of specific information about the earth can be gleaned from this splitting of the electromagnetic spectrum into separate parts?; (3) Why is blue-green light information desired?; (4) Why would red light information be selected?; and (5) Why multispectral and not something else?

First, multispectral remote sensing obtains different kinds of information from the same image area. It is done to gain as much information from a remote location as can be acquired by imaging systems. It is done to acquire and select the spectral signatures of objects for object identification.

By splitting the multispectral scene into discrete segments, it is possible, for example, to use black and white infrared imagery which best shows water features and land-water boundaries. Conversely, by using normal black and white film and a green filter, information about water depths and turbidity characteristics can be obtained. While both of these examples are aimed at gleaning information about water features, the first one utilizes black and white infrared film with a deep red filter to show land-water boundaries but *not* water depth. The second example

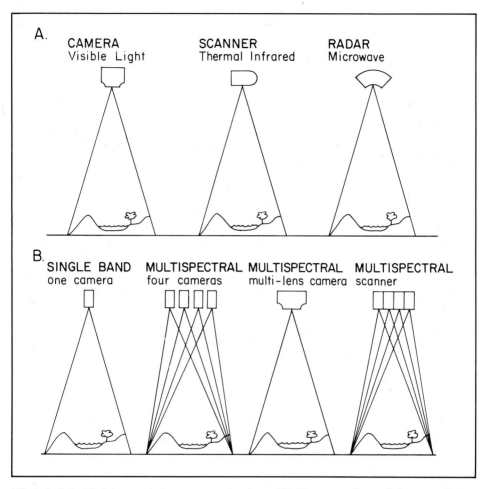

Figure 4.1. Multispectral remote sensing. (A) The sensing of the same target by different sensors representing different regions of the electromagnetic spectrum. (B) The sensing of the same target by different sensors used in multispectral remote sensing.

provides better information about water depths and sedimentation but cannot show the land-water interface as well as the first.

Why is information from blue-green and red light spectral regions important? The remote sensing of blue-green light information involves those features which display moisture characteristics, such as soil moisture, atmospheric moisture, sediment laden waters, and shallow water depths. Data coming from red light information include sharper contrasts between forests and crops, and between all forms of vegetation and cleared land, curves, roads, airports, and railroads. In short, red light information provides sharp contrasts and enables the interpreter to distinguish cultural landscapes from physical features.

Why use multispectral remote sensing devices and not something else? A single camera with no optical filter and a film that is sensitive to broad wavelength regions, such as black and white panchromatic film that is sensitive to *all* colors perceived by the human eye, may yield imagery which is too broad and too generalized. Such a system may be inadequate to produce the discrete spectral signatures which provide good object-to-background contrasts for proper object identification. Water features, soils, and vegetation in this example could possibly display similar gray tones and thus would be difficult to distinguish from each other. The alternative is to turn to multispectral remote sensing devices to provide a number of narrow spectral wavelength slices. With the proper film/filter combinations in cameras, or through the use of multispectral scanners, such elements as water features, soils, and vegetation can be identified and distinguished from each other.

The application of this principle is clearly shown in the imagery examples in figure 4.2 which illustrate the variety of information obtainable in the remote sensing of a single target area with a multilens, multispectral camera. Furthermore, the imagery from Landsat in figure 4.3 illustrates the principle of multispectral data from a satellite viewpoint in which bands 4, 5, 6, and 7 represent energy coming from the blue-green ($.5\mu$ to $.6\mu$), red ($.6\mu$ to $.7\mu$), red to near infrared ($.7\mu$ to $.8\mu$), and infrared ($.8\mu$ to 1.1μ) portions of the electromagnetic spectrum.

How Many Multispectral Slices?

Now that the concepts of multi—meaning more than one—and spectral—meaning parts of the magnetic spectrum—have been identified, the next question is how many multispectral slices are considered effective? In our American culture *more* is usually thought to mean *better,* but in multispectral remote sensing this is only partially true. The more spectral slices which are produced, the more information is available to an interpreter. This may mean 18 to 26 different spectral slices. There is, however, one drawback. The human interpreter usually is limited to not more than three sets of multispectral imagery which can be effectively interpreted and utilized at a given time. Consciously or unconsciously aware of his selection, the interpreter will select the three best sets from a group of four or more.

It is not redundant, however, to obtain more than three sets of imagery of an area, knowing that the human interpreter will choose only three. The three spectral regions may not be the same ones necessary for different targets. As the number of differing targets increases, it becomes necessary to increase the number of spectral bands to obtain the best identifiable signatures for each target. As the interpreter shifts to a different kind of target, he may require a different three-band set of imagery for the best interpretation possible.

In the following sections the current equipment which is used in the acquisition and interpretations of multispectral imagery will be examined.

Multispectral Equipment

The kinds of equipment or "hardware" in multispectral remote sensing include cameras, films, photographic filters, multispectral scanners, and multispectral color additive viewers. Conventional multispectral remote sensing is based upon a selection of film/filter combinations in a battery of cameras. Initially, multispectral remote sensing is thought of as a selection of black and white film types with different colored optical filters to screen out all but one discrete part of the visible portion of the spectrum for each camera. By simply using color film, however, the investigator is actually performing multispectral remote sensing as well. Color film contains three emulsions—yellow, magenta, and cyan, each with its own spectral range.

1. Red

2. Blue

4. Infrared

3. Green

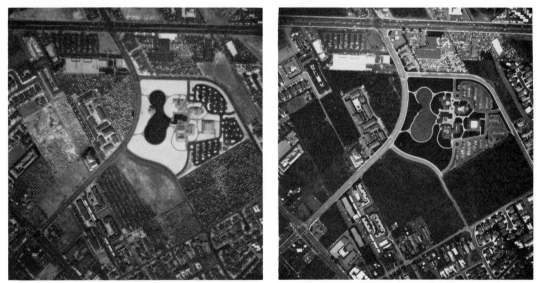

Figure 4.2. Multispectral imagery. Frames 1, 2, 3 are black and white film filtered by red, blue, and green filters. Frame 4 is a black and white infrared image filtered by a deep red 89B filter. (Courtesy of I²S—International Imaging Systems Division of Stanford Technology Corporation)

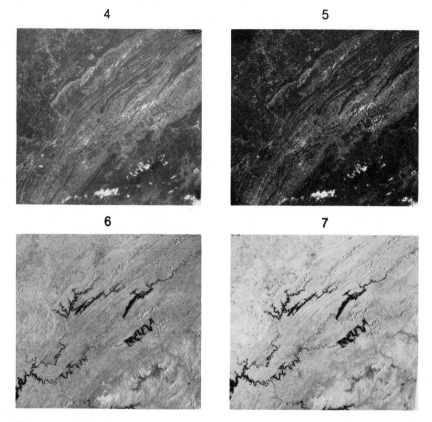

Figure 4.3. Multispectral Landsat imagery of eastern Tennessee. Band 4 represents blue-green light, band 5 represents red light, and bands 6 and 7 represent infrared energy. (EROS Data Center)

When viewing a color image, the eye sees three color-coded images superimposed—a multispectral scene. Color film, therefore, can turn any camera into a multispectral system.

Cameras

The simplest conventional multispectral camera system is based on two cameras with different film/filter combinations mounted in a single aircraft. The system can be as simple as two Kodak Instamatics or a pair of inexpensive 35 mm cameras; however, four camera installations are more common. These multicamera arrays, variously known as "quad clusters" or "quadri-camera," usually include four 70 mm Hasselblad cameras or four motorized Nikon 35 mm cameras. Almost any brand will do, but the concept is the same—use four cameras with four different film/filter combinations, mount them together with a common film advance and shutter release mechanism, and an effective homebuilt multispectral camera system is produced.

More advanced multispectral camera systems are the multilens cameras (table 4.1). In the

TABLE 4.1
Multispectral Aerial Cameras

Camera	Nine Lens	Spectral	Mark I	Aero I	MPF
Manufacturer	Itek	Spectral data	I²S	Dot Products., Inc.	Itek
Format (cm)	5.7 × 5.7	8.9 × 8.9	8.9 × 8.9	5.7 × 5.7	5.6 × 5.7
Film length (m)	76	76	76	120	30
Frames/rolls	325	188	300	470	NA
Number of lenses	9	4	4	4	6
Focal length (mm)	152	150, 100	150, 100	150, 100	150
f/no. range	2.4-16	2.8-16	2.8-16	2.8-16	2.8-16
Wavelength range (μm)	9 bands	4 bands	4 bands	4 bands	0.5-0.9 (4 bands) 0.4-0.7, 0.5-0.9
Filters	Wratten	Forty available	0.4-0.7 W47B; 0.47-0.59 W58; 0.59-0.69 W25; 0.74+ W88		Special
Can be used with normal color film?	Yes	Yes	Yes	Yes	Yes (band 1)
Can be used with IR color film?	Yes	Yes	Yes	Yes	Yes (band 1)
Shutter speed (sec.)	1/30-1/120	1/25-1/133 1/150-1/135 1/350-1/800	1/150-1/350(A) 1/350-1/800(B)	1/150-1/350(A) 1/350-1/800(B)	0.0025, 0.005, 0.01
Shutter type	Focal plane	Focal plane	Focal plane (2 types: A, B)	Focal plane (2 types: A, B)	Rotary, intralens
Weight (kg) without film and mount	23	43	26	– – –	57
Data annotation	Band no. & flight detector	NA	NA	– – –	Card digital
Intervalometer	– – –	Yes	External	External	Yes
Other	Fiducials	Yes	Modified K-22 camera with 4 lens, 24-cm film	Modified K-22	Grid reseau 70 mm film

early 1960s ITEK Corporation developed a camera which had nine lenses mounted together on a single camera body. Three separate rolls of film each supplied a set of three lenses. After exposure, the film was advanced three exposures ahead to provide a new unexposed strip of film for each of the lenses. Over the past fifteen years, however, ITEK Corporation, International Imaging Systems Division of Stanford Technology Corporation (I²S), and Spectral Data Corporation, among others, have advanced the technology of multilens cameras. ITEK's Multispectral Photographic Camera (MPC) was originally developed for NASA's Skylab program. The camera has six separate lenses and six separate 70 mm film magazines, drives, and controls.

The multispectral camera MARK I from I²S (International Imaging Systems Division of Stanford Technology Corp.) has four lenses mounted on a single camera body. The camera records four different spectral images, each 3.5 inches square, on a single 9 × 9 inch roll of film. Although the camera has but one 250 foot roll of film in it at any time, the multispectral capabilities come from the use of different filters placed over the four lenses (fig. 4.4). The Spectral Data multilens camera which is similar to the one from I²S has four lenses mounted on a single camera body. In this system photographs are made in four spectral bands ranging from .36μ to .92μ (fig. 4.5). In addition, both camera systems are combined with color additive viewers which accept multispectral images. These images are registered and projected through colored filters for image analysis in the laboratory.

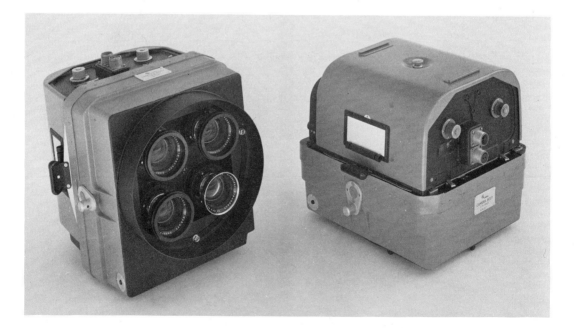

Figure 4.4. Multispectral camera. I²S Mark I multilens camera. (Courtesy of I²S—International Imaging Systems Division of Stanford Technology Corporation)

Figure 4.5. The Model 10 Multispectral Camera System. Courtesy of Spectral Data Corporation)

Films

The choice of films for multispectral remote sensing includes normal black and white and normal color films in the visible portion of the spectrum and black and white infrared and color infrared films for the visible and near infrared portions of the spectrum. The complete knowledge of brand names and emulsion numbers is not totally necessary to a basic introduction to multispectral remote sensing; however, table 4.2 lists a selection of most film types and their general applications.

The films marked with an asterisk (*) in the table are the ones most commonly used in remote sensing. The choice of films depends upon the number of cameras to be used and the specific needs of the user. In flying multispectral research missions for landscape change and landuse analysis in one study area, film needs consisted of:

1. A normal black and white film (Kodak Panatomic X 3400)
2. A normal color reversal film (Kodak Ektachrome 2448)
3. A color infrared film (Kodak Aerochrome Infrared 2443)
4. Either a black and white infrared film (Kodak Infrared Aerographic 2424) or a normal black and white film with a different colored filter

TABLE 4.2
Aerial Films

Sensitivity	Film Name	Film Number	Characteristics and Applications
Black and white panchromatic	Kodak Plus-X Aerographic	2402	Medium sensitivity, medium speed for aerial mapping and reconnaissance.
	Kodak Tri-X Aerographic	2403	High sensitivity, high speed for aerial mapping and reconnaissance under low light conditions. Grainy.
	Kodak Double-X Aerographic	2405	Medium to high sensitivity, medium to high speed for aerial mapping.
	*Kodak Panatomic-X Aerial	3400	Medium sensitivity, medium to low speed, high contrast. Good for high reconnaissance. Suitable for small negative format cameras such as 35 mm and 70 mm. Fine grain.
	Kodak Plus-X Aerial	3401	Medium sensitivity, medium sensitivity high-contrast, fine grain for reconnaissance.
	Kodak High Definition Aerial	3414	Low sensitivity, slow speed, thin base but high definition—fine grain for high altitude reconnaissance.
	Kodak Plus-X Aerecon	8401	Medium sensitivity, medium speed for low altitude reconnaissance.
Infrared (IR)	*Kodak Infrared Aerographic	2424	Black and white infrared for haze penetration. For surface water, soil moisture, forest surveys, multispectral photography.
	*Kodak Aerochrome Infrared	2443	Color infrared. False color. For forest surveys, camouflage detection, agricultural surveys, surface water features.
	Kodak Aerochrome Infrared	3443	Color infrared. False color for forest surveys, camouflage detection, agricultural surveys, surface water. High spool capacity and minimum storage space.

TABLE 4.2 (continued)

Sensitivity	Film Name	Film Number	Characteristics and Applications
Color	Kodak Aerocolor Negative	2445	High sensitivity, high speed color negative film for print making. For mapping and reconnaissance.
	Kodak Ektachrome MS Aerographic	2448	Normal color. Color reversal film for medium altitude mapping and reconnaissance.
	Anscochrome D/200	D/200	Normal color, color reversal film. High speed for general remote sensing.
	Anscochrome D/500	D/500	Normal color. High speed color reversal film suitable for low light conditions. Can be processed in negative form. For general remote sensing.
	Kodak Water Penetration Aerial Color	SO-224	A two color (blue-green) sensitive film with spectral sensitivity in the shorter blue-green wavelengths. For water penetration and bathymetric studies.

Source: Manufacture's literature and Eastman Kodak Company, *Kodak Data for Aerial Photography* #M-29 (Rochester, New York: Eastman Kodak Co., 1971).
*Films most commonly used in remote sensing.

Minimum needs of the writer have consisted of two cameras with a normal color film and a color infrared film. The black and white films have been used to obtain multispectral information, as well as prints for publication. The normal color and color infrared films have been primarily for photo interpretation in the laboratory. The color infrared film provides more information about general surface water features, land-water boundaries, and vegetation types. However, normal color film is used as a natural color record of the scene for purposes of comparison with the other film types and for color and ground truth.

Filters

Photographic filters are optical glass, or gelatin surfaces, mounted in front of a camera lens to prevent unwanted spectral energy such as blue haze, or green or red light energy from reaching the film. In essence, filters absorb or block out certain spectral light while allowing others to pass through the lens and ultimately to the film.

The best rationale for optical filters in photography can be found in the following quotation:

> Light is a prime mover in all photography . . . and in the majority of color and black and white pictures the light reflected from the subject is colored. Because the perception of a color by the human eye does not always agree with the perception of this same color by film, a filter is often essential to the effective rendition of a subject in the final print.[2]

Table 4.3 provides a selected list of aerial photographic filters and their functions. Those which are used in conventional multispectral remote sensing in conjunction with black and white film are marked by an asterisk.

TABLE 4.3
Aerial Camera Filters

CORRECTION FILTERS

Filter	Wratten Number	Color	Will Weaken the Intensity of:
K1	6	Light yellow	Blue light
K2	8	Yellow	Blue light
X1	11	Light green	Blue and red light
X2	13	Green	Blue and red light

CONTRAST FILTERS

Filter	Wratten Number	Color	Will Absorb
A	25*	Red	Blue and green light
B	58*	Green	Blue and red light
C	47*	Blue	Red and green light
F	29	Deep red	Blue and green light
N	61	Deep green	Blue and red light
G	15	Orange amber	Blue light

HAZE PENETRATION FILTERS

Filter	Wratten Number	Color	Will Absorb
89B	89B*	Deep red or black	All colors except IR
25A	25A	Red	UV, violet, blue, green
Minus blue	12	Deep yellow	UV, violet, blue
Aero 1	3	Light yellow	UV, violet, blue
Aero 2		Yellow	UV, violet, and some blue

*Those which are used in conventional multispectral remote sensing in conjunction with black and white film.

Film/Filter Combinations

What is a good film/filter combination? The multispectral film/filter combinations in figure 4.2 may serve as an example. All four images were taken with an I'S Mark I multilens camera.

Frame 1 represents an image produced by normal black and white film with a red (25A) filter, which allows only red light to reach the film. For the most part, vegetated areas are in dark tones while cleared bare earth and concrete and building surfaces are light toned.

Frame 2 was imaged on black and white film with a blue (47B) filter which allows only blue light to reach the film. One of the characteristics of this film/filter combination is the ability to display water in various tones of gray depending on the sediment load and water clarity. The "figure 8"-shaped pond in the center of the image is quite discernable.

Frame 3 was produced on normal black and white film with a green (57A) filter which allows only green light to reach the film. It, too, can be applied to water quality and clarity detection.

Frame 4 illustrates a black and white infrared film image with an 89B filter. Near infrared radiation energy was reaching the film. Water is almost universally black as illustrated by the pond in the center of the image. Green living vegetation is highly reflective of infrared radiation and this appears as bright white to light gray tones. The grassy lawns surrounding the pond and the grass islands and median strips in the parking lot adjacent to the building complex in the center of the image attest to this fact. Furthermore, the groves of trees across the road west and south of the building complex are reflecting infrared radiation.

Study the "figure 8" feature in the center of each image. In frame 1, the object is black, in frame 2 it is light grey with some light-toned swirls in it. In frame 3, it is gray but not as light as in the blue image of frame 2. Finally, in frame 4 the object is totally black. Notice how all black tones do not represent the same surface features. Perhaps the one image which identifies the "figure 8" feature as a pond is the infrared image; however, to learn something about its water characteristic, the "blue" image in frame 2 must be consulted. In both the red and infrared images (1 and 4), the water body is black. In the blue and green images (2 and 3) the pond is gray with suspended sediment or the bottom appearing in light tones.

Study the parking lot just east of the pond and large building complex in the center. Can you select the imagery which best registers grassy median strips and landscaped islands in the parking lot? You should pick frame 4, the infrared image. Why? Healthy vegetation on infrared black and white appears in a white tone. The large areas of white tones which surround the "figure 8" pond are grass, and thus the very bright white areas in and around the parking lot are grassy median strips and islands as well. Furthermore, the bright white tones representing green living vegetation in frame 4, the infrared band, are shown as very dark tones in frame 1, and red filtered scene.

As you analyze these images, you may see more information than has been introduced here. By comparative anaylsis you can realize the significance of multispectral remote sensing.

Multispectral Scanners

To the uninitiated, the group of sensing devices called multispectral scanners appears to be a formidable mass of machinery. The more widely used scanners are for sensing the infrared portion of the electromagnetic spectrum. These are heat sensing devices producing imagery much differently interpreted from the conventional photographic part of the spectrum. Although there are some scanners which operate exclusively in the mid to far infrared portions of the spectrum, there are others which span ultraviolet, visible, and infrared as well. Others operate only in the visible portion of the spectrum. Multispectral scanners have a wide latitude over the electromagnetic spectrum—much wider than any of the photographic film/filter systems.

Despite the great range of spectral coverage provided by multispectral scanners, only one will be emphasized—the Multispectral Scanner (MSS) found on board the Landsat 1 and 2 earth resources satellites.

The Landsat System

NASA's earth resources technology satellites Landsat System (formerly called ERTS) orbits the earth 14 times per day in a north to south polar orbit. Global coverage is completed by each satellite every 18 days. Because the two satellites are 180° out of phase with other, the Landsat System can sense any given point on the earth between 82° north latitude and 82° south latitude every nine days. For this reason, Landsat is a timely, sequential system capable of producing images of any part of the earth in 9 day intervals.

Sensors and Imagery

Sensors on Landsat consist of a Return Beam Vidicon (RBV) and the four channel Multispectral Scanner (MSS). The MSS scanner systems have been reliably scanning and imaging the earth with outstanding clarity and have produced several hundred thousand image-frames to date.

The MSS scans the earth in a sweeping motion. Optical energy—that is, light and infrared radiation coming from the earth's surface—is sensed simultaneously by an array of six detectors in four spectral bands designated as bands 4, 5, 6, and 7. Similar to viewing the earth through different colored windows, band 4 senses blue-green light ($.5\mu$ to.6μ), band 5 senses red light. ($.6\mu$ to .7μ), and bands 6 ($.7\mu$ to .8μ) and 7 ($.8\mu$ to 1.1μ) sense infrared radiation (fig. 4.3).

Although the spectral data are detected as colors, they are transmitted as a sequence of black and white dots to ground receiving stations. The data points are recorded first on computer tapes. Later they are transferred to image processing equipment where the initial black and white images are produced.

Imagery from band 4, representing blue-green light, appears as a hazy, often foggy scene because the band 4 sensor is sensitive to the blue-green light reflected by atmospheric haze and moisture. Band 5, sensing red light, displays sharp contrasts between cleared land and vegetation cover. Band 5 has the appearance of "normal" black and white photographs. Forests are dark, cleared lands and cities are white, water surfaces appear in shades of gray to white. Bands 6 and 7, the infrared sensing bands, both render water features sharply in black, cities in very dark gray, and the surface roughness of landforms in tones of gray. Although both have remarkable haze penetration, band 7 is preferred over band 6 because its contrasts are sharper.

Color imagery from Landsat is produced exclusively on the ground. Light is projected through three bands of the black and white transparent images (bands 4, 5, and 7) and then through colored filters onto color film. The result is a color infrared composite image which renders green living vegetation into shades of red and pink, water into various shades of blue, cleared lands into shades of yellow, tan, or white, and dense urban areas into shades of dark blue-grey to black. The color infrared composite imagery as well as each band of the black and white are available in either prints or transparencies.

Multispectral Viewers

A counterpart to the production of multispectral imagery is the analysis and interpretation of imagery. The primary function of multispectral viewers or color additive viewers is to project the imagery in such a way that different spectral images can be viewed either separately or in combination with other images of the same target (figs. 4.6 and 4.7). By combining images, the interpreter can exclude unwanted information while obtaining needed information in the photo analysis. Multispectral color additive viewers also have the capability to add colors to the projected black and white images through the use of colored filters. By projecting different spectral scenes of the same target through different colored filters in the viewer, a reconstructed color scene is produced.

Among the viewers currently on the market are those built by Spectral Data Corporation and International Imaging Systems (I²S) Division of Stanford Technology Corporation. For the sake of brevity the following discussion applies to the I²S Mini-Addcol Viewer (fig. 4.7). The Mini-Addcol Viewer is a four channel optical projector designed for the color enhancement of black and white imagery. It accepts either 9 1/2 inch roll film transparencies from the I²S Mark I camera or 70 mm cut film transparencies from Landsat.

The machine uses a light source, lens, filter, and mirror system to project the image onto a screen. Imagery is loaded into the machine either on spools for roll film or in a 70 mm cut film holder which holds four image frames. The 70 mm cut film holder has been especially designed

Figure 4.6. Multispectral Viewer. Spectral Data's Series 60/70 Additive Color Viewer (Courtesy of Spectral Data Corporation)

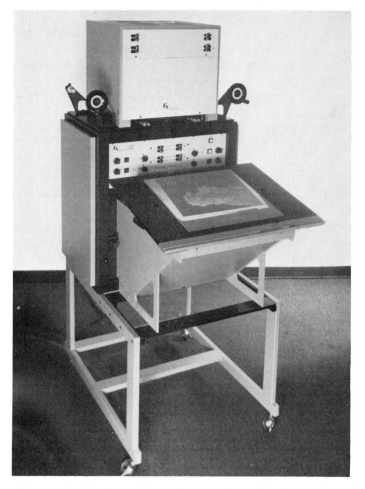

Figure 4.7. Multispectral Viewer. The I²S Mini-Addcol Color Additive Viewer. (Courtesy of I²S—International Imaging System Division of Stanford Technology Corporation)

to accept the four bands of imagery from Landsat (band 4, 5, 6, 7). The imagery fits between the light source of four 500 watt projection bulbs and the main optical system. The optical system consists of four Schneider Comparon 150 mm f/5.6 lenses, four filters for each lens (blue, green, red, clear), a front-surface mirror to project the image onto a screen, and the viewing screen itself. The machine has controls for filter selection, illumination intensity, and lamp, lens, and image positioning.

To obtain color additive scenes the four individual images must be aligned together, or registered, to form a single image. Once the scenes have been registered, each of the four images can be projected through one of the four filter colors—red, green, blue, clear. The range of color combinations is great, enabling the user to construct an extensive array of color images for different interpretation and mapping purposes. For an infrared color composite image from Landsat imagery, the best image/filter combination is band 4 projected through a blue filter, band 5 through a green filter, and band 7 through a red filter.

The color image is reflected by a mirror onto a 9″ × 9″ viewing screen where the user can hand transcribe the data to map form by using plastic or tracing paper overlays. Also the image on the screen can be photographed with a 35 mm camera; printing heads are optionally available for generating more accurate photographic copies of any screen presentation at full size. An optional screen attachment can be used to enlarge a Landsat image to a scale of 1:500,000 on the viewer screen.

In summary, the I²S color additive viewer provides a simple and effective means for analyzing and evaluating multispectral imagery. With the I²S it is possible to color enhance any 70 mm black and white transparencies or multispectral 9 1/2 inch film to produce an unusual range of multispectral combinations for image interpretation.

Multispectral Applications in Terrain and Geologic Analysis

Multispectral remote sensing as applied to geologic analysis has yielded results from geobotanical investigations and studies of lineations in fracture zones. Different vegetation types can be associated with different terrain and geologic formations. For example, in southern forests, cedar glens can be found in limestone areas while pines are associated with sandstone materials. It has been difficult, however, to determine clear-cut differences among types of rock from conventional multispectral photography.

In the Apollo 9 SO-65 experiment of NASA, multispectral imagery taken by four cameras of a test site in New Mexico and another in Alabama yielded entirely different kinds of information. For the New Mexico test site, the three black and white film/filter combinations (Panatomic X with a green 58 filter, Panatomic X with a red 25A filter, and Infrared Aerographic with a deep red 89B filter) showed little or no differences between surface rock types. In fact, for the New Mexico site, the multispectral black and white imagery in the three cameras offered no real advantage over the color infrared imagery in the fourth camera. On the contrary, for the Alabama site, geologic structure, sharp differences between decidious and coniferous evergreen vegetation, open bare earth and rock surfaces, and open water features were clearly discernable on the three sets of multispectral black and white imagery.[3] From this early example of multispectral analysis it was demonstrated that terrain and vegetative cover can determine the usefulness and limitations of black and white multispectral imagery for geologic applications.

For mapping geologic lineations—fault lines and fracture zones, multispectral imagery especially from Landsat has yielded a wealth of information from many parts of the world. In a recent U.S. Geological Survey Professional Paper, examples of lineation mapping from Landsat imagery included: the Jordan Rift Valley of the Middle East; Western Brooks Range in Alaska; Los Angeles and the Salinas Valley of California; Northern Arizona; the Andes Mountains in Peru, Boliva, Chile; earthquake zones in Nicaragua; and a newly discovered hundred-mile-long lineament in Tennessee.[4] The mapping of geologic lineations in such places as these gives information about the structural geology of earthquake and volcanic zones, water resources, and potential mineral exploration sites.

Multispectral Applications to Forestry

Experimental applications of multispectral remote sensing to forestry have been in existence since the 1940s. More recently, multilens cameras and color additive viewers have been very effective in vegetation/terrain classification and tree species identification. For example, an interpretation accuracy experiment by Lauer, Benson, and Hay showed that coniferous vegetation could be correctly identified 88 percent of the time on black and white images projected through red, blue, and green filters. The images spectrally covered $.53\mu$, $.62\mu$, and $.75\mu$, respectively, and gave better interpretation results than those obtained either from single band black and white imagery, normal color, or color infrared imagery.[5]

The band 5 imagery from Landsat is being used to distinguish forestland from cleared land surfaces. Furthermore, it has been particularly useful in mapping the general forest cover for large regions, such as state-wide forest inventories. Perhaps the most exciting aspect of Landsat is the temporal (time) dimension. With its eighteen-day cycle of coverage (nine days with both satellites), color composite imagery of Landsat enables the investigator to interpret and monitor the seasonal changes of forest types. This is the effect on the leaf coloration of deciduous trees of the progressive "green wave" in springtime and the fall "brown wave." Interpreters not only have the ability to monitor the seasonal changes of the nation's forests, but they can also monitor the progress of the seasons.

Multispectral Applications to Surface Water and Marine Environments

Surface water possesses properties which are ideally suited to its detection, monitoring, and mapping from multispectral remote sensing. In this section three applications will be examined: flood inundation mapping, wetlands mapping, and water depth analysis and measurement.

Flood Inundation Mapping

Perhaps one of the more difficult tasks in remote sensing is to distinguish floodwaters from bare soil in the visible portion of the electromagnetic spectrum. During flood stages, sediment laden water has the appearance of bare earth. The contrasts between dark blue water against brown earth in nonflood conditions disappear when muddy waters cover soil surfaces. Furthermore, in shallow waters, bottom sediments are difficult to distinguish from bare earth surfaces. Such problems can be solved by the use of infrared imagery.

With black and white infrared imagery, all exposed water surfaces appear in black tones on the imagery, no matter how shallow or sediment laden they appear. On color infrared imagery, water surfaces can range from black to dark blue for clear water to lighter tones of blue for sediment laden waters. Water presents an absorptive condition for infrared energy, promoting high tonal contrast between water (dark = absorptive) and land (light = reflective). Moreover, as floodwaters recede, wet soils continue to be darker than dry soils. The detection and mapping of floods and flood affected areas are greatly facilitated by the use of black and white or color infrared imagery.

An example of flood mapping can be seen in figure 4.8, which shows a portion of the Mississippi River in western Tennessee during a preflood period in October, 1972, and during one of the worst floods in recent history in May, 1973. Both are Landsat 1, band 7 imagery and illustrate the significance of Landsat for its broad aerial coverage and the value of infrared imagery for flood inundation mapping.

A.   B.

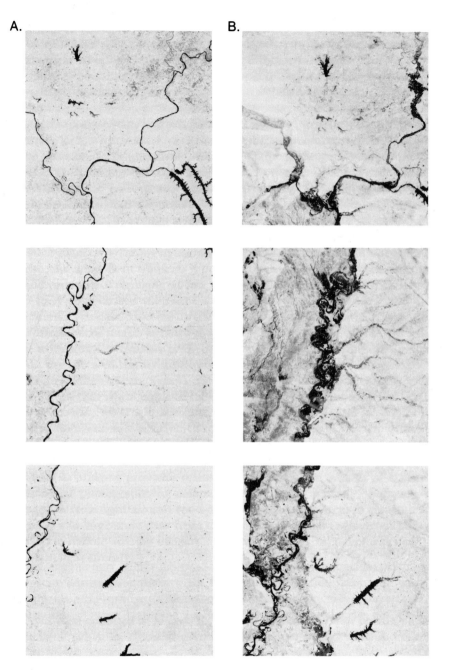

Figure 4.8. Landsat band 7 imagery of the Mississippi River between Cairo, Illinois, and Memphis, Tennessee. Images at left represent a low water period in October, 1972. Images at right represent flood inundations in May, 1973. (EROS Data Center)

Wetlands Mapping

The mapping of wetlands conveniently takes much the same course of action as flood inundation mapping. The water surface signature is sought, but in this case it is modified by the presence of vegetation. Wetlands are low-lying, naturally flooded, vegetated surfaces, either with trees in swamps or grasses in marshes. In either case, it is the underlying presence of water which identifies the feature as wetland.

Just as in flood mapping, infrared imagery is used in wetlands mapping. In the U.S. Geological Survey Paper 929, *ERTS-1, A New Window on Our Planet,* seven papers are concerned with wetlands classification and mapping. The areas presented range from coastal marshes in Virginia, South Carolina, and Georgia to the Dismal Swamp of North Carolina and Virginia and the Florida Everglades.

In a wetlands mapping project in western Tennessee for NASA, several forms of multispectral imagery have been used: (1) MSS bands 4, 5, 6, 7, and color composite imagery from Landsat; (2) high altitude color infrared from a NASA U-2 overflight at 19,819 meters (65,000 feet); and (3) midaltitude color infrared imagery from a 3,658 meter (12,000 foot) altitude. The purpose of the project is to determine the reliability of Landsat data for the wetlands mapping of large areas. Based on preliminary findings at the time of writing, it has been found that the Landsat infrared imagery (bands 6, 7, and the color composite) are very reliable for wetlands mapping in western Tennessee.

Water Depth Analysis and Measurement

Multispectral applications to water depth analysis and measurement (bathymetry) have wide utility, including near-shore navigational charting for peacetime navigation or to assist the military in landing troops. Such fields as oceanography, hydrology, limnology, and even archeology require information about water depths, water quality, and submerged archeological sites. Multispectral remote sensing has been applied to these problems and has provided significant results. It has also led to the development of new water penetration film and better film/filter combinations.

Contrary to the use of infrared imagery for flood and wetlands studies, in bathymetric applications imagery from the blue-green visible portion of the electromagnetic spectrum yields the most information about water depths. Among the more promising film developments has been the Kodak Water Penetration Film SO-224, a high speed, two-color film with spectral sensitivities in two narrow spectral regions. One is centered at $.48\mu$ in the blue-green part of the spectrum, and the other is at $.55\mu$ in the green portion. During processing, a magenta (light purple) dye is formed in the blue-green sensitive layer of the film while a green dye is formed in the green sensitive layer. The film is filtered with a wratten 4 light yellow filter.[6] The resulting imagery is a combination of green and light purple color differences which represent bottom configurations and depth variations to depths of 9.14 to 22.86 meters (30 to 75 feet) in clear water.

Multispectral Applications to Agriculture and Soils

Multispectral remote sensing as applied to agriculture and soils is widely used for crop surveys, plant vigor and stress analysis and mapping, soil types, and soil moisture. In this last section two quite different applications will be examined: (1) crop vigor and stress analysis and (2) the analysis of bare earth/plowed soil areas.

Crop Vigor and Stress Analysis

Crop vigor and stress analysis is one of the more common and more dramatic applications of multispectral remote sensing to agriculture. The health or vigor of a crop is mirrored by the amount of infrared radiation the crop reflects to the sensor. Healthy broadleaf vegetation is highly reflective of infrared radiation and thus appears in bright gray tones on black and white infrared imagery and very bright red on color infrared film. Diseased vegetation of the same plant species is less reflective and appears dark in tone on black and white infrared and dark red to brown on color infrared imagery.

Table 4.4 shows four black and white film/filter combinations applied to an analysis of healthy and diseased alfalfa—one of the nation's leading livestock feeds. By reading the table vertically, light and dark tone differences between healthy and diseased alfalfa can be distinguished in three of the four film/filter sets. Only the black and white panchromatic film with a 47B blue filter fails to detect a difference. By reading the table horizontally, healthy alfalfa has the same dark tone signature for the three normal black and white films but has a light tone signature on black and white infrared. Diseased alfalfa has a variety of tone signatures among the normal films; however, the normal black and white film with the 47B blue filter has a dark signature like the black and white infrared with a 89B red filter. By studying the table the significant differences made by multispectral imagery in detecting diseased crops can be seen.

Analysis of Bare Earth/Plowed Soil Areas

One of the more unusual and interesting multispectral applications to soils is the study of plowed fields from Landsat band 5 imagery in the negative format. On band 5 black and white *positive* prints—the ones which can be purchased from the EROS Data Center, Sioux Falls, S.D.—forest cover is represented in dark tones while plowed fields and bare soil surfaces are in white tones. Reversing the tones by photographic contact printing causes the forests to appear in light tones and the plowed fields to appear in deep black tones on a negative print. The negative imagery in figure 4.9 provides a high contrast mapping medium from which black dots and blocks of plowed fields can be detected and mapped.

Perceived as aggregates, the dark tones in figure 4.9 form photomorphic regions of similar tones. The value of the photomorphic region is that like-tones can be assumed to reflect similar

TABLE 4.4
Multispectral (Four-Band) Signatures of
Healthy and Diseased Alfalfa

	Normal B & W Panchromatic 47B (blue) Filter	Normal B & W Panchromatic 61 (green) Filter	Normal B & W Panchromatic 25A (red) Filter	B & W Infrared 89B (dark red) Filter
Healthy alfalfa	Dark	Dark	Dark	Light
Diseased alfalfa	Dark	Light	Very light	Dark

Source: Robert N. Colwell, "Uses and Limitations of Multispectral Remote Sensing" *Proceedings of the Fourth Symposium on Remote Sensing of the Environment* (Ann Arbor: University of Michigan, Institute of Science and Technology, 1966), p. 88.

Agricultural Landscape Change

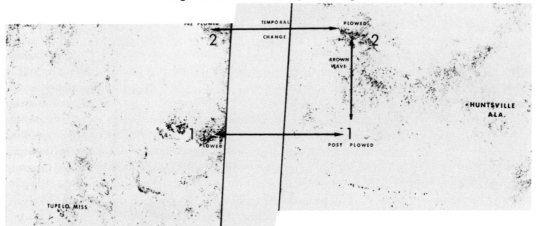

4 May 1973   21 May 1973

Figure 4.9. Negative prints of Landsat band imagery. Agricultural landscape changes can be detected based on changes in plowed field signatures in northern Alabama and south central Tennessee. (EROS Data Center)

landscape characteristics. In this case the dark tones together constitute a photomorphic region of plowed earth features. Analysis of Landsat imagery taken on successive cycle dates for an area reveals the capabilities of detecting plowed fields and agricultural crop differences.

Figure 4.9 shows two negative band 5 prints for May 4 and May 21, 1973, in south central Tennessee and northern Alabama. The temporal distance is only one Landsat cycle apart, yet significant changes can be detected. In region 1, the Muscle Shoals-Florence, Alabama, area, there is a direct transformation from dark plowed earth signatures for May 4 into lighter tones for May 21. This represents a change from plowed conditions into an initial flourishing or greening of the cotton crop. Northward in region 2, the Lawrenceburg, Tennessee, area, the field signatures of light tones for May 4 indicate only minor areas of activity. Plowing has just begun for a small area of cotton, whereas the majority of fields remain undisturbed. However, by May 21, dark tones appear, indicating a freshly plowed condition for the planting of the major crop—soybeans.

As indicated by the arrows on the two images, a temporal change can be detected between the two dates of May 4 and May 21 in terms of plowing signature changes. Region 1 changes from plowed to post-plowed early crop signatures, whereas region 2 changes from pre-plowed to plowed signatures. Furthermore, a brown wave effect can be detected spatially in a south-to-north movement. Unlike a seasonal brown wave, this wave represents the northward migration of plowing practices as a response to a variable crop and plowing calendar.[7]

NOTES

1. William A. Fischer et al., "History of Remote Sensing," in *Manual of Remote Sensing,* vol. 1, ed. Robert G. Reeves (Falls Church, Va.: American Society of Photogrammetry, 1976), pp. 37-39.
2. "Vivitar: A Guide to Filters" (Santa Monica, Calif.: Ponder and Best, 1974). A trade brochure.
3. P. D. Lowman, Jr., *Apollo 9 Multispectral Photography: Geologic Analysis,* pub. X-644-69-423 (Greenbelt, Md.: NASA Goddard Space Flight Center, 1969).
4. Richard S. Williams, Jr. and William D. Carter, eds., *ERTS-I, A New Window on Our Planet,* U.S. Geological Survey Paper 929 (Washington, D.C.: Government Printing Office, 1976).
5. D. T. Lauer, A. S. Benson, and C. M. May, "Multiband Photography—Forestry and Agricultural Applications," *Proceedings of the ASP-ACSM 1971 Fall Meetings* 71-348, pp. 531-53.
6. M. R. Specht, D. Needler, and N. L. Fritz, "New Color Film for Water Penetration Photography," *Photogrammetric Engineering* 39 (1973):359-69.
7. John B. Rehder, "The Uses of ERTS-I Imagery in the Analysis of Landscape Change," in *Remote Sensing of Earth Resources,* vol. 3, ed. F. Shahrokhi (Tullahoma, Tenn.: University of Tennessee Space Institute, 1974), pp. 573-86.

5

Ground Level Observations of Signal Contrast from Various Electromagnetic Spectra

Ray Lougeay

State University College of Arts and Sciences, Geneseo, New York

Introduction

The term remote sensing often carries the connotation of exotic electronic instrumentation which produces unfamiliar picturelike images understood only by a few with extensive training in photogrammetry, image interpretation, and physics. At times the space-age jargon which surrounds remote sensing belies its simplicity and leads geographers and other scientists to believe remote sensing image interpretation is a field for physicists and engineers. This is not the case, and, therefore, it is not necessary to have extensive background in quantum physics in order to understand the basic imaging process and be able to appreciate the usefulness of remote sensing imagery. Actually, the process of remote sensing is something everyone does on a day to day basis. A brief introduction to simple image interpretation, using normal photographs or snapshots, will greatly enhance an ability to understand and interpret more exotic remote sensing imagery.

Three remote sensing systems commonly employed for geographical applications are photography (visible and near infrared), electro-optical scanning (visible, near infrared, and thermal infrared), and microwave systems (much longer "radio" wavelengths) (see chapter 2). Great strides in electromagnetic remote sensing have, within the past half decade, made this imagery readily available. Remote sensing imagery now appears in modern textbooks and classrooms as a natural addition to the more classic maps and photographs. Landsat imagery (plate 1) is a good example of imagery with useful applications to many environmental problems, once the user gains familiarity with the image format.

Developing Familiarity with Imagery and Imaging Processes

Remote sensing is a mapping tool. The electronic and optical systems which produce remote sensing imagery record reflected or emitted electromagnetic energy (such as sunlight or radiated "heat energy") and display spatial patterns of the energy in a maplike format. Varying patterns of gray tones are seen on a black and white photograph. These represent differing amounts of sunlight reflected from various surface materials at the ground level, here referred to as "signal contrast." A black and white photograph is simply a mapped record of the reflectance values (i.e., albedo) of various surfaces within the field of view. Albedo is defined as the percentage of incident solar energy reflected away from a surface. Surfaces with high albedo, such as a new metal roof or snow, reflect more energy back to the film in the camera and produce lighter patterns on the printed positive photograph. Dark surfaces, such as asphalt or green vegetation, reflect smaller amounts of sunlight in the visible spectral wavelengths and produce dark patterns on the positive photograph.

Patterns on a photograph do not seem unfamiliar because the image is produced in much the same way as images produced by the human eye. The imaged patterns parallel the human visual experience. The simple act of seeing with the human eye, common to all sighted people, is a remote sensing process familiar to all; therefore, when presented with imagery produced in much the same way as "seeing," the interpretation process is almost automatic.

Color photography is a common teaching medium in almost all geography classes. These color photographs are rather sophisticated remote sensing images because they not only record varying albedo values within the field of view, but also split the reflected sunlight into various spectral bands of visible light. The color dyes of the film respond to different wavelengths of visible light, producing patterns of blue, green, and red. Thus, the standard color photograph is a multispectral image, usually designed to simulate the human visual experience.

There are some wavelengths of solar energy, such as the near infrared, which cannot be seen with the human eye. Certain films are designed to record patterns of reflected sunlight in these invisible wavelength bands. For some applications, phenomena can be detected which are not readily apparent in the visible spectral band. An example would be crop distress due to drought or disease. Healthy, green vegetation is highly reflective, but distressed vegetation quickly displays reduced reflectivity in the near infrared wavelengths. In both the visible and near infrared spectral bands, however, the imaging system is recording patterns of reflected solar energy. A given object can be detected on the photographic image if it reflects sunlight at different amounts or wavelengths from its surroundings and background. This signal contrast distinguishes the target on the final image.

"Hands-on" experience with the actual electromagnetic energy patterns which produce the imaged tonal patterns greatly enhances the perception of the imaging phenomenon. Being able to measure some of these energy fluxes in the field is most helpful. This can be done with cameras, perhaps using various film and filter combinations. Also, radiometers can be used to measure absolute amounts of incident and reflected sunlight or thermally emitted infrared radiation (figs. 5.1 and 5.2). These signal contrasts, observed in the field, can be compared with remote sensing imagery (fig. 5.3).

Changing Observational Perspective

Ground level photographs are probably the most familiar types of remotely sensed images of scenes and events. These images are mostly oblique "snapshots" and so are within the natural spatial scale and observational perspective. Once it is realized that pictures, or images, are produced by signal contrast and energy patterns, the perspective can be modified to a higher platform where there is a broader field of view and work is with smaller map scales covering a

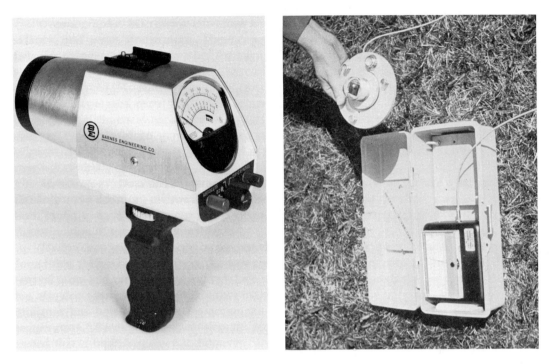

Figure 5.1. (Left) The Barnes PRT-10 Infrared Thermometer is a radiometer which measures the radiant emittance of a target surface in the spectral band between 6.5 and 20 microns. This emitted energy is directly proportional to the fourth power of the surface temperatures. (Courtesy of Barnes Engineering Company)

Figure 5.2. (Right) A small solar radiometer can be hand held. The electrical current generated by the sensor, when exposed to sunlight, is measured on the meter in the box. (Photograph by author)

larger area. The observational perspective can be changed to a view from above by ascending to the top of a hill or a tall building, or by flying in an airborne or satellite vehicle. Working with smaller scale imagery will also give experience with varying resolution levels, because at ever smaller scales the detail for a given point becomes less clear.

At the extreme of small scale imagery are satellite images where patterns of signal contrast are sensed and transmitted to the surface of the earth by sophisticated electronic processes. The resultant image, however, is still a map of varying target energy values. In the visible wavelength bands this energy is reflected sunlight, the same as in an optically produced photograph. In the thermal infrared and microwave spectral bands, signal contrasts represent electromagnetic energy radiated away from the surface as a function of varying surface temperatures and materials. Daily weather satellite images, which often appear on television forecasts, are familiar examples. The light and dark patterns on this extremely small scale imagery represent varying albedo values of clouds and the earth surface below.

Interpretation of weather satellite images is dependent upon association. Meteorologists are interested most in air mass temperature, density, and humidity values. They also know that

Figure 5.3. A Barnes PRT-10L infrared radiometer is being used to scan across the distant fields and forest canopy. These observations will record temperature differences between the fields and the forest canopy. (Photograph by author)

various cloud patterns can be associated with certain air mass phenomena and frontal conditions. Here ground truth is used to enhance the interpretation of remote sensing imagery. The network of weather observation stations provides data representing point samples. These data are used in conjunction with the overall two-dimensional view of the imagery. This allows interpolation between ground-based point samples, and enhances the understanding of atmospheric dynamics and forecast meteorology.

The reader should not be shocked by the strong patterns of false color, extremely small map scale, and vertical perspective of satellite and other remote sensing images. It is now understood why the color tones are unlike the visual experience. It is not necessary to discuss extensively the imaging systems themselves. The process of "making" the image inside the camera or electronic systems is less important at this stage than an understanding of the types of electromagnetic energy being sensed and the production of signal contrasts. It becomes easier to apply remote sensing imagery to solutions of geographical problems as more experience is gained with energy profiles, ground level phenomena, and mapping from the imagery.

Usually remote sensing systems are employed to provide rapid data collection of a large sample population, providing a broad scale survey. For these purposes the remote sensing system is usually mounted on an aerial or satellite platform, where the sensor can respond to electromagnetic energy reflected and/or emitted from the ground surface. The resultant imagery provides information about the subject matter without the expense of laborious and time-consuming ground level data collection.

"Ground Truth"

Accurate interpretation of the remote sensing imagery and accurate analysis of the subject matter is dependent upon an understanding of the relationship between imaged patterns of electromagnetic energy and the ground level phenomena under study. This understanding is provided by ground truth or field checking. The term "ground truth" is commonly used in reference to ground level observations, compatible with the airborne remote sensing observations, which are designed to enhance the understanding, interpretation, and analysis of the subject matter contained within the remote sensing imagery.

All researchers have their own specialties. It is not possible in this discussion of ground truth problems to indicate exactly what must be observed at ground level to make remote sensing systems a useful tool to each specialty. Instead, specific techniques of ground truth measurement and problems of scale, sampling, and logistics will be considered to aid in future work involving remote sensing surveys.

Most remote sensing systems and imagery deal with the same "stuff," that is, electromagnetic energy. No matter what the subject matter on the ground, all remote sensing systems discussed in this chapter provide a final product which is literally a map of spatial variation in electromagnetic energy reflected or emitted toward the sensing system. The key to truly understanding the subject matter imaged on remote sensing imagery is not only to understand the subject but also to understand the processes which produce the spatial variation in electromagnetic energy sensed, and subsequently imaged, by the remote sensing system.

An example would be high altitude aerial photography used to map the area of snowpack in a total study area of 2500 square kilometers (965 square miles). Hydrologists need an accurate assessment of the amount of snow held in the mountains which will melt and produce the spring runoff and water supply. To accomplish such a survey on the ground is a very laborious task. If all of the white (i.e., high albedo) areas on the imagery are assumed to be snow surfaces, it is quite easy to determine the percentage of study area covered by snow. An understanding of surface albedo values may, however, reveal that other surfaces may have solar reflectance properties similar to snow. Clouds and buildings with white shingle roofs may both appear as white areas on the photographs. Ground level observation may estimate cloud cover and record surfaces with albedo values similar to that of snow. This information would increase the accuracy of photo interpretation, or at least provide some measure of error present in the survey.

Spatial variation of energy intensity is also involved where thermal remote sensing is used to map vegetative associations in a rural area. Various vegetation associations may exhibit differing surface temperatures due to plant structure, soil moisture, and density foliage. Since the thermal characteristics of various soils and vegetation surfaces will differ, the diurnal pattern of surface heating and cooling will not be constant. Imagery from different times of day, therefore, may show very different thermal emittance patterns. Ground level observations of actual surface temperature and thermal emittance at sample locations would be necessary for accurate analysis of the remote sensing imagery.

Ground Level Observations Directly Related to Intensity
of Transmitted Electromagnetic Energy

Detection and identification of remotely sensed subject matter relies upon a signal contrast resulting in image contrast on the imagery, that is, patterns of light and dark or color on the imagery, produced from various intensities of electromagnetic energy striking the sensor as it views the study area. The following are examples of various signal contrasts: (1) temperature difference between two currents of water detected by thermal remote sensing, where these surfaces emit differing amounts of radiant energy due to their differing surface temperatures; (2) tone difference between various types of sea ice, as detected by aerial photography, where these ice surfaces reflect differing amounts of solar energy due to their differing surface albedo values. Information at ground level is transferred to the imaging system by electromagnetic energy. To truly understand the relationship between ground level phenomena and the remote sensing imagery, the ground truth observer must monitor signal contrasts within the study area in spectral wavelength bands comparable to those of the airborne or satellite remote sensing system.

Signal contrasts are directly related to surface characteristics. Photographic signal contrasts reflect variations in surface albedo. Electromagnetic energy leaving the surface in these wavelengths bands (i.e., 400-1,000 nm) is reflected sunlight:

$$\text{albedo} = \frac{R\uparrow}{R\downarrow} \times 100 = \% \qquad (1)$$

where,

$R\downarrow$ = total amount of solar radiation incident upon the surface

$R\uparrow$ = the amount of solar radiation which is reflected away from the surface.

Thermal infrared signal contrasts reflect variations in surface temperature. The nature of the ground truth needed for a given study is dependent upon the objectives of the study and the surfaces to be imaged. In general, the lesser the thermal contrast between target surfaces the greater the need for more extensive ground truth. When thermal contrasts are small there may be few periods of the day when one can safely predict that airborne imagery will be able to show a contrast between these surfaces.

Infrared radiant energy emitted from the ground surface is expressed by the equation:

$$M = e\sigma T^4 \qquad (2)$$

where,

M   = the radiant emittance in Langley (ly) min^{-1}(one ly min^{-1} = 0.698 kJ m^{-2} s^{-1})

e   = the emissivity of surface material

σ = the Stefan-Boltzmann Constant (8.132×10^{-11})

T   = the absolute surface temperature.

Since σ is a constant, only differences in surface temperature or surface emissivity can produce changes in surface radiant emittance power and apparent thermal contrasts. A relatively small emissivity difference can produce thermal contrasts on the order of a degree or two between various surfaces, even where surface temperatures may be equal. With a knowledge of the emissivities of the surface materials under study, or of the thermal contrasts which are sufficient to mask the effects of emissivity differences, reliance can be placed on ground truth measurements of surface temperature to indicate thermal contrast which could be depicted on infrared imagery. In this case surface temperature probes, a contact pyrometer, or simple thermometers could yield valuable ground truth data at a minimum of financial expense.

The question of emissivity and the problems of accurately measuring surface temperature can all be eliminated by the use of an infrared radiometer. The radiometer responds to infrared radiation emitted from the ground surface in the same manner as the airborne infrared detector. The radiometer integrates all surface temperatures within its field of view. This is important when dealing with a complex surface, such as vegetation over exposed soil, or coarse gravel where there are both sunlit and shady areas. On these surfaces, temperatures will vary markedly and it is difficult to aggregate contact temperature measurements. The radiometer, however, does this automatically. The surface temperature data which are plotted on a diurnal curve could provide information on the time of day when thermal contrasts would be such that airborne imagery could detect the surface phenomena of interest.

Infrared thermometers, or radiometers, sense surface signal contrasts in the thermal infrared spectral wavelength bands (1,000-20,000 nm). Some of these radiometers can be filtered to narrower spectral bands to better simulate the airborne system. The use of an infrared thermometer to remotely obtain surface temperatures assumes that the material of the surface in question radiates as a blackbody radiator (i.e., emissivity approaches unity). If the surface material does not radiate as a blackbody radiator and the emissivity of this substance does not approach unity, the observed radiant temperature will be somewhat less than the absolute temperature of the surface. Errors, therefore, may be inherent in recording absolute surface temperatures unless a measure is available of the emissivity of various surface materials in question. Emissivity is not important to the question of radiant thermal contrast, however, it is important in obtaining the absolute temperature of these surfaces.

The surface temperature and the resultant radiant temperatures of an environmental surface are direct products of the various components of the energy balance. The total energy balance, or heat balance, of a surface is expressed by the following equation (in this case provision has been made for ice or snow on the surface):

$$\pm R \pm S \pm E \pm I \pm P - M = 0$$

where,

R　= the all-wave net radiation budget
S　= heat energy used in convective heating of the atmosphere—sensible heat flux
E　= latent heat transferred by evaporation or condensation— latent heat flux
I　= energy used in the conductive heating of surface and subsurface material
P　= heat energy transferred by precipitation (direct, not latent), when applicable
M　= energy used in the melting of snow or ice (latent heat of fusion), when applicable.

The primary components of the energy balance are the latent, sensible, and soil heat fluxes. The energy for the three heat fluxes is derived from the net radiation at the soil surface, with solar insolation as the original source of heat.

While the energy balance approach to thermal remote sensing is perhaps the most accurate way of understanding the processes producing signal contrast, it is also the most laborious and most costly. Usually it is only possible to monitor the total energy balance at a few key sample points.

Radar signal contrasts reflect variations in surface roughness and the electrical properties of surface materials. The surface characteristics which determine the amount of radar energy scattered back to the radar sensing system produce signal contrasts. These surface characteristics are the roughness of the surface and the electrical properties of the surface materials.

Ground Level Observations of the General Environment and
Phenomena Indirectly Related to Signal Strength

Usually the primary subject matter of a remote sensing study is not, in itself, a problem of electromagnetic radiation. It is, nevertheless, closely related to the signal contrasts discussed in the previous section. Ground level observations of the primary subject of the study are needed to enhance the interpretation of remote sensing imagery. The types of ground truth observations needed for any given study involving remote sensor imagery or data depend primarily upon the problem under investigation. Obviously, the ground data collected for an urban heat island study will differ somewhat from that need in a rural land use survey, or an analysis of watershed hydrology. Since remote sensing is most often employed to enhance ground survey capabilities by covering more area in a shorter time period, a discussion of all possible research problems employing remote sensing would be similar to a listing of all possible geographical research.

The most useful characteristic of remote sensing imagery is that it provides a means of data collection over the whole study area. Often ground observations must rely upon a collection of point samples to represent the total population of features or events under study. Remote sensing allows the investigator to interpolate between the point samples, providing a technique for the geographer to switch from point samples to information with two or three dimensional quality (i.e., the contiguous area, or region). The "point to area" interpolation capability of remote sensing systems provides a powerful tool of special interest to geographers. In light of these capabilities, remote sensing lends itself to almost any applied geographical problem. Most ground level observations needed in the study of any specific problem are those which would normally be used if remote sensing were unavailable. It is not possible in this short space to adequately discuss all ground level observation techniques not specifically related to the remote sensing system, because this would lead to a discussion of all field geography! There are, however, certain general considerations of scale and sampling techniques, plus knowledge gained from past experience, included in subsequent sections of this chapter which will help in the planning of future remote sensing ground truth operations.

Ground level observations are used to describe the general environment and surroundings of the study area, plus providing measurement of the specific phenomena under study. The more extensive and intensive the ground truth is the better the sample will be. The greater the amount of knowledge observed at ground level the better will be the understanding of related remote sensing imagery. This will also produce greater accuracy within the study.

The investigator is cautioned not to be afraid to "overkill" on ground truth collection. These observations will always be useful, and they will enhance the study. It is always worth the extra time and effort to provide as much ground truth information as possible. If something should occur to produce remote sensing imagery of a quality inferior to that originally expected, then more intensive ground truth observations would be necessary in order to glean needed information from the degraded imagery. Some of the obvious types of information and data sources to be collected are: existing photographs; maps and topographic sheets; existing data and studies of the area; a general mapping of the study area for location and surface characteristics; and extensive photography of the general area and specific phenomena under study. Modification of hypotheses and operational definitions must be made in the field and reasons for these changes recorded at the time.

In general, investigators are urged to record what is in the study area so they can interpret the remote sensing imagery. Sometimes this means intensive mapping of small areas with large scale imagery, such as aerial photography of one city center; at other times it means extensive generalization of existing published data, such as satellite imagery of a total continent.

Temporal and Spatial Scale Problems

Most remote sensing systems collect information about the subject almost instantaneously (i.e., at the speed of a camera shutter or in the length of time it takes the aircraft or satellite to pass over the study area). It is difficult to provide ground level observations this quickly. At best, a few instantaneous observations can be made at key ground control points coincident with the time of the airborne imagery. Even when ground truth observations are made on the same day as airborne imaging, some parameters of the environment will change within the time of observation. Examples of these rapidly changing phenomena are climatic phenomena, such as shadow and sunlight, wind and waves, and cloud cover. Also, stream flood stage, tides, and traffic patterns change more rapidly than the speed of ground level observation. Other parameters of the environment, such as topography and cultural features, change at much longer time intervals.

Sometimes ground truth observations are conducted at the same time as the remote sensing imagery is being flown. This usually involves a research team in the field at the time of the overflight. This type of ground truth involves monitoring, all environmental parameters and signal reflectance and/or emittance contrasts simultaneously, or almost simultaneously. Ground truth of this type is most costly since it can involve extensive manpower and expensive instrumentation with rapid data collection capabilities. Often this will include the cost of providing, or contracting, for your own airborne remote sensing system.

The data which are collected simultaneously with the overflight of an airborne system provide the most comparable and accurate ground truth for imagery interpretation. The remote sensing system and imagery can be custom designed to a specific study, and the investigator often can choose the type of system, spectral bands, flight time, and environmental conditions for the flight. The remote sensing system, therefore, will be designed for the study at hand, and will probably provide the best possible results.

When ground observations are conducted concurrently with airborne remote sensing, an opportunity is provided for greater control of scale and identification of sample points. This can be accomplished in a study area by placing special markers at strategic locations. These markers will produce strong signal contrasts with the surrounding environment. The markers can be surveyed to provide exact location and identification of points on the remote sensing imagery. This is especially useful with some scanning systems and oblique photos where the imagery may be skewed spatially relative to the true map. Sample markers for various remote sensing systems might be: large cloth panels of contrasting color for the photographic system; flares, fires, or hot targets for the thermal infrared system; and metallic reflectors, especially three-corner reflectors, for radar systems. Some natural targets within the study area may be included as control markers, such as road intersections, buildings, bodies of water, or steel structures and junk yards for radar. Control markers may also provide some measure of imagery resolution if size (photographic), temperature (thermal), or electrical properties and roughness (radar) are recorded.

If imagery is available for a given study area, it can be obtained for only the cost of reproduction. In this case ground truth observations are made at some other time than that of the imagery. Examples of existing imagery are: aerial photography from governmental agencies; Landsat and Skylab satellite imagery; and weather satellite imagery. Even in these cases one may still need ground level observations of the study area for accurate interpretation of the imagery. It must be realized that changes may have occurred since the time of the imagery (e.g., seasonal patterns, diurnal patterns, and physical movement of subject). Although available imagery is less costly, changes may have occurred in subject matter, which produces a lower level of accuracy than when the scene is imaged more nearly to the time of ground truth.

A way to aid in the estimation of environmental conditions is to attempt to simulate conditions present at the time of the imagery. A popular method of simulation is to choose a comparable local time and day for ground truth so that conditions will be similar to those sensed by the remote sensing system.

Computer simulation models are also useful in estimating environmental conditions. Sometimes environmental parameters which are not readily measured at ground level can be studied by simulating conditions present at the time of imagery with computer simulation models. Climatic simulation models have been successfully used to enhance ground truth for thermal remote sensing by predicting the environmental conditions necessary to produce certain surface temperature contrasts. Hydrologic models have been useful in the prediction of water levels in lakes and streams. Also simulation modeling is used with data collected at the time of remote sensing overflight to obtain information about environmental parameters which cannot be measured in the field.

It is common to use remote sensing techniques to measure change through time by obtaining two sets of imagery taken at different times. Aerial photography can be used to show rates of surface flow on glaciers by observing a known point on the surface of a glacier. Also observations of the advance and retreat of glacier termini can be obtained in this manner. Aerial photography or thermal infrared scanning can be employed to survey traffic patterns at various times of day (or night) to provide some estimates of diurnal traffic patterns. Satellite imagery is often used to monitor the shift in cloud patterns associated with the travel of storm centers across the surface of the earth.

Standing on the ground in a study area gives a much different view than from the airborne or satellite platform. Finite point sampling at ground level can often be lost in the spatial generalization of small scale imagery. A comparison of the ground observer's surroundings to satellite imagery may show that the scene which the ground observer is able to discern from one point is lost in one grain of image resolution. This scale problem can persist even with relatively large-scale remote sensing imagery. For example, the aerial view of a forest reveals a surface below; however, is the surface forest canopy, underbrush, leaf litter, mineral soil, or something else? The ground truth observer must have knowledge of the remote sensing system and how the image signal will be produced by the type of surface or, perhaps, surface layers. For these reasons it is often helpful to use cameras and radiometer at, or just above, the surface since they integrate the scene in a manner similar to that of the airborne imaging system.

The problem of choosing a representative sample is too large a subject to treat adequately here. A few general considerations, however, follow. The remote sensing system provides imagery, or information, at a two or three dimensional scale. Usually ground level observations are only point samples. Perhaps it is best to have a few very intensively studied plots within the total study area which, when viewed on the imagery, become control points. A few intermediate points can be monitored less intensively, filling in some of the gaps between the "first order central points." The accuracy of image interpretation will be expected to decrease with increasing distance from ground level control points. The imagery, however, provides data between the control points and extends the observer's knowledge to a broad regional view of the total study area. Other methods of ground level data collection may allow a series of transects to be laid out, criss-crossing the study area. Data collected along these transects may prove useful when the ground observer attempts to distinguish boundaries between surface types (i.e., signal contrast). Also, if a network of transects is dense enough it is possible to get some indication of the density and variation of changing surface types and phenomena within the study area.

Ground level observations should be designed with some idea of the airborne system's resolution in mind. The data collected on the ground should be somewhat compatible with the resolution of the airborne imagery. This applies both spatially and temporally. A single point in time or space will not be representative. On the other hand, a ground truth network which is too dense may do the job well, but will be very expensive in time and money, and may simply duplicate that which the airborne system is providing much more economically. Ground truth must be so designed that it will enhance the interpretation of the imagery and accomplish the task of data collection, while still realizing resolution, accuracy, and generalization characteristics of the remote sensing system.

It is often helpful to have large scale remote sensing data, or imagery, obtained from an altitude between the study area surface and the remote sensing platform. The intermediate level imagery is useful in bridging problems related to spatial scale. Examples of intermediate level imagery are: aerial photographs to compare with satellite imagery; oblique and vertical photographs taken from light aircraft (or a hill above the study area) to compare with remote sensing imagery; and large scale data from thermal scanners and radiometers for comparison with smaller scale remote sensing imagery. Thermal scanners or radiometers can be ground based or flown in light aircraft.

Other Considerations

The type of data collected at ground level may depend upon how certain phenomena are being defined in the study, and the sensitivity of given remote sensing systems. For this reason it is necessary to develop operational definitions. Consider a statewide survey of "wetlands," defining "wetland" as any flooded surface. Once problems of temporal scale (tides and stream levels) have been overcome it is not difficult to survey control points and determine the strand line, i.e., boundary between dry land and water surface. Thermal remote sensing flown on a clear night when the land surface would be cooled, providing good thermal contrast between land and water, would produce fairly accurate results. If, however, "wetlands" are defined by vegetation associations, the ground observers will be working with a totally different set of observations. In this case they are mapping vegetation, and multispectral photography and thermal scanning may provide the best accuracy. On the other hand, a system in the microwave region may provide the general reconnaissance accuracy needed for much less money.

A second example of the need for operational definitions might be an aerial survey attempting to identify city traffic problems. If traffic problems are defined as density of cars per block (e.g., a bottleneck), then vehicles must be counted at given points. If, however, traffic patterns are defined by time-distance, the speed of travel must be observed, and a very different system of remote sensing and ground truth is needed.

In some cases it is impossible to image phenomena under study. In these cases, observations of associated phenomena affected by the subject of the study must be relied upon. This may be true at ground level and in the remote sensing imagery. The ground truth observer and image interpreter must understand the association between the actual subject of the study and the imaged phenomena. Weather satellite imagery is a good example of using associated phenomena to get at the actual subject matter, the primary subject matter in this case being air mass characteristics. Cloud patterns and perhaps surface temperature are imaged by the satellite. The interpreter must then associate these with air mass temperature, humidity, and density. Weather satellite imagery, when compared to ground level weather observation station records, is a good example of using remote sensing to interpolate between point samples, thus providing an overall areal pattern.

In general, as signal contrasts or physical size of the phenomena under study approach the resolution limits of the remote sensing system, greater emphasis must be placed upon ground level observations. In cases where the subject is readily identified on the remote sensing imagery a minimum of ground truth may be involved in a given study. When it is difficult to distinguish and identify the subject on the remote sensing imagery, emphasis must be placed upon accurate ground level observations to provide clues to image interpretation. When successful interpretation of controlled sample points is accomplished, the remote sensing imagery can be used to provide a broad survey of the total study area.

6

Microwave Remote Sensing

Floyd M. Henderson
State University of New York at Albany

James W. Merchant, Jr.
University of Kansas

Introduction

The microwave portion of the electromagnetic spectrum encompasses those wavelengths longer than those of thermal infrared.

It is generally considered that electromagnetic energy having wavelengths between 0.1 cm and 200 cm falls in the microwave region. Since wavelength and frequency are interdependent, the microwave region may be referred to in terms of energy having frequencies of about 100 MHz to 50,000 MHz. Microwave energy, like thermal energy, is invisible to the human eye. A number of remote sensing systems, however, have been developed to detect, measure, and display microwave energy emitted or reflected by land, water, and atmospheric targets. These include both passive sensors (e.g., microwave radiometers) and active systems (e.g., radars).

Most microwave remote sensors operate in one portion, or band, of the microwave region (fig. 6.1). During World War II these bands were assigned letter designations for security purposes. The letter identifiers have remained a part of the remote sensing vocabulary and are used frequently when discussing microwave remote sensing.

Remote sensing in the microwave region complements remote sensing in the optical and thermal portions of the electromagnetic spectrum. Microwave data will be found most useful when used in conjunction with collateral information and imagery. Sensors operating at the long microwave wavelengths do, however, possess many unique characteristics, including the ability to operate day or night and the ability to obtain quantitative data on certain conditions which require sensing beneath the terrain surface (e.g., soil moisture).

This chapter deals with the elements of microwave remote sensing, including the acquisition, interpretation, and application of such data. Because radar is, by far, the most commonly encountered microwave remote sensing system, emphasis is placed upon its operation and use. Microwave radiometers are discussed briefly.

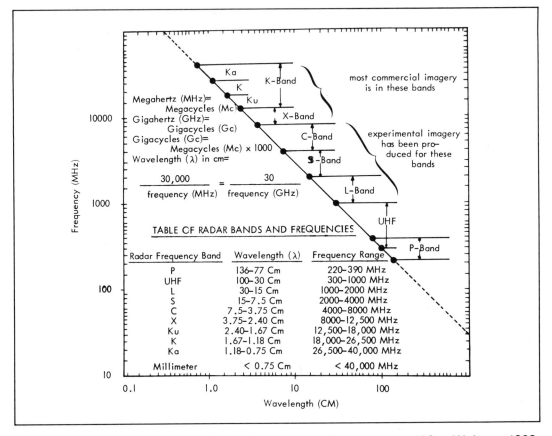

Figure 6.1. Microwave portion of the electromagnetic spectrum. (After Walters, 1968, courtesy of University of Kansas CRINC)

Microwave Radiometry

Background

Microwave radiometers are passive sensors which are designed to detect, measure, and display the natural microwave energy reflected or emitted by earth surface features and the atmosphere. Originally microwave radiometers were developed for use in radio astronomy; however, they are applied now to a variety of earth resources monitoring tasks.

All objects in the natural environment emit or reflect microwave energy. The amount of microwave energy emission is small relative to emission of thermal energy, but it, nevertheless, can be detected and measured, and can provide valuable information about the environment. The magnitude of microwave emission is generally determined by the temperature of the target and the emissivity of the target material. Reflected and scattered microwave energy from other external sources, such as the atmosphere, also may be sensed (fig. 6.2). Some data may actually arise from internal characteristics of the target.[1]

The emissive properties of materials are very important in microwave radiometry. *Emissivity* may be defined as an inherent characteristic of a material which is a measure of the

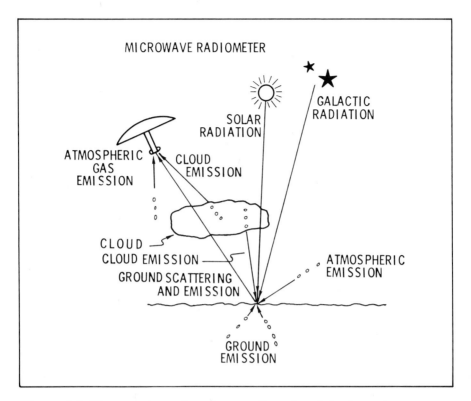

Figure 6.2. The passive microwave radiometer detects and measures microwave energy which occurs naturally in the environment. (Courtesy of Dr. R. K. Moore, University of Kansas CRINC)

ease by which it will give up energy through radiation. If there are two objects at the same temperature in the same environment, with one of the objects having a higher emissivity than the other, the one with the higher emissivity will radiate more strongly than its counterpart. The microwave radiometer measures these differences in the microwave portion of the spectrum in terms of the apparent temperature or "brightness temperature" of the target.[2] This is not the same as the target's true temperature, but is a measure of the total emissive characteristics of the target, its immediate environment, and the intervening atmosphere between target and sensor.

System Characteristics

The microwave radiometer consists essentially of a very sensitive receiving antenna which may be mounted in either a stationary or scanning mode, an amplifier which strengthens the received microwave signal, and a means for recording and/or displaying acquired data. Most passive microwave sensors operate at wavelengths of 0.1 mm to 3.0 cm.[3] At these wavelengths the microwave signal is not seriously attenuated by clouds, moderate rain, or other atmospheric phenomena, and an acceptable resolution is obtained. (Attentuation in remote sensing means a decrease in the flux density the farther a sensor—or antenna in radar—is from the source.) These wavelengths represent a compromise. At longer wavelengths somewhat better atmospheric penetration would be achieved but at a cost of poor resolution. The sensor may, of course, operate at night as well as during the day.

A typical microwave radiometer often employs a reception technique developed for radio astronomy. This receiver compares the signal received by the microwave antenna (antenna temperature or brightness temperature) with a calibrated reference temperature by switching back and forth between the two. The result is essentially a linear measurement of the differences between the brightness temperature and the calibrated reference. The brightness temperature is not the actual temperature of the target but is related to the emissivity and the real target temperature. Other factors which influence the brightness temperature include the polarization of the received radiation, the surface roughness of the target, the angle at which the object is viewed (*angle of incidence*), and the internal noise of the radiometer system.[4]

The detected and amplified brightness temperature is recorded on magnetic tape as a signal proportional to the received energy. Depending upon whether the sensor is operating in a profiling or a scanning mode, the data may then be displayed as either a graph or image, respectively.

If the sensor is mounted in a stationary fashion so that data are collected for only a trace along the surface directly below the aircraft, then the sensor is said to be operating in a profiling mode. The resulting data may be displayed as a graph of brightness temperatures versus distance (fig. 6.3).

Scanning radiometers are designed to collect energy from a broader swath beneath the aerial or satellite platform. The scanner senses along a path transverse to the direction of vehicle movement (fig. 6.4). The resulting data can be presented in an image form where brightness temperatures are represented as various gray tones or colors (fig. 6.5).

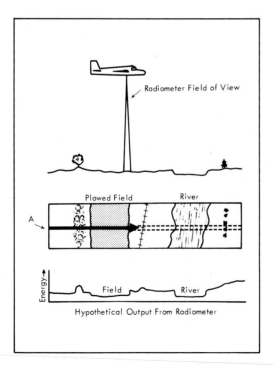

Figure 6.3. Passive microwave radiometer operating in profiling mode. Observe the area sensed along the ground trace (A). (After Scherz and Stevens, 1970; courtesy of authors and University of Wisconsin Institute for Environmental Studies.)

Figure 6.4. Line scanning mode of operation for a radiometer. (After Scherz and Stevens, 1970; courtesy of authors and University of Wisconsin Institute for Environmental Studies.)

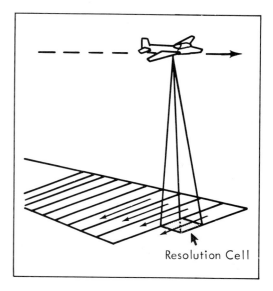

Resolution Cell

MICROWAVE RADIOMETER DATA—NIGHT AND DAY COMPARISON

NIGHT—2500 FEET

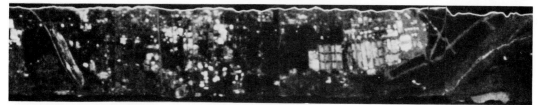

DAY—2500 FEET

AERIAL PHOTOGRAPH

Figure 6.5. Microwave radiometer imagery of San Bernardino, California, compared to conventional aerial photograph. (Courtesy of Naval Weapons Center, China Lake, Calif.)

Image Data Characteristics and Interpretation

Images obtained from microwave radiometers tend to exhibit poor spatial resolution. The spatial resolution of the radiometer is a function of antenna size, distance to the target, and operating wavelengths. Altering any one of these will change the resolution. Because the microwave signal is relatively weak, the radiometer must be designed with great sensitivity and either must "look" at an area for a long period of time, or must collect data for a relatively large field of view (*resolution cell*). Radiometers operated from aerial and space platforms are not able to view any one area very long; therefore, data must be collected and averaged over rather large resolution cells.[5] This is reflected in the rather coarse resolution of the imagery. Since these resolution cells are usually rectangular, the imagery may appear to have a "blocky" structure.

At microwave frequencies the correlation between actual temperature and intensity of signals received is not as good as in the thermal infrared region, but the interpreter can tell a great deal about emissive characteristics of the surface. The emissivities of various materials differ widely at microwave frequencies. Consequently, the amount of energy radiated by objects having the same actual temperature may differ considerably. Boundaries between materials having different emissivities are readily apparent.

Ice, for example, has an emissivity approximately twice that of water in the microwave region. If ice and water with the same temperatures are imaged with a microwave radiometer, the ice will radiate twice as much energy as the water. The water will appear radiometrically "colder" than the ice.[6] On passive microwave imagery these differences in brightness temperature are represented by various tones of gray or by colors.

Capabilities and Limitations

Microwave radiometers provide a unique means to obtain data on surface temperatures and emissive characteristics. They can operate day or night and almost independently of weather conditions. The passive microwave sensor can obtain information on materials which, although they may have the same actual temperature, are quite different. In addition, passive microwave detectors provide a means of obtaining data on subsurface phenomena, such as soil moisture, in some instances.

Present limitations in the use of microwave radiometers include generally poor spatial resolution and the difficulties an inexperienced user may have in interpreting imagery. Nevertheless, research is continuing to improve the imagery hardware, the applications of the data are being defined. Several of these are summarized below.[7]

Applications

Although microwave radiometers generally are considered to be experimental systems, applications have been demonstrated in many areas. Passive microwave sensors have been mounted aboard both aircraft and space platforms. A 1.55 cm (19.35 GHz) radiometer was carried aboard the Nimbus 5 weather satellite, and, more recently, two radiometers (one operating at 13.9 GHz, the other at 1.4 GHz) were utilized from the Skylab manned space laboratory. Laboratories such as those of the Naval Weapons Center (China Lake, Calif.), the Jet Propulsion Laboratory (Pasadena, Calif.), and the National Aeronautics and Space Administration continue to investigate the application of data derived through microwave radiometry.

Oceanography. Microwave emissions have been shown to vary with sea state (roughness), salinity, and water temperature. These types of observations can be invaluable to shipping, fisheries management, and weather forecasting. In certain instances pollution can be detected and mapped. Oil slicks, in particular, have been effectively monitored. Thermal pollution may also be observed. As stated previously, ice and water possess different emissivities and can readily be separated even when their temperatures are similar. Ice itself shows brightness temperature differences according to its surface roughness and composition. Ice of different ages and types, therefore, may be classified and mapped.[8] This may be accomplished either day or night, during polar night (winter), and in spite of inclement weather. Implications for shipping will certainly be great.

Meteorology. Data on sea state can provide much information on surface wind speeds over the oceans. In addition, microwave radiometers which operate at the shorter wavelengths have been found to have applications in the measurement of atmospheric levels of water vapor and the estimation of temperature profiles.

Hydrology. Snow packs can be separated from their environment. In addition, it has been found that the emissivity of snow decreases with increasing water content. Information on extent of snow and moisture content can be a valuable aid in the prediction of runoff. Research has also demonstrated that soil moisture mapping may become operational. An increase in soil moisture has been shown to be accompanied by a decrease in emissivity with brightness temperature.[9] Because longer wavelengths provide deeper surface penetration, it has been suggested that a multifrequency radiometer can produce a profile of soil moisture. Agriculture would benefit from more adequate knowledge of soil moisture distributions.

Land Use and Land Cover Mapping. Several investigations have shown that microwave radiometry is an effective means of terrain and land use mapping, though the technique cannot compete with many other sensors except in cases where inclement weather poses a problem.[10] Geologistis have noted that some rock types appear to be separable on passive microwave imagery, and some success has been obtained in delineating lineaments based primarily upon soil moisture differences. Passive microwave data probably will find its greatest value as a complement to other information obtained from remote sensing and/or ground investigations.

Radar

History and Background

The acronym "radar" was derived by two U.S. Naval officers from the words, *RA*dio *Detection A*nd *R*anging. Although some work with radio waves had been performed as early as the 1900s, the possibility of using them to detect distant objects was not seriously explored until 1922 when the military conducted research in transmitting high frequency radio beams. Employed first as a navigation and air defense aid, refinements and further development continued prior to and through World War II by the armed forces. Early radar systems transmitted from ground platforms to airborne targets and, in some instances, from airborne platforms to ground targets. During this period radar engineers noticed "noise or ground clutter" when monitoring hard targets (i.e., planes and ships) and attempted to eliminate it. Fortunately, it was soon realized that this "noise" was in reality a crude image of the terrain.

The first nonmilitary application and report of this "ground clutter" was made in 1948 by Lt. H. P. Smith using radar scope photographs from a Plan Position Indicator (PPI), a radar screen familiar to many people. Typically it has a faint line that sweeps around a circular screen like a spoke on a slowly turning wheel. Planes or other recorded images are seen as phosphorescent green "blips" on a dark background. When Smith compared these radar images with existing maps and charts of northwestern Greenland, he observed that the radar images contained detailed landform information exceeding that available from the maps. Smith also recognized the similarity of the PPI display to a map and the relationship between the intensity or brightness of the signal return from the ground and the types and configurations of landforms. From this work he concluded such images might have potential for terrain studies.

Following World War II, interest in radar declined with the reduction of military intelligence activities, and no appreciable advances were made until the early 1950s. At that time the need to track missiles and satellites led to the development of more complex components. Simultaneously, the development of supersonic manned bombers precipitated the need for highly accurate navigational and bombing radars. Since the mid-1950s, radar technology has been continuously refined. Shortly thereafter the civilian scientific community began investigating radar imagery systems to define possible nonmilitary applications. Thus, in comparison to some of the more traditional systems, such as aerial photography, radar is a relatively new remote sensing tool.

System Characteristics

Radar is an active remote sensing system. The word "active" means that it sends out its own microwave signal and measures the strength of the signal as it is reflected from objects. Passive sensors, on the other hand, sense only the energy provided by nature. Most radar systems operate at wavelengths between 1 mm and 11 cm. In simplest terms, radar is a system which uses microwaves instead of light waves to image an object. This is accomplished by sending out short powerful pulses of microwave-frequency energy at regular intervals in a certain direction. When these pulses strike, or illuminate, a target, they are reflected back to the source (receiver), whereupon the time and distance to the target and the intensity of reflection can be measured and displayed.

A simple ground to air radar system will illustrate the radar operation. A transmitter on the ground emits a burst of microwave-frequency electromagnetic energy. When the signal is interrupted by an object, such as an aircraft, it is reflected back to the receiver. The velocity of electromagnetic waves in free space is 3×10^8 meters/second (approximately 1,000 feet/microsecond). To detect an object one nautical mile away (1,853 meters or 6,080 feet), therefore, it will require slightly more than 12 microseconds for the transmitted energy to reach the target and return.

If the transmitter were in continuous operation, it would be very difficult to separate the transmitted and reflected signals. A common practice, therefore, is to transmit a short pulse of energy at delayed intervals. This enables one transmitted pulse to be reflected to the receiver before another is transmitted. Thus, the transmitter and receiver do not operate simultaneously, and signal confusion is avoided. Moreover, the same antenna can be used for both transmitting and receiving.

There are many possible and complex transmitter-signal configurations depending upon the intended use of the radar system. For geographers and other scientists interested in studying environmental phenomena on the earth, the most important is a radar system referred to as side-looking. Side-looking radars scan a path to the side of the flight path of the aircraft carrying the radar system. Since the antenna looks at right angles to the direction of flight, this system is referred to by the acronym SLAR for Side-Looking Airborne Radar.

An area being imaged by a SLAR system is depicted in figure 6.6. The transmitter generates short bursts of microwave energy. These pulses are propagated into space at right angles to the flight path by means of a directional antenna (A), and radiate from the antenna as a block of energy (B) at the velocity of light. As can be seen, the energy pulse is confined to a narrow path on the ground. At any one instant the terrain illuminated by a transmitted pulse is limited in the range direction (direction at right angles to the flight path) by pulse duration and in the azimuth direction (direction along the flight path) by the beam width of the antenna. The radar resolution will be primarily determined by these two factors. If a terrain feature intercepting a signal is illuminated at point (a), a fraction of the transmitted energy will be reflected back

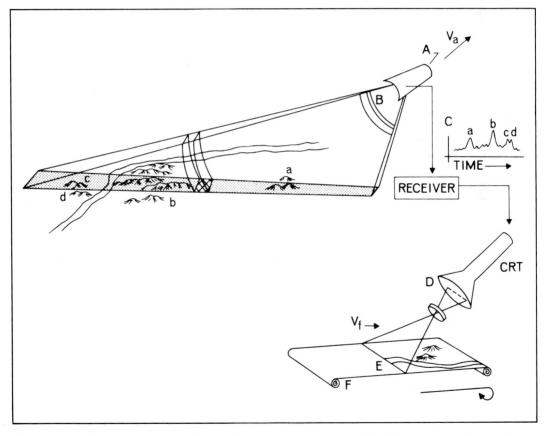

Figure 6.6. Real aperture side-looking airborne radar (SLAR) system. (From Mac-Donald, Lewis, and Wing, 1971; Modified from Westinghouse Electric Corp. diagram; reprinted from the *Geological Society of America Bulletin* by permission of the Geological Society of America.)

to the antenna. An object at point (b) will also be illuminated by the pulse, but at a later time because it is further from the antenna. Reflected energy from point (b) will thus be received by the antenna later. The same applies to features (c) and (d). The reflected energy received by the antenna from points (a), (b), (c), and (d) is converted by the receiver into video electrical signals of varying intensity depending on the shape, composition, size, and distance from the antenna of objects.

The antenna is repositioned literally at the speed of the aircraft (Va). Each pulse transmitted returns signals to the antenna from the terrain features within the beamwidth. These returning signals (echoes) are then converted into a time-amplitude video signal which is imaged across the face of a cathode ray tube (CRT) and onto film (F) as a scan line (E). The strength of the CRT scan line varies as a function of the strength of the echoes. By moving the photographic film past the CRT display line at a velocity (V_f) proportional to the velocity of the aircraft, an image of the terrain is recorded on the film as a continuous strip (fig. 6.7).

This type of radar system is referred to as a real aperture, noncoherent, or brute force system. The system records the signals on film as soon as they return, and the beamwidth of the signal is a function of the length of the antenna and the wavelength of the radar signal. That is,

$$\text{beamwidth} = \frac{\text{wavelength}}{\text{antenna length}}$$

Since the narrower the beamwidth the better the resolution, it is necessary to shorten the wavelength or lengthen the antenna if improved detectability and resolving power are desired.

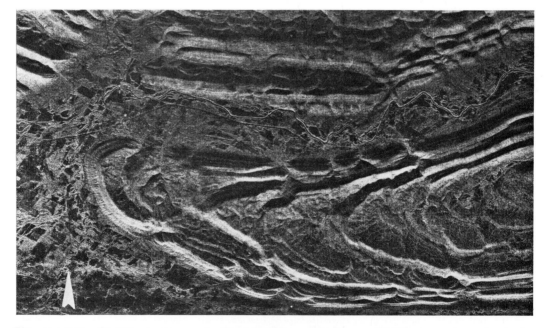

Figure 6.7. SLAR image of Tuskahoma syncline, Oklahoma, acquired by Westinghouse K-band SLAR system. Look direction is indicated by arrow. Note shadowing. (Courtesy of Westinghouse Electric Corp.)

As the wavelength becomes shorter, however, the radar is subject to greater attenuation from clouds and particles in the atmosphere—similar to the way blue wavelength light is more easily scattered than red wavelength light in the visible portion of the electromagnetic spectrum. On the other hand, if the antenna becomes too long, it physically cannot be carried by the aircraft. In a brute force system the beamwidth is constant; that is, it is narrower near the aircraft than it is farther away. Consequently, resolution is better closer to the aircraft. Because of these limitations, this radar system is of optimum use when short-range, low-level operations are desired. Some problems emerge when images of large areas from high aircraft altitudes or even from space are desired.[11]

Synthetic aperture or coherent SLAR systems have been designed to overcome the restrictions imposed by real aperture systems. Synthetic aperture antennas produce a narrow effective beam without requiring a long actual antenna or short wavelengths. A physically short antenna transmits and receives pulses at regular intervals along the flight path. These received signals are not transmitted directly to the CRT and a photographic film strip as is the case in real aperture systems. Instead, the signals are stored and then added electronically to produce the effect of a physically long antenna. For example, the actual antenna may be only three feet in length, but the process may create a synthetic antenna of 200-300 feet since the synthetic antenna length is determined by the length of time (distance) the aircraft is receiving signals from an object.

In a real aperture system only a single pulse is transmitted, received, and recorded as a line of the image. For a synthetic aperture system, each target is viewed and imaged a number of times by the antenna, producing a large number of return pulses. These phase histories of pulses are stored on a data film until all the signals have been received. Subsequently, the numerous views of a target, as well as the record of signal intensity, are electronically processed and combined (sometimes even after the plane has landed). These composite signals are then transmitted across the face of the CRT on a film strip.

Synthetic antennas obviously are much more complicated than real antennas because of the extremely complex operations involved. By utilizing the system, however, improved resolution can be achieved without employing shorter wavelengths which are more susceptible to atmospheric attenuation. Synthetic antennas are desirable when both high-altitude, long-range operations and short-range operations are required of the same radar.[12]

Characteristics of SLAR Imagery

Near, Mid, and Far Range

Near, mid, and far range are terms used to describe locations on radar imagery in relation to distance from the imaging aircraft. Although precise formulas and calculations exist for determining these locations, they are beyond the scope of this text. In general the radar image can be divided into three segments perpendicular to the flight path. That portion closest to the aircraft is considered near range; the middle third is midrange; and the section farthest from the antenna is far range.

Slant Range, Ground Range

Depending on the design, SLAR imagery may be recorded in either a slant range or a ground range presentation. These formats are illustrated in figure 6.8 where h = height of imagery aircraft, S_r = slant range, and G_r = ground range.

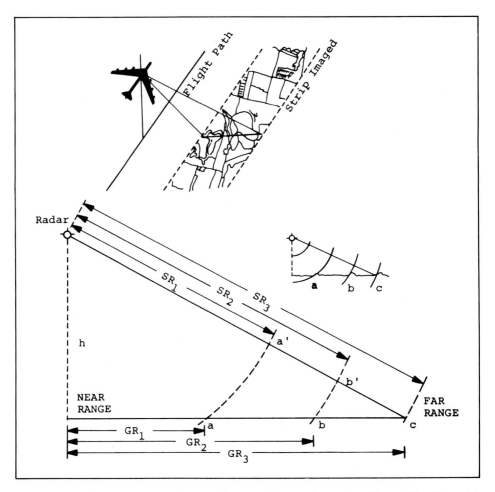

Figure 6.8. Slant range and ground range image formats. (From MacDonald and Waite, 1971a, reprinted from *Modern Geology* by permission of Gordon and Breach, Science Publishers, Inc.)

Radar records targets with respect to their direct line distance from antenna to the target, that is, the slant range. Since the slant range sweeps are linear, the spacing between returning signals is directly proportional to the time interval between the imaged terrain objects. The distortions that result from a slant range presentation are depicted in figure 6.9. Note the compression of scale in the near range. Objects a, b, and c represent objects of equal size located at increasing distances from the SLAR antenna (i.e., a = b = c). Compression is greatest at a, the object closest to the aircraft, and decreases outward; thus, on a slant range presentation A < B < C and hyperbolic distortions result. By applying a hyperbolic correction for this distortion (an electronic process performed in the CRT), a ground range presentation can be obtained, and the scale relationship between image and ground range presentation becomes linear (fig. 6.9). This correction process becomes increasingly more difficult as the terrain becomes rougher.[13]

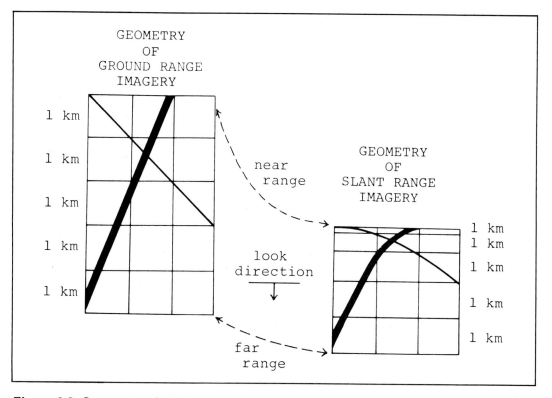

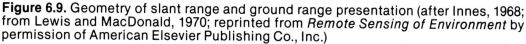

Figure 6.9. Geometry of slant range and ground range presentation (after Innes, 1968; from Lewis and MacDonald, 1970; reprinted from *Remote Sensing of Environment* by permission of American Elsevier Publishing Co., Inc.)

Ground range presentations are often derived when the work requires planimetrically correct results. Two points should be made in this regard. First, whereas the distortion on a vertical photograph increases radially from the center to the edges, the distortion displacement of a slant range radar image is a function of distance from the radar antenna. Second, by utilizing a ground range presentation, it is possible to approximate the appearance of a vertical photograph, but the image will contain shadows similar to an oblique photograph.

Incidence Angle and Depression Angle

The angle of incidence is that angle formed at the point of contact with the ground by a transmitted beam of radar energy and the perpendicular of the incident surface. This can be seen as angle θ in figure 6.10. The angle between a line from the transmitter to a point on the ground, and a horizontal line intercepting the transmitter is the depression angle. As illustrated in figure 6.10, the geometry of SLAR systems is such that along the swath width (near to far range) of an imaged area there is a continuous change in the angle of incidence. When homogeneous flat terrain is imaged and the depression angle along the flight path is constant, the angle of incidence at a given range will remain constant. Natural surfaces, however, typically contain variations of slope which in turn can alter the incidence angle. For level terrain the angle of incidence is the complement of the depression angle.[14]

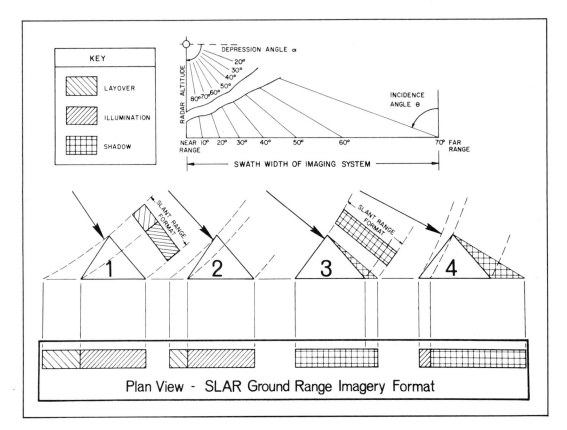

Figure 6.10. Relationship between radar layover and shadowing, terrain slope, and radar depression angle. (From MacDonald and Waite, 1971a; reprinted from *Modern Geology* by permission of Gordon and Breach, Science Publishers, Inc.)

If terrain slopes are inclined at an angle toward the imaging radar and the effective angle of incidence decreases to a point where the angle of incidence equals the angle of slope, vertical incidence (maximum reflection or brightest return) results. Conversely, if the terrain slopes away from the impinging signal to a point where signal grazing results, maximum shadows (dark or black areas on the image) are observed. These angles of terrain illumination are somewhat analogous to the sun's position in vertical aerial photography. When the sun is at its zenith, shadows are at a minimum, but shadowing increases as the sun approaches the horizon. Rugged terrain slopes in the near range of a SLAR image, therefore, may be illuminated while the same type of slope in the far range may be in complete shadow. Such data have been used to compute landform slopes from SLAR imagery obtained over rugged, inaccessible areas.

Radar Foreshortening

Radar foreshortening is a phenomenon encountered when imaging uneven terrain. When slope measurements are made at different incidence angles there is a variation in the length of equal terrain slopes as seen by radar. That is, the length of time a slope is illuminated by the radar pulses determines how the terrain slope will appear. Thus, the length of the terrain slope, and by implication its appearance on the imagery, is a function of the incidence angle and the terrain slope.

Distances between terrain features on radar imagery are recorded as a function of the time required for the microwave energy to traverse this distance. Time intervals between equally spaced radar energy pulses as they intercept terrain features at various distances or ranges from the antenna are illustrated in figure 6.10. The actual lengths of the fore (toward radar) and back (away from radar) slopes of features 1, 2, 3, and 4 are equal. On radar imagery, however, the fore slope of feature 2 will appear shorter than the back slope, since radar uses time as a discriminant. Note the change in slope length with changing incidence angle. Terrain features sloping toward the radar will appear shortened on the imagery as opposed to those sloping away.[15] While this characteristic may appear to be a handicap, it can be put to good use. If the range position of an object is known, a fairly accurate measurement of terrain slope can be ascertained by incorporating the foreshortening characteristic. This method may not be the most desirable; however, it may nevertheless provide data where none previously existed.

Radar Layover

Returning or reflected radar signals are directly proportional to the time interval (distance) between the transmitter and the imaged terrain object. Radar layover is an extreme case of relief displacement which is often encountered when steep slopes are present in the near range. In such instances the tops of terrain features on the imagery appear to lean toward the aircraft. Radar layover is a result of the impinging radar beam intercepting the top of a terrain feature before it comes in contact with the base. Such a case is shown in figure 6.10. The first point detected by the wavefront generated from the antenna is the top of feature 1; this point will be recorded first. As the wavefront progresses across the terrain, the base of feature 1 is less than the distance to the base. With feature 3, both points are recorded simultaneously, even though there is an actual distance difference on the ground. Terrain feature 4 is imaged in a normal fashion since the slant range distance to the base is less than to the top. When layover occurs, the top of a terrain feature, such as a hill or mountain, is recorded before the base, and the object will appear to lean toward the antenna.[16]

Radar Shadow

Since radar is an active sensor, it provides its own illumination; it acts as a "sun" when imaging terrain. Consequently, in areas of even moderate relief, the side of the terrain feature farthest from the antenna may be in shadow. Some of the shadow characteristics associated with SLAR systems are illustrated in figure 6.10. The shadow of terrain features of equal height increases from the near to far range (i.e., the angle of incidence increases). These cases are evident in the illustration: (1) the backslope is fully illuminated producing no shadow; (2) the backslope is partially illuminated producing a condition referred to as "grazing;" and (3) the backslope is concealed from the radar beam producing a full shadow or "no return area." Shadow areas are most often found in the far range at low depression angles.[17] As with foreshortening, the analysis of radar shadow in relation to range and angle of incidence can be utilized in slope measurements to group terrain units into classes. Since radar shadows may be controlled through careful mission planning, they may be used to deliberately highlight certain terrain features.

Polarization

Polarization refers to the orientation of the electric and magnetic components of a radar signal which is transmitted to and received from the ground (fig. 6.11). Specifically, signals are transmitted and received in either a horizontal or vertical plane, producing "like-polarization" images (HH—sent horizontal, received horizontal; or VV—sent vertical, received vertical) and "cross-polarized" images (HV—sent horizontal, received vertical; or VH—sent vertical, received horizontal). In a basic SLAR configuration, this electronic signal is usually transmitted and received horizontally, producing a "like-polarized image. Many SLAR systems, however, are "dual" or "multiple polarization" systems. That is, they transmit and receive signals in two or more polarizations of the same area. The result is similar to that of a multiband camera system in that many images of an area are produced, each containing slightly different information about the environment. In the case of radar, the impinging signal interacts with the terrain feature and is depolarized and rotated to varying degrees, but only the reflected horizontal and/or vertical signals that return in the direction of the antenna are received and recorded. Comparison of the multifrequency imagery can be a substantial aid in interpretation (fig. 6.12).

Multifrequency Radar

Early SLAR systems operated at single wavelengths, and most commercial systems still do. This is analogous to a camera system equipped with a filter that would only allow wavelengths of .55 mm to pass through and expose the film. Obviously, such a procedure severely limits the amount of potential information available. As a result, radar systems are being designed to

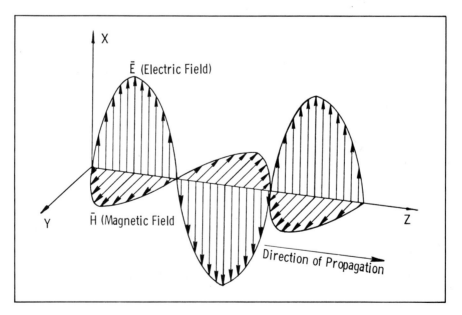

Figure 6.11. Polarization refers to the orientation of the electric and magnetic components of the electromagnetic wave. (From Dellwig, MacDonald, and Waite, 1973, courtesy of University of Kansas CRINC)

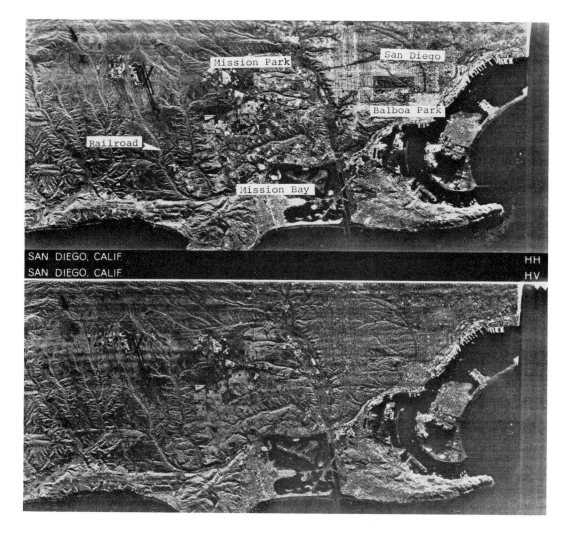

Figure 6.12. Use of multipolarized imagery in interpretation. Imagery was acquired by Westinghouse K-band system. (From Lewis, 1973; imagery courtesy of Westinghouse Electric Corp.)

send and receive signals at many frequencies. Such a multi-frequency system operating at M wavelength and employing N multi-polarizations would yield MN distinct images of the terrain. Examining and identifying objects based on variations in spectral as well as polarization responses quickly expands the number of interpretation techniques and information obtainable from the imagery. One of the problems, however, in such a system is designing and operating the tremendously complex equipment needed to keep all the signals separated and properly recorded.

Radar Stereoscopy

Relief displacement is an inherent characteristic of SLAR systems. If the object is above the datum it will lean toward the radar; and away from the radar if the object is below the datum. This is the opposite of the displacement of optical camera systems. When an object is imaged twice—from different look directions or from different heights, the latter producing different angles of incidence, radar parallax can be measured and stereoscopy observed. The characteristics of radar systems, however, make stereo interpretation very difficult in comparison to aerial photographs. Fore slopes may be illuminated to different degrees by changing altitudes and may also change appearance or shape. Back slopes may be in shadow in one case and become foreshortened fore slopes with a new look direction. Radar images taken from different directions will result in effects similar to taking aerial photographs in early morning and again in late afternoon. For these reasons, same-side, same-height coverage, with 60 percent side lap is preferred for resource evaluation and geoscientific analysis.

Radar Interpretation

Many geographers, as well as other scientists, are familiar with aerial photograph interpretation techniques (see chapter 3). For those persons considering radar as an educational or research tool, it is significant to note that SLAR imagery is very similar in appearance to black and white aerial photography. Consequently, the basic interpretive techniques developed for photographic analysis are applicable to radar systems. Analyses of tone, texture, shape, and pattern are recognition elements which contribute to interpretation of geographic data from radar imagery as well as from aerial photography. It should be remembered, however, that radar is an active sensor and the *causes* of tone, texture, shape, and pattern variation are different from those associated with passive sensors operating in this portion of the electromagnetic spectrum. In this section the characteristics and properties of radar imagery are discussed as they relate to imagery interpretation.

Surface Roughness

Surface roughness is the most important factor in determining the strength of the returning signal and thus the image tone. Roughness of the terrain, such as forest, grasses, bare ground, soil particle size, and water, is relative to wavelength of the radar signal and the incidence angle. Surfaces with microrelief less than a wavelength wide will appear smooth (little or no return); surfaces with roughness equal to or greater than a wavelength will appear brighter. A surface which may appear smooth at one wavelength may appear quite rough at a longer or shorter wavelength. Also, a surface that appears rough in the near range may appear smooth at angles near grazing because of the angle at which the signal impinges on the objects (the angle of incidence). Examples of surfaces which give little or no return and appear black on a radar image are water, paved roads and highways, and airport runways. Forests, brush, grasses, and rocky terrain tend to appear as various shades of gray depending on the wavelength of the system and incidence angle (fig. 6.13).

Returning radar signals to a degree may come from and be influenced by subsurface features. In vegetation this means the signal may penetrate through many of the leaves and branches so that the return signal is a composite of signals coming from the surface visible to the eye and from regions within the plant and/or beneath the vegetation canopy. In soils this

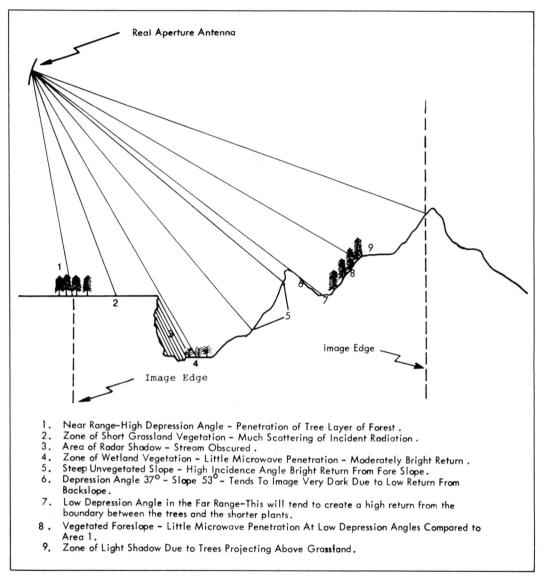

Figure 6.13. Influences of surface features on radar return imagery. (From Hardy, 1972, courtesy of University of Kansas CRINC)

1. Near Range–High Depression Angle – Penetration of Tree Layer of Forest .
2. Zone of Short Grassland Vegetation – Much Scattering of Incident Radiation .
3. Area of Radar Shadow – Stream Obscured .
4. Zone of Wetland Vegetation – Little Microwave Penetration – Moderately Bright Return .
5. Steep Unvegetated Slope – High Incidence Angle Bright Return From Fore Slope.
6. Depression Angle 37° – Slope 53° – Tends To Image Very Dark Due to Low Return From Backslope.
7. Low Depression Angle in the Far Range–This will tend to create a high return from the boundary between the trees and the shorter plants.
8 . Vegetated Foreslope – Little Microwave Penetration At Low Depression Angles Compared to Area 1.
9. Zone of Light Shadow Due to Trees Projecting Above Grassland .

means that not only the surface roughness, but the characteristics of the near subsoil govern the signal return. When vegetation is sufficiently sparse or dry, or the wavelength long enough, the return signal from the vegetated surface includes both vegetation effects and soil effects. The degree of signal penetration depends upon various factors. In general, however, the signal penetration increases with increasing moisture content and longer wavelength. Thus, a 1 cm radar would image only the upper layers of a forest or a cornfield, while a 30 cm radar would image a combination of leaves and trunks in a forest, and of leaves, stalks, and perhaps even soil in a cornfield. If the cornfield were dry, the return from the 30 cm radar would be determined primarily by soil effects. On the other hand, if the plants contain much moisture (as at the height of the growing season), even the 30 cm radar return may be due primarily to the plants themselves.[18]

In the microwave region surface roughness must be in terms of wavelength. Surfaces that would appear very rough to the human eye can be very smooth at centimeter wavelengths. Small surfaces, however, such as leaves, that appear smooth at optical wavelengths may be only a fraction of a wavelength across at centimeter wavelengths. They thus scatter incident microwave energy uniformly in all directions, but reflect light in only one predominant direction. A plowed field may appear very rough at centimeter wavelengths, but smooth at a meter wavelength. The same field, therefore, might appear either bright or dark on an optical image, bright on an image made wtih one centimeter wavelength radar, and dark on a one meter wavelength SLAR image. Complete understanding of these geometric effects on radar return is still the subject of considerable research.[19]

Although at first glance a SLAR image appears much the same as a black and white air photo, the user must be aware of the differences between them. Shadows, for example, appear on photos as a function of camera and sun position. SLAR provides its own illumination and shadows are always created on the side of the object away from the radar position. Operating at a very low incidence angle the SLAR image has the appearance of a photo taken from a vertical camera at low sun elevation. However, since SLAR is a ranging sensor, the compression of the far range at low oblique viewing angles does not occur as in a camera.

The SLAR user also must be cognizant of the fact that the radar image represents reflected microwave energy and not optical wavelengths. Objects that appear identical in an aerial photograph may appear quite differently at microwave frequencies. This represents one of the advantages of using SLAR to complement other sensors. Yet SLAR image interpreters can use many of the same tools as aerial photograph interpreters if the aforementioned differences between the two image types are kept in mind. Standard photograph interpretation features, such as shape, pattern, size, tone, and shadow, although modified, can be used in interpreting SLAR imagery, but uses should be aware of the differences between the SLAR image and the aerial photograph. Coiner and Dellwig have noted some of the similarities and differences between aerial photograph and radar image interpretation (fig. 6.14).[20]

Tone

Tone may be defined as the intensity of signal backscatter, converted to a video signal and recorded on photographic film as shades of gray from black to white. Tone on radar imagery is influenced by the backscatter, or signal return, as a function of terrain geometry (micro and macro), environmental elements (e.g., vegetation types), surface roughness, and incidence angle. Tonal variations can be observed on radar imagery and analyses of them can contribute much to geographic interpretation. For example, black, or no return areas, on an image may be water bodies or areas of radar shadow; white or light gray may indicate settlements, houses, cities, or crop types; and shifts in gray tone may delineate soil types, soil moisture, vegetation communities, and land use. It should be noted that the tone of an object also can vary on radar imagery as a function of polarization, wavelength of the signal, and location of the image (i.e.,

INTERPRETATION CRITERIA

SLAR IMAGE *PHOTOGRAPH*

TEXTURE—

Also strongly influenced by system and target parameters--

TEXTURE = or ≠ TEXTURE

SHADOW and ILLUMINATION—

SYSTEM CONTROLLED   NATURE CONTROLLED

SHAPE—

If objects are smaller than resolution cell size--

The more the size of the object exceeds the size of the resolution cell, the more nearly--

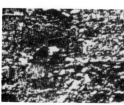

SHAPE ≠ SHAPE

SHAPE = SHAPE

PATTERN—

Although Individual objects lose shape--

PATTERN = PATTERN

TONE—

Sensing roughness and dielectric constant and influenced by wave-length, aspect angle, and resolution--

TONE ≠ TONE

Figure 6.14. Comparison between image attributes of shape, pattern, tone, texture, size and distance, and shape and illumination as they influence the interpretation of aerial photography and SLAR imagery. (From Coiner and Dellwig, 1973, courtesy of University of Kansas CRINC)

near or far range). Note in figure 6.14 that image tone may vary between images acquired with different bands in the microwave region. Imagery taken in the P and K bands is illustrated.

Texture

Texture may be defined as the degree of roughness or nonhomogeneity (smoothness) on an image. Although difficult to quantify, the qualitative attributes of coarse, medium, or fine texture can lend much to the ability to discriminate among phenomena on radar imagery. A forested area may appear speckled and groves of trees as small bumps. Texture has been an aid in inferring rock types, differences in soils, crop types, and urban land use features.

The texture of a feature also may vary in reference to the wavelength of the radar system. For any system, however, there are three relevant types of textures on an image—micro, meso, and macro texture. Micro texture is related to electronic characteristics of the radar signal and is on the order of and related to the resolution cell. Micro texture results from the appearance of several resolution cells which produce light and dark areas on the image. For example, differences in meso texture often provide clues to such vegetation entities as marsh, pine forest, or burn areas. Macro texture is that observed when gross variations in the geomorphology or geology of the terrain are noted. Micro texture tends to be random because it is more a function of the system than the terrain. Meso and macro textures, on the other hand, exhibit a spatial orientation related to and dependent upon the terrain feature being imaged.

Shape

Shape is defined as the geometric configuration or spatial form of an object or objects. Cultural features on the landscape are generally more regular in shape than natural elements. Knowledge of the shape of objects such as field patterns, road networks, reservoirs, alluvial fans, or stream types are useful when interpreting natural features.

In the interpretation of medium and large scale aerial photographs, the shapes of objects are primary factors in their identification. The geometry of individual scene objects is expressed in terms of differences in their shape (e.g., object A is round, object B is square; therefore, object A is not object B). This reliance on shape is reinforced by the object's familiarity to the photograph interpreter. On aerial photographs objects tend to retain their natural shapes and thus are often recognizable on that basis.[21]

A scene consists of an aggregate of individual scene units. For instance, a building, a smokestack, a cooling tower, and a transformer yard are individual scene units which, when aggregated, are interpreted as a power station. On commercially available SLAR imagery, the basic scene unit often loses its shape because of resolution and microwave reflection properties of the individual scene objects. For example, in an image of an airport, the support facilities, administrative buildings, and terminals may be agglomerated into clusters of high returns (bright areas). The functions of these facilities thus can only be deduced from their positions relative to the runways and each other. Suppression of individual scene units tends to emphasize regional shapes and structures. In an urban scene, when a small scale photograph and a SLAR image are compared (fig. 6.14), the overall urban region is more easily defined on the SLAR image. The merger of basic scene unit shapes tends to enhance the regional units, such as central business districts and residential areas.[22]

Pattern

Pattern is defined as the repetitive spacing of units or arrangement of terrain features which, in combination, aid in the identification of a scene or scene elements. For example, the regular spacing of houses, roads, yards, and automobiles can lead to numerous inferences about suburbs on an aerial photograph.[23] Settlement patterns indicate a certain cultural background or agricultural system. Since the acquisition scale of most radar imagery is on the order of 1:150,000 or smaller, detailed terrain patterns are not generally detectable. A pattern of image textures from which broader, more general observations are made are seen. This is true of urban and other cultural features, as well as geologic and geomorphic elements such as lineaments, lithology, soil patterns, drainage patterns, or vegetative patterns.

Distortions

The distortions inherent in SLAR systems have already been discussed concerning layover, foreshortening, and range. There are other geometric distortions, however, which may appear on an image and precipitate a loss of resolution or change of scale, such as those related to aircraft motion. Short term distortions are caused by a change in the position or attitude of the aircraft. On the image film these may appear as a smear or other obvious anomaly. Distortions that become apparent when attempts are made to mosaic strips of imagery or match a strip of imagery to a map are of long term nature. Drifting of the aircraft from the straight ground track can result in a circulinear geometry. As the aircraft attitude and speed change, expansion or compression in the azimuth (along track) scale. The distortion apparent on a radar image is a combination of short and long term anomolies in aircraft motion.

As is the case in aerial photography, these distortions are minimized as much as possible in mission planning and actual flight. Both physical control and recorder compensation techniques are employed to mitigate the problem. Only when space platform radar systems are developed, however, can any major abatement of this problem reasonably be expected (similar to that which occurs when satellite and aerial photography are compared).

Radar Mosaics

Although the construction techniques of radar mosaics have improved greatly in recent years, a number of problems still exist. This is particularly true when ground control is lacking or inadequate and/or a short range format is employed. The problems encountered when assembling slant range imagery into mosaics are similar to those found in assembling mosaics with oblique aerial photographs. On slant range imagery the far range is the most geometrically correct in the range scale. In the near range there are problems of radar foreshortening and layover. Some areas will be omitted, others will be redundant or offset if sufficient ground control is lacking. While these images may prove sufficient for broad scale general surveys, they are too inaccurate for planimetric mapping or detailed analysis. Use of ground range presentation does alleviate many of the problems inherent in the slant range format but presents problems in mountainous areas in that radar layover is maximized.[24]

One of the principal advances in assembling radar mosaics is the incorporation of Landsat imagery. When Landsat imagery is used in combination with SLAR imagery, costs are reduced and accuracy improved. The Landsat imagery provides an economical, reliable base for geographical and positional control, particularly at scales near 1:250,000. Consequently, fewer radar ground control points are required to achieve an acceptable product, and less time is needed to assemble the radar mosaic.

Interpretation Keys and Procedures

Interpretation keys similar in style and format to those employed in aerial photograph interpretation have been developed for use with SLAR imagery. These may take the form of descriptive clues, such as "rectangular field pattern," "a series of white dots in linear array," or "a mottled appearance along streambeds." More specific keys may incorporate variations observed from near to far range, different polarizations, multipolarization, or time of year. Chips or samples of radar imagery and/or training sets may be given to assist interpretation. When extremely detailed analysis is required, a matrix (fig. 6.15) may be designed to examine and delineate the relationship of multiple parameters of an entity, such as soil type, crops, or vegetation. For example, a comparison of the tone and texture variation on HV polarization may be matched with tone and texture appearance on HH polarization to provide specific guides in identifying certain vegetation species. As another approach, a dichotomous key (fig. 6.16) may be designed in which the interpreter need answer only "yes" or "no" to a list of questions that proceed from general to specific, detailed statements.[25]

Such qualitative methods provide valuable tools in image analyses; however, some problems are so convoluted or involved that they are beyond the scope of such procedures. In these cases, machines assist the user in interpreting imagery. The film may be digitized on a computer, electronically enhanced, color combined, or scanned by densitometers to identify shifts in gray level or texture changes too subtle for the human eye to discern. Or, once the machine has been given the proper guidelines, it can be used to provide the data much faster than would be possible by a person examining the study area. In short, most of the qualitative and quantitative procedures utilized in analyses of other forms of remote sensing imagery can be and are employed in interpreting radar imagery, but the user must adapt them with the unique traits of this active sensor in mind.

Capabilities and Limitations

Many characteristics of radar imagery are advantageous to specific problems, such as subsurface penetration and ability to assign and change polarization. While the utility of such functions cannot be denied, a discussion of such uses is beyond the scope of this text.

The unique capabilities of radar most often cited are: (1) weather and time of day independence and (2) synoptic view. Because radar is an active sensor it can be flown any time day or night. The wavelengths at which radar systems operate allow the signals to illuminate surfaces regardless of most weather conditions. Areas that are perpetually cloud-covered or have rainy environments, therefore, can be imaged virtually any time of year or day in a relatively short time period. The synoptic view afforded by radar imagery allows large extensive areas to be imaged quickly on a relatively small quantity of film. Moreover, the film is a continuous strip and not a series of individual photographs. Both the time needed to assemble mosaics and identify imagery and the space needed to store imagery are much less than with normal aerial photography.

Any remote sensing system not only has advantages but also disadvantages, and radar is no exception. No sensor is a panacea to the needs of all potential users, in comparison to many

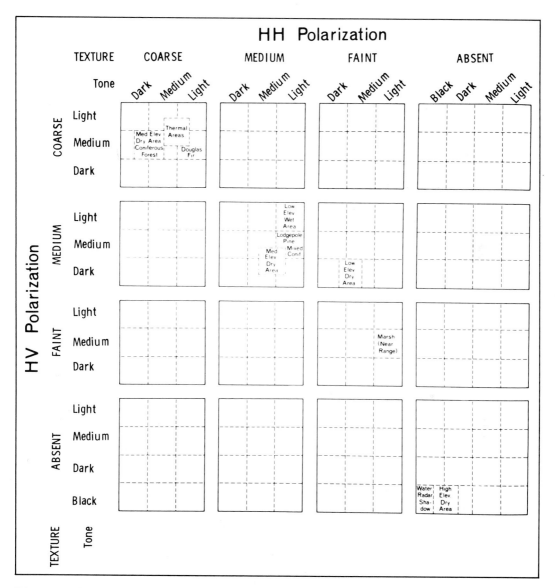

Figure 6.15. The matrix key utilizes attributes of tone and texture in both like and cross polarizations. This example was prepared by Norman E. Hardy for interpretation of vegetation patterns in Yellowstone National Park. (From Coiner and Morain, 1971. Copyright 1971 by the American Society of Photogrammetry. Reprinted with permission.)

other systems. Radar imagery is relatively expensive to acquire. The scale and resolution of unclassified SLAR systems are less than that possible with aerial cameras and conventional film. For many needs, therefore, conventional photography is more economical and provides as much or more data as radar imagery.

```
A     Field is light gray to white on HH ---------------------------- Go to B
A'    Field is not light gray to white on HH -------------------------- Go to D

B     Field gray tone shifts from light gray/white HH to medium gray HV-  Go to C
B'    Field gray tone shifts HV lighter than HH  ------------------------------ cut alfalfa

C     Field gray tone on HV homogeneous  ------------------------------------ sugar beets; or wheat >3"
C'    Field gray tone on HV not homogeneous  ---------------------------------- fallow

D     Field has medium to dark gray tone on HH  --------------------- Go to E
D'    Field has very dark gray tone on HH  -------------------------------------- recently tilled

E     Field gray tone is homogeneous  ----------------------------------- Go to F
E'    Field gray tone is not homogeneous ------------------------------ Go to I

F     Field has lineations parallel to long axis -------------------------------- maturing alfalfa
F'    Field does not have lineations  --------------------------------- Go to G

G     Field has medium coarse texture  ----------------------------------------- grain sorghum (rows ⊥ flight line)
G'    Field does not have medium coarse texture --------------------- Go to H

H     Field has same gray tone on HH and HV ---------------------------------- wheat >3"
H'    Field has moderate gray tone shift HH to HV ----------------------------- alfalfa >12"

I     Field has a cultivation pattern observable -------------------------------- emergent wheat
I'    Field does not have cultivation pattern observable but ------------------ mature corn
      displays pronounced boundary shadowing
```

Figure 6.16. This example of a dichotomous interpretation key was developed for identification of crops in southwest Kansas. Attributes of polarization, tone, texture, and pattern are used. (From Coiner and Morain, 1971. Copyright 1971 by the American Society of Photogrammetry. Reprinted with permission.)

The reader must be aware of a few significant qualities. First, radar is a relatively new sensor to the scientific community. The potential of multiple polarization, space platform radar, and multiple wavelength systems if virtually unknown. Second, if data are needed within a specified time frame, radar may be the only sensor useable. Radar can operate at night, "see through clouds," and provide data at desirable scale and format for resource inventory or analysis. Third, when used in combination with other sensors (e.g., Landsat), the utility of radar and the quality of data are improved by several magnitudes. This is generally the case with any combination of remote sensing systems, but observations point to the fact that radar can and does make a contribution alone or in combination with other sensors. With that in mind, a brief overview of applications of radar imagery follows.[26]

Applications

Landscape analyses. Large areas can be inventoried quickly owing to the synoptic view and small scale afforded by radar imagery. Land use maps (scale 1:250,000 and smaller) can be created at a Level II of generalization. For state and federal agencies desiring to monitor land use changes and trends, this amount of detail is satisfactory and cost beneficial. As mentioned above, mosaics, stereographic viewing, and enlarging enhance the advantages and interpretability of radar.

Natural resources. Radar imagery can be used for regionalizing landforms, as well as for identifying individual landform features. Also, relative relief and slope can be measured by calculating radar shadow, foreshortening, and layover.[27] Since fracture zones, rock-type associations, and structure can be observed on imagery, radar provides a useful tool in identifying areas for petroleum and mineral exploration.[28] By studying the surface geology and terrain configuration, faults and lineaments characteristic of known resource areas have been used to map heretofore unknown natural resources in such areas as Panama, the Amazon basin, Nicaragua, and Indonesia.

Water resources. Ground water has been located by using techniques similar to those employed for mineral exploration. Because of the unique appearance of water on radar imagery and radar's all-weather capability, radar imagery is a very useful tool for flood mapping. Real time data on water levels can be acquired when other sensors are hampered by clouds or foul weather.

Other applications include studying wind patterns over water, monitoring lake level changes and spring thaw of ice-covered lakes, and detecting water bodies. Also, radar can prove useful in detecting tidal zone features, such as mud flats, surf zones, and shell reefs of coastal wetlands. Such cultural features as canals, jetties, or oil platforms are evident on radar images. Drainage basin features (basin area, network length, basin perimeter) can be delineated with radar imagery to a level of detail found on a 1:24,000 topographic map, but multiple-look angles are necessary, especially in mountainous terrain where radar shadow is an important factor. Knowledge of radar return from snow is limited, but it is known that old snow appears different from new snow. Inventorying the amount and location of snow and ice is critical for navigation, water supply, and irrigation planning. Also, radar has shown potential in monitoring oil spills, but little work has been conducted to date regarding other aspects of water pollution and water quality.

Agriculture. The most important advantage of radar imagery in agricultural analysis is its all-weather capability. Timeliness in gathering data at specific stages in crop growth is critical for acreage and field estimates. Monitoring overgrazing and erosion assessments are related applications. Since the radar return is a function of plant structure, plant and soil moisture, and amount of bare ground visible, work in this area is quite complex. Some crops are readily identified, such as corn and sugar beets, but others are more difficult to image on radar because of topography, field size, and planting methods.[29] With the development of multiple polarization and multiple frequency radar systems, advances in this area can be expected. Presently, work is being performed relating to detection of insect infestation, disease, flooding and storm damage, and maintenance of range quality. The most significant contribution of radar imagery for agriculture and range surveys appears to be the ability to detect crop-water phenomena and total leaf area. Radar imagery can provide data needed to produce reconnaissance maps and shifting cultivation patterns.

Vegetation. Radar imagery is very sensitive to variations in the density of natural plant communities. Broad vegetation structures are identifiable on multipolarized imagery; however, there are many variables that affect the radar return signal, and all plants are not equally identifiable. Results of studies to date indicate that radar imagery might have a role in the following: preparing small scale regional maps of vegetation types (especially when there is a marked difference in the structure of the plants); delineating vegetation zones that vary with elevation; delineating large burn patterns from forest fires and some regrowth areas; locating timberlines; and supplementing small-scale, low-resolution photography where more detailed vegetation texture differences are absent or weak.[30]

Soils. Radar sensing of soils has progressed slowly because of the complex relationship of soil, soil moisture, and radar signal return. The radar signal is influenced by the soil type and composition, size of particles, and moisture in the soil.[31] Each of these factors will vary with the wavelength, polarization, resolution, look angle, and depression angle of the system. Since most soils have at least some vegetation growing in them, the interpreter also must determine if the signal is being influenced by nonsoil parameters and to what degree. For example, a soil that might appear smooth at one wavelength, might appear rough or coarse textured at another. Radar imagery will not replace photography in mapping detailed or general soil surveys. Radar, however, probably can provide unique data on texture, moisture, and texture-related soil phenomena with continued research.[32]

Cultural features. With current commercial radar systems (resolution about 15 meters) major cultural features, such as roads, railroads, field patterns, and built-up areas, can be detected. For most urban areas and cultural features, however, photography can provide more information and do it more economically than can radar imagery. Although not the preferred sensor, it should be remembered that in many cases radar may be the only sensor available or practical (e.g., cloud-covered or rainy environments such as the tropics). Gross land use categories can be delimited on radar imagery, but the return from individual buildings or groups of buildings will vary owing to the problem of scintillation. Corners of buildings and building materials produce a varying signal which often "balloons" and obscures surrounding features on the imagery. This does not imply that radar imagery is of little value in mapping land use, but only reminds the reader of the complexity of an active sensor. For example, bridges and powerlines parallel to the flight path show up better on like polarized imagery, but if the same feature is at any other angle relative to the flight path, cross polarized imagery is superior for detection. Studies to date have shown that radar can be used to estimate populations of urban areas and to map land use.[33] As work progresses and radar systems improve, radar imagery will provide a useful tool and complement to photography.

Overview

The above discussion has provided a brief glimpse of radar capabilities and limitations in relation to geographic investigations. It is readily apparent that for many applications radar seems to be inferior to other sensing systems. In this regard the reader must recall that radar is a relatively new sensor. Little work has been done in radar image interpretation as compared with some older, more established systems. Radar is unique in that it is an active sensor. As such, interpreters are faced with an additional group of variables to consider in their analyses.

Although the role of radar imagery as a tool for geographic investigation will increase as more work is conducted to understand the signal/environment relationship, radar imagery even now is an operational tool in many instances. Brazil, Venezuela, and Indonesia, among others, have had their entire countries imaged by commercial radar systems. Only with this imagery have they been able to inventory their mineral, petroleum, timber, and other natural resources. Since these countries are large and parts are often cloud covered, it was not until radar imagery was obtained that scientists were able to map these countries. In the process it was discovered that new or previously mislocated mountain chains and river systems could be accurately located. General land use maps were created to monitor and plan future growth. Obviously, radar is a valuable tool whose worth will multiply in the future.

NOTES

1. M. L. Bryan, "Microwave Remote Sensing," in *Remote Sensing Short Course Notes,* ed. M. L. Bryan (Washington, D.C.: Association of American Geographers, Committee on Remote Sensing, 1976).
2. D. Harper, *Eye in the Sky: Introduction to Remote Sensing* (Montreal, Canada: Multiscience Publications, 1976).
3. M. R. Holter, "Imaging with Non-Photographic Sensors," in *Remote Sensing with Special Reference to Agriculture and Forestry* (Washington, D.C.: National Academy of Sciences, 1970).
4. Bryan, "Microwave Remote Sensing."
5. J. P. Scherz and A. R. Stevens, *An Introduction to Remote Sensing for Environmental Monitoring* (Madison, Wis.: University of Wisconsin Institute for Environmental Studies, 1970).
6. Harper, *Eye in the Sky.*
7. For more detail on passive microwave applications, see, for example, R. G. Reeves, ed., *Manual of Remote Sensing* (Falls Church, Va.: American Society of Photogrammetry, 1975); and E. Schanda, "Passive Microwave Sensing," in *Handbook of Remote Sensing Techniques,* ed. E. Schanda (New York: Springer-Verlag, 1976).
8. W. Bradford and M. Plaster, "Microwave Radiometry and Imagery," in *Handbook of Remote Sensing Techniques,* ed. W. Bradford (Orpington, England: Technology Reports Centre, 1973).
9. Ibid.
10. J. Hooper and R. Moore, "The Remote Sensing Capabilities of Microwave Radiometry," in *Proceedings of the First PanAmerican Symposium on Remote Sensing* (Panama City, Panama: Inter- American Geodetic Survey, 1973).
11. L. F. Dellwig, H. C. MacDonald, and W. P. Waite, *Radar Remote Sensing for Geoscientists* (Lawrence, Kans.: University of Kansas Center for Research, 1973).
12. Ibid.
13. A. J. Lewis and H. C. MacDonald, "Interpretive and Mosaicking Problems of SLAR Imagery," *Remote Sensing of Environment* 1 (1970):231-36.
14. H. C. MacDonald and W. P. Waite, "Optimum Radar Depression Angles for Geological Analysis," *Modern Geology* 2 (1971):179-93.
15. Ibid.
16. Ibid.
17. Ibid.
18. L. F. Dellwig et al., *Use of Radar Images in Terrain Analysis: An Annotated Bibliography,* Technical Report 288-2 (Lawrence, Kans.: University of Kansas Center for Research, 1975).
19. Ibid.
20. J. Coiner and L. F. Dellwig, "Similarities and Differences in the Interpretation of Air Photos and SLAR Imagery," in *Radar Remote Sensing for Geoscientists,* eds. L. Dellwig, H. MacDonald, and W. Waite (Lawrence, Kans.: University of Kansas Center for Research, 1973).
21. Ibid.
22. Ibid.
23. Ibid.
24. Lewis and MacDonald, "Interpretive and Mosaicking Problems of SLAR."
25. J. C. Coiner, *SLAR Image Interpretation Keys for Geographic Analysis,* Technical Report 177-19 (Lawrence, Kans.: University of Kansas Center for Research, 1972).
26. For further details see R. Mathews, ed., *Active Microwave Workshop Report,* SP-376 (Washington, D.C.: National Aeronautics and Space Administration, 1975); or R. Reeves, ed., *Manual of Remote Sensing* (Falls Church, Va.: American Society of Photogrammetry, 1975).
27. A. J. Lewis, *Geomorphic Evaluation of Radar Imagery of Southeastern Panama and Northwestern Colombia,* Technical Report 133-18 (Lawrence, Kans.: University of Kansas Center for Research, 1971).
28. L. F. Dellwig, H. C. MacDonald, and J. W. Kirk, "The Potential of Radar in Geological Exploration," in *Proceedings of the Fifth Symposium on Remote Sensing of Environment* (Ann Arbor, Mich.: University of Michigan, 1968).
29. See, for example, D. S. Simonett, J. Eaglemen, et al., *The Potential of Radar as a Remote Sensor in Agriculture: 1. A Study with K-Band Imagery in Western Kansas,* Technical Report 61-21 (Lawrence, Kans.: University of Kansas Center for Research, 1967); and D. Schwarz and F. Caspall, "The Use of Radar in the Discrimination and

Identification of Agricultural Land Use,'' in *Proceedings of the Fifth Symposium on Remote Sensing* (Ann Arbor, Mich.: University of Michigan, 1968).

30. See S. Morain and D. Simonett, "K-Band Radar in Vegetation Mapping," *Photogrammetric Engineering* 33 (1967):730-40; or N. E. Hardy, *Interpretation of Side Looking Airborne Radar Vegetation Patterns: Yellowstone National Park,* Technical Report 177-24 (Lawrence, Kans.: University of Kansas Center for Research, 1972).

31. S. A. Morain and J. B. Campbell, "Radar Theory Applied to Reconnaissance Soil Surveys," *Soil Science Society of America Proceedings* 38 (1974):818-26.

32. H. C. MacDonald and W. P. Waite, "Soil Moisture Detection with Imaging Radars," *Water Resources Research* 7 (1971):100-10; and J. Cihlar and F. Ulaby, *Microwave Remote Sensing of Soil Water Content,* Technical Report 264-6 (Lawrence, Kans.: University of Kansas Center for Research, 1975).

33. M. L. Bryan, "Interpretation of an Urban Scene Using Multi-channel Radar Imagery," *Remote Sensing of Environment* 4 (1975):49-66; and F. M. Henderson, "Radar for Small-Scale Land-Use Mapping," *Photogrammetric Engineering and Remote Sensing* 41 (1975):307-19.

7

Gemini and Apollo Programs and Imagery

L. Alan Eyre
University of the West Indies, Jamaica

The Gemini and Apollo Programs

The Satellite Era Begins

From October 4, 1957, when the Soviet Union startled the world by putting Sputnik I into orbit about 900 kilometers (560 miles) above the earth, until the splashdown of Apollo IX in March, 1969, eight hundred known spacecraft were rocketed from the earth. It can be asserted confidently that from these satellites at least a million photographs of earth were taken, returned by various retrieval methods, and recovered for use.

Most of the photographs were small scale images of meteorological interest taken from Tiros satellites and military-sponsored photography from the Russian Cosmos reconnaissance satellites and their United States counterparts. More than a thousand photographs of earth were taken by Mercury and Gemini astronauts and two thousand more during the Apollo missions. The intrinsic interest and technical quality of this unique bank of imagery have enabled it to retain practical utility as an educational tool in the earth sciences. It is mainly in such a role that those aspects of Gemini and Apollo missions which were concerned with imaging the earth are examined in this chapter.

The Impact of Gemini and Apollo

It is most important for students of the earth to recognize the great influence that Gemini and Apollo photography had on scientists. It excited them and fired their imagination. There was a widespread realization that orbiting platforms of one type or another offer tremendous potential for imaging the earth and monitoring the constant physical and man-made changes that occur on, above, and below its land and water surfaces. It is true that Gemini and Apollo photographs formed no more than a small sample of possible global coverage, but it is, nevertheless, an exciting sample.

Important examples of the scientific analysis and interpretation of the more interesting of these photographs, such as plate 4, are considered in this chapter. They were landmarks in the field of remote sensing, showing in a small way what could be achieved on a larger scale whenever more extensive coverage became available. These classic studies in the 1960s were an input into the decision to launch long-life spacecraft specifically for imaging the earth, and so paved the way for Landsat and Skylab in the 1970s.

The Importance of Orbital Track and Altitude

Of the twelve Gemini missions, eight returned with usable photographs of earth, ranging in number from 25 in Gemini III to 250 in Gemini VI. All of these photographs, and those of the Apollo series which followed, are of locations on earth which fall within or close to a zone from 33 °N to 33 °S latitude. The reason for this derives from the launch and orbit characteristics of the manned spacecraft from which the photographs were taken.

The planned orbit of any artificial satellite may be tailored to its functions, but also it has to take into account the physical forces involved, particularly gravity. A satellite circling at an altitude of 1,000 kilometers (621 miles), therefore, is almost friction-free and may remain in orbit for many years. A satellite orbiting at an altitude of 36,000 kilometers (22,370 miles) makes one revolution every 24 hours like the earth beneath it; therefore, it appears to remain over the same spot on earth and is called *geostationary*. A spacecraft, particularly if manned, may be sent into what is known as a *parking orbit,* usually close to the minimum altitude above earth for free orbit to be possible, i.e., 160 to 270 kilometers (99 to 168 miles). From a parking orbit, the spacecraft may be directed and powered to a higher orbit, to lunar or planetary flight, or to re-entry. Most Gemini and Apollo photographs were taken during parking orbits.

Attainment of orbit depends principally upon:

1. Site of launch (latitude)
2. Launch angle
3. Weight and velocity of spacecraft
4. Gravitational forces of earth and other bodies

To achieve and maintain a near-circular orbit at an altitude of 200 kilometers (124 miles), a spacecraft must maintain a horizontal velocity of 7.9 km/sec. From Cape Kennedy Space Center, Florida (28 ° 30 ′N), the Gemini and Apollo spacecraft were launched in an easterly trajectory, thereby deriving from the earth's own eastward motion about its axis an added component of acceleration. The launch angle was such that each orbit formed a plane passing through the center of the earth and making an angle of 33 degrees with the equator. As a result, only the zone from 33 °N to 33 °S latitude could be photographed vertically. Photographs of areas to the north and south of this zone could only be taken at high oblique angles.

A fact of general interest is that monitoring of and communication with the Gemini and Apollo spacecraft and astronauts required the largest and most complex computer facility ever made, capable of handling 80 billion computations daily, in addition to a worldwide network of optical, radar, and radio tracking stations.

Image Targetting

Special maps of the orbital zone were prepared before each mission so that the astronauts would be able to pinpoint suitable targets for their hand-held photographic experiments. Even so, on one mission the crew initially misidentified the irrigated Nile delta as a large lave flow! On the other hand, Conrad's excellent eyesight picked out individual houses, aircraft landing, and, on one occasion, a moving border patrol vehicle in the United States Southwest—all from almost 200 kilometers in space!

To maximize the interpretation and use of the photography, particularly large scale imagery of the arid regions, such as plate 4, it is desirable to know:

1. Position of the satellite and the subsatellite point on earth
2. Altitude of the satellite (i.e., of the camera)
3. Direction in which the camera was pointing
4. Date and time of the photograph

The dating and timing of Gemini and Apollo photography has distinct peculiarities. Each completed orbit formed an arc of 384 degrees, 24 degrees more than a circle, since the earth had rotated 24 degrees during the orbit. Also, as the astronauts passed over localities with different time zones, orbiting the earth twenty-four times a day, they took pictures of places whose local times were hours or even a day before other places that they had already photographed. For this reason all satellite imagery has to be referred to Greenwich mean time (GMT) rather than local time.

Image Quality

Photographic Equipment

The overall quality of the Gemini and Apollo imagery was a landmark in applied technology. This was largely because of the equipment used, which approached the ultimate in Swedish engineering and German optics.

The principal camera used was the Hasselblad 500C, described by its makers, Viktor Hasselblad Aktiebolag of Goteborg, as the most comprehensive and versatile camera system made in the 70 mm × 70 mm format (fig. 7.1). It is a single-lens reflex camera with interchangeable lenses, magazines, rangefinders, film-winders, and filters. Each of the three superb Zeiss lenses has a diaphragm shutter with both automatic and manual apertures and automatic depth of field indicator and exposure value scale.

The Zeiss Planar 80 mm lens was the most frequently used of the lenses, especially for vertical photographs. The Zeiss Biogon 38 mm super-wide-angle lens was used (on a Hasselblad SWA camera) to take extensive oblique photographs, especially when it was desired to extend the field north or south of the orbital zone. The Zeiss Sonnar 250 mm telephoto lens yielded relatively large scale vertical photographs in very sharp detail with the least distortion due to the curvature of the earth.

It is a useful approximation that *vertical* photographs taken from an orbital altitude of 200 kilometers with the Zeiss Planar 80 mm lens cover a ground area of 50,000 square kilometers (19,305 square miles); with the Sonnar 250 mm telephoto lens they cover 3,000 square kilometers (1,158 square miles); while many of the oblique photographs taken with the Biogon lens include from 50,000 to more than 1,000,000 square kilometers (19,305-386,100 square miles) of ground area within one frame.

Film Characteristics

In the early Gemini missions most of the photographs were obliques. SO-368 Ektachrome 70 mm color film, sometimes with a haze filter, was used throughout the Gemini program.

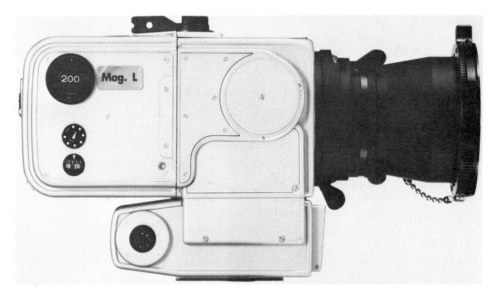

Figure 7.1. The Hasselblad camera which was used by Gemini and Apollo astronauts. (Courtesy of Viktor Hasselblad Aktiebolag)

From Gemini VII Borman and Lovell took the first color infrared ("false color") images. Ektachrome Infrared Aerial film reacts to green, red, and infrared rather than to blue, green, and red as does ordinary color film. When developed, healthy vegetation appears bright red because of the high reflectivity of the chlorophylls for infrared radiation; while dead vegetation appears gray or gray-green.

In unmanned Apollo VI, a single camera mounted in a hatch window used high resolution S-121 Ektachrome film with a minus-blue filter. This unit took a series of vertical photographs with 60 percent stereoscopic overlap. The same film was used in the later manned missions, with consequent improved clarity and resolution. Even with the minus-blue filter, the prevailing blueness of the earth, particularly in humid regions, reduced image quality throughout the program and was never fully overcome from a technical photographic point of view.

Apollo IX

Of all the Gemini-Apollo series, Apollo IX was the mission in which sensing of the earth was most emphasized. It was also the last major project of earth photography before ERTS-1 (Landsat 1) was launched in July, 1972. In the interim, attention was diverted to lunar objectives.

In Apollo IX, astronauts McDivitt, Scott, and Schweickart took many hand-held vertical photographs using a minus-blue filter and color infrared film. In addition, an array of four Hasselblad cameras was hatch-mounted and took simultaneous multispectral vertical photographs of selected areas, including some with stereo overlap, particularly in Southwestern United States. This was named the SO-65 experiment. A good example of Apollo IX photography is the set of frames AS9-26-3700-ABCD covering part of the Imperial Valley, California, and Yuma, Arizona (plate 4 and fig. 7.2 A, B, and C).

Figure 7.2. Multispectral photograph frames from the SO-65 experiment of Apollo IX. The imaged scene is of the Imperial Valley of California and adjacent parts of Mexico. (A) Band B using black and white type 3400 film with a 58B green filter. (B) Band C using black and white infrared SO-246 film with a 89B filter. (C) Band D using black and white type 3400 film with a 25A red filter. Observe the large alluvial fans which have formed along the San Andreas fault. (NASA, supplied by Technology Application Center)

Technical Limitations

When photographs taken by astronauts first began to be published widely, they created excitement and many optimistic claims concerning their utility. It soon became apparent that for most applications Gemini and Apollo imagery has technical limitations. The number of cloud-free frames is only a small proportion of the whole. Bits of cloud cover are often an irritating hindrance to interpretation, obscuring important parts of the desired scene. One unexpected result of the imagery as a whole, however, was an upward revision in estimates of the cloud cover of the earth both globally and in specific regions. Other limitations are related to characteristics of the atmosphere which affect color and tone. Still others are optical deformations due to camera tilt, film, emulsion, spacecraft motion, and various aspects of photographic technology.[1]

Image Resolution and Interpretation

It cannot be overemphasized that analysis and interpretation of space photographs depend essentially upon the degree of image resolution achieved. This, in turn, depends upon the quality of the transparency or print used and the sophistication of the equipment available for interpretation. Such equipment ranges from the slide projector and screen to densitometers and expensive scanning devices.

It is not worth skimping on the basic equipment such as projector, screen, transferscope, and photo-enlarger. Quality is essential in these items if the maximum use is to be made of any space photography, particularly those in the small scale range of the Gemini-Apollo series.

A clearing house for the Gemini-Apollo photography has been established at the Technology Application Center, University of New Mexico, Albuquerque. The original transparencies are filed at the Johnson Space Center in Houston, Texas. Users may find wide variation in the brilliance and definition available to them in the third generation transparency they purchase. For example, this author found that even using a 35 mm reduction obtained from a 70 mm second-generation positive transparency and projecting with a Hasselblad slide projector onto a reverse image screen, frame S-65-45768 of mainland China imaged in August, 1965, revealed the urban area of Shanghai in considerable detail with a network of new canals in the surrounding countryside. Other investigators, however, said that they could see nothing of the kind on their copies of the same frame. Conversely, others have been able to pick out on certain frames roads, railways, and some other lineations which this author has not been able to discern on the same frames by the best techniques available to him.

Summarizing the Gemini and Apollo photography in general, its main characteristics can be described as follows:

1. It is limited to low latitudes (approximately 33°N to 33°S).
2. Clarity is greatest over the earth's arid lands.
3. In humid areas the color film has a distinct bluish cast, confirming the optical quality of the earth as a "blue planet."
4. With the exception of the limited telephoto vertical photographs, most frames are on very small scales.
5. Areal coverage globally is very spotty, offering small samples. The only two large contiguous areas imaged in detail are a band from California to Texas and a smaller region of the central Sahara in Libya and Algeria.

6. Given its limitations of scale and distribution, selected frames have proved to offer considerable scope for interpretation with a wide range of methods and under the aegis of several diverse scientific disciplines. This is true especially in respect to the SO-65 multispectral and the stereoscopic coverage.

The Interpretation of Gemini and Apollo Imagery

Basic Procedures for Assessing the Potential

When the Gemini, and more particularly the Apollo, photography was becoming available rapidly in 1968 and 1969, a sober assessment of its utility was an urgent necessity. It was important to transform the first ecstatic whistles of admiration at the new dimension to the realm of practical application. As Robert N. Colwell asked half-rhetorically: What result is there to be besides man on the moon for a fifty billion dollar space program?[2]

Colwell offered a basic procedure for assessing the ways in which orbital photography could be of use to the environmentalist and the earth scientist. He stressed the need for both competent analysts and proper equipment. It is very easy to *imagine* things on a photograph showing a million square kilometers of earth. On the other hand, it is also very easy to miss something which is there. For this reason, competent map-reading skills are essential to proper, accurate interpretation and use of any imagery from space, indeed of any kind of remotely sensed image of the earth. Colwell also emphasized the importance of obtaining ground truth checks to ensure accurate interpretations.

Published Output

Photo-anthologies. Four types of published output emerged immediately from the Gemini and Apollo missions. Those in the first category can be described as photo-anthologies or collections of space photographs. These are usually printed in color on high-grade glossy paper, with brief or detailed general interpretations, or sometimes simply captions, to each selected frame (refer to chapter 1). These photo-anthologies served a purpose by introducing space photographs to the profession and to the public.

Photo-mosaics. The second type of early output from the missions consisted of attempts to utilize Gemini-Apollo imagery as if it were aerial photography; that is, to make a photo-mosaic and map observable features over a wide area of earth space. One example of this class is the photo-mosaic of western Peru prepared by J. McKallor.[3] This photo-mosaic was studied intensively over several years by geomorphologists in the field in Peru and Bolivia with interesting results. From the imagery there were identified for the first time important concentric geomorphic structures in the Peruvian and Bolivian Andes which are genetically related to abundant mineral deposits. One of these structures in the vicinity of Cuzco proved to be associated with two hundred occurrances of minerals and was named Gemini VI in honor of the astronauts.

"Land Use in the Southwestern United States from Gemini and Apollo Imagery," by Norman Thrower was published as a map supplement to the *Annals* of the Association of American Geographers, vol. 60, no. 1, 1970. This effort may perhaps be viewed as something of a cartographic landmark, a prelude to even better things to come. Thrower was able to identify and map ten classifications of land use on the broadest scale. One distinctly intriguing category on his map is "Oilfields," the symmetric well pattern in the Permian Basin and elsewhere in the western rangelands providing a useful and characteristic signature.

User guides: suggested applications. The third type of output emerging from the Gemini-Apollo missions may perhaps be described as user guides: books and articles suggesting the various applications to which the photography could be put and giving illustrated examples. As might be expected, there was quite an outpouring of these between 1968 and 1972, led, in number of contributions at least, by the geologist P. D. Lowman, Jr., who was closely involved in the photographic aspect of the missions from the early 1960s.

Very broadly, the authors of the users guides showed that Gemini and Apollo photography, although having limited coverage in aerial terms, did provide valuable input into the following ten fields of scientific research. Each will be considered later in this chapter.

1. Terrain analysis and geomorphology
2. Vulcanology
3. Multidisciplinary regional planning
4. Agriculture and land utilization
5. Forestry
6. Hydrology
7. Pollution studies
8. Marine and coastal studies
9. Urban studies
10. Meteorology

Techniques of use. The fourth type of output, which followed a little later, presented, explained, and assessed a number of techniques which could be used to give an added depth and sophistication to the analysis of Gemini and Apollo photography. Brief discussions of a few of the most significant of these techniques follows.

Selected Techniques of Use

Color/tonal enhancement. Tonal enhancement techniques were quickly applied as Gemini photographs became available. These use photographic methods such as special developing procedures to accentuate specific selected portions of the tonal or color range. Such techniques improve readability greatly, and make possible further instrumented analysis using densitometers and other measuring devices.[4]

Color coding. This somewhat similar technique was developed by scientists at Philco-Ford, Palo Alto, California. It entailed dividing the color tonal range in a particular Gemini photograph into twelve individual tones and isolating these one at a time. The twelve were color separated and assigned arbitrary colors as codes for further identification. The picture was then reconstituted with the code colors replacing the original tones. In this way bathymetric depths on the Bahamas Bank could be shown with surprising accuracy. On the published color-coded photograph, different colors show water depths of two meters, two to three meters, three to five meters, and deeper than five meters.

Autographic theme extraction. Basically, this technique automatically utilizes density and tonal differentials in the imagery to extract significant phenomena. Based on experiments in data discrimination instrumentally, a system has been developed at the EROS Data Center which automatically extracts specific "themes" from orbital imagery. The process is officially defined as *autographic theme extraction.* Apollo IX multispectral imagery has been dissected by this process so that such specific spatial components as "deep snow," "light snow," "clear water," "turbid water," and other features can be imaged in isolation and made available for accurate quantification.

Television waveform analysis. Under the U.S. Geological Survey (USGS) Geographic Applications Program, a technique was developed at Florida Atlantic University to transform the color photograph through linked projection and television systems into a form capable of being stored and analyzed by computer. Gemini and Apollo transparencies were projected on a special screen and scanned by television camera. An attached waveform analyzer enabled the characteristic of each television scan line to be examined separately and digitalized for computer processing.

The multiband approach. The Apollo IX SO-65 experiment (figure 7.2 and plate 4) in which four images were made simultaneously assumed that appropriate techniques would be applied to analyze the data obtained. The important factor here is the means—mainly instrumental—to discriminate between the image characteristics on the four bands and, on the basis of such physical properties as reflectance, emissivity, and relative radiance, to provide criteria for accurate identification of ground data.[5]

Orthophotomapping. Apollo IX vertical images can be combined with line images to make a photomap. A true orthophotomap needs sophisticated instrumentation to correct deformations, but reasonably accurate photomaps can be made by combining basic photography and cartography, using a transferscope if one is available.

Fourier optical transform analyzer. Typical of several highly sophisticated methods of research beyond elementary applications is the use of a Fourier transform analyzer technique which employs a laser optical system. Using this advanced technique, it was possible to measure the wavelength and direction of ocean swells near New Guinea produced by Typhoon Gloria east of Taiwan and seen from over 3,000 kilometers (1,864 miles) away.

Contributions to Terrain Analysis and Geomorphology

The majority of published remote sensing studies utilizing Gemini and Apollo photographs have been in the field of terrain analysis and geomorphology. This is probably because the scale, camera angle, resolution, and photographic properties admirably suit this type of analysis. Both NASA and the USGS recognized the potential of the imagery for this purpose, and the number of NASA special papers, USGS monographs, conference papers, and journal articles which utilized Gemini and Apollo photographs for terrain analysis and geomorphology grew steadily after 1966.

Striking Landforms

Photographically, the most striking scenes of the Gemini-Apollo series are those of the landforms of the central Sahara. The numerous frames taken of this region, both verticals and obliques, have contributed greatly to knowledge of this beautiful, forbidding, but resource-rich region. One of the best examples is that of the Algeria-Libya border. The desert landforms of the Ubari sand sea, the Cambro-Ordovician rocks of the Hoggar massif, and the huge canyons and mesas of the massive Acacus sandstones, with their dendritic drainage patterns clearly inherited from the period when the Sahara had a humid climate, were photographed in sharp detail. (Rock drawings of giraffes and other savanna animals are ground-level relics of the same period.) Figure 7.3 shows a detailed Sahara landscape south of Colomb-Bechar, Algeria. This is

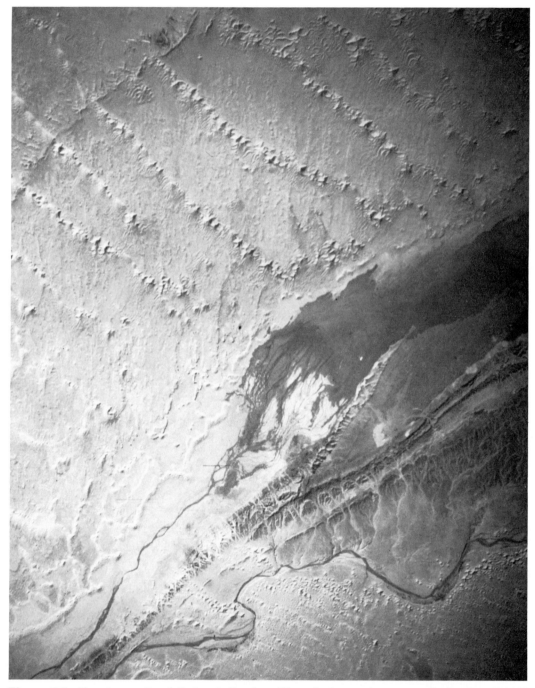

Figure 7.3. Algeria south of Colomb-Bechar. The photograph was made by Gemini VII astronauts in December, 1965. This nearly vertical photograph was made with a 250 mm telephoto lens. (NASA, supplied by Technology Application Center)

a telephoto vertical frame taken in 1965 by Borman and Lovell from Gemini VII at an altitude of 250 kilometers (155 miles). The area is quite small for a Gemini frame, only 2,500 square kilometers (965 square miles). Dune landforms, rock outcrops, wadi structure, and oases are strikingly evident.

It is hard to say whether the utility of the pre-Landsat orbital photography to terrain analysis and geology was realized in practical terms, but some very interesting investigations in many areas certainly added a new and useful dimension to terrain interpretation and to the search for economic resources from the earth.

Plate Tectonics

The Gemini and Apollo perspectives, particularly the oblique photographs, supported the newly revived concepts of continental drift and plate tectonics. Studies in tectonically mobile areas, such as Baja California and the Red Sea, revealed numerous transcurrent faults and other lineaments not previously suspected. These photographs highlighted ways in which plate movements were expressed in landforms. For example, the huge alluvial fans generated by strike-slip movement on the San Andreas fault are illustrated in figure 7.2.

A Classic Study in Landform Analysis

The interpretation of Monem Abdel-Gawad of Gemini frame S-66-54664 is a classic of its kind (fig. 7.4).[6] The interpretation is summarized below.

The displacement of two major shear zones or wrench faults either side of the Red Sea—indicates that Arabia has not only rotated and separated from Africa, but also has moved 150 kilometers (93 miles) northward along the rift of the Gulf of Aqaba. A reconstruction of the area as it was in the Miocene Period before this movement began brings the Maqna block of Saudi Arabia into a position south of Sinai and the Gulf of Suez. This is within the area of Miocene oil-yielding sediments extending along both sides of the Gulf of Suez, where at least eighteen oilfields are known.

On the basis of the reconstruction, oil-gas deposits would be expected to occur in the Maqna block. The expectation proved to be justified when hydrocarbons were discovered later at depth in the Maqna block. A number of west-northwest to east-southeast trending faults were first identified by Abdel-Gawad on the same Gemini imagery.

These parallel faults were shown to be closely related to the same oilfields along the Gulf of Suez. By careful examination of the orbital photography, these subtle landform features could be traced for hundreds of kilometers westward across the Egyptian desert toward and even beyond the Nile where they had not been previously suspected.

Alluvial Morphology

Deltaic and alluvial plain morphology is very strikingly shown from orbital altitudes, and scars and other relict features often are more evident than from nearer the ground. Macro-scale water features—which are often critical to understanding geomorphic processes—of such regions appear with clarity on the infrared and false color frames of the Apollo IX multispectral series.

Arid Landforms

The striking detail on Apollo photographs of playa basins in the southwestern part of the United States attracted the attention of geomorphologists and hydrologists. There are many

Figure 7.4. This photograph shows the northern portion of the Red Sea and the lands bordering it (Egypt, Saudi Arabia, Israel, and Jordan). The photograph was made by astronauts aboard Gemini XI in September, 1966. (NASA, supplied by Technology Application Center)

frames available showing landscapes in southern Arizona which can be used as a base for interesting and instructive studies of arid landforms within the United States.

Contributions to Vulcanology

The Gemini-Apollo imagery has provided two kinds of input for vulcanology. On many frames for many areas which are difficult of access, arid lands, lava flows, and other volcanic landforms are visible with eye-catching clarity. They can be mapped with ease using a transferscope or similar device. It has been possible to distinguish between aa and pahoehoe lava in the Sierra del Pinacate of Mexico, and, using orbital photography, the morphology of volcanic landscapes has been investigated in Namibia, Somalia, the Sahara, Arizona, and Australia.

The second type of input has come from the infrared frames of the United States Southwest. These indicated that the thermal characteristics of the ground in some volcanic regions could be remotely sensed and could lead to an identification of "hot spots," i.e., areas of radioactivity and low-level vulcanicity.

Multidisciplinary Regional Planning

The Arizona Regional Ecological Test was an attempt to focus multidisciplinary analysis on a specific site. An Apollo IX multispectral mosaic was made covering 86,000 square kilometers (33,205 square miles) in the southern part of the state and examined by a team of specialists from twelve of the major earth-atmosphere sciences. Information on twenty-seven categories was extracted by intensive instrumented and visual analysis, including the categories of cloud, water, snow, pollution, geology, urban, and crops. Including almost all aspects of the total environmental base, the whole was viewed as a unique combined inventory, and as a planning tool for a region with relatively limited natural resources.

On one vertical frame, AS9-26-3801 (fig. 7.5), a small area (Maricopa Test Area) was selected for larger scale interpretation. This area, near Phoenix, Arizona, is probably the one place on earth which has been examined most closely from space. When color infrared is used, the spatial organization of this small region is clearly expressed, and it can be mapped and quantified with the aid of such ground truth data as urban, suburban, and rural settlement; along with irrigated cropland, rangeland, watershed, drainage, and the network of transportation that links them all. Areas of fragile ecology can be identified—those likely to suffer environmental stress under unsuitable development.

Agriculture and Land Utilization

The obvious potential of the Gemini imagery for agricultural research and land use inventory purposes was recognized quickly, and later it was developed in a number of NASA studies. On sequential Gemini photographs field acreages of alfalfa, cereals, and fallow were computed in parts of Arizona, with some essential aid from ground and near-ground obliques. Later, techniques were proposed for reliable identification of agricultural types. As early as 1967, a global mapping of vegetation from orbital photography was being suggested. Following the analysis of land use at the Maricopa Test Site, the utility and cost saving of photography from

Figure 7.5. Phoenix, Arizona, and surrounding areas imaged by Apollo IX astronauts in March, 1969, from an altitude of 204 kilometers (127 miles). The land use in this area was estinated to be 5 percent urban, 20 percent agricultural, 43 percent rangeland, and 24 percent mountains. (Data from Oran W. Nicks, *This Island Earth,* p. 115. NASA Photography)

orbital altitudes in regional agricultural surveys were acknowledged. Recognizing the vast bank of orbital data which was then available or expected, methods of mapping this type of data by computer were undertaken.

All investigators stress that, with so many different green plants under cultivation, great care must be taken to insure accurate identification. This is an important caution for eager persons attempting to interpret space photographs without experience. In many cases, a poor type of one crop which appears on a color infrared photograph may actually be a totally different crop when identified on the ground, and vice versa. The rich reds on the California side of the international boundary in the Imperial Valley on frame AS9-SCI-3152 (plate 4) indicate the United States emphasis on alfalfa, while the duller shades of red on the Mexican side reflect cotton, which is more popular there. The boundary may be a signature indicating different crop choices rather than better or worse agricultural methods or production.

Plate 4 also demonstrated the key role of water management in the modern agricultural economy. Sharp discontinuities occur where people have turned an arid region into highly productive cropland. Trunk canals transfer water from undeveloped or useless land to fields being used for irrigation agriculture.

Forestry

Economic Forestry Survey Aided by Apollo IX Imagery

Apollo IX photographs saved thousands of flights and attendant costs in a forest inventory survey covering 20,000 square kilometers (7,722 square miles) in Arkansas, Louisiana, and Mississippi.[7] Survey by air alone entailed the analysis of 35,000 pictures every time an updating survey was required. Discriminating use of selected Apollo IX frames, however, supplied the following information:

1. Changes since the previous survey from forest to cropland and vice versa
2. Identification of marginal areas of forest not worth larger scale aerial survey
3. Identification of complex areas of forest where detailed aerial surveying would be desirable

As a result, only 2,000 vertical aerial photographs were needed. One of the frames utilized, figure 7.6, shows the Ouachita River in flood.

The first step in this particular forestry inventory survey was to superimpose a grid upon a print of each Apollo positive. Each grid square selected as warranting aerial cover was enlarged 16 times. An aerial photo-mosaic at a scale of 1:60,000 was made of this square, and a grid was superimposed. A further selection to a scale of 1:12,000 followed. Finally, a sampling was made to select small grid squares at a scale of 1:2,000 for detailed bole counts and measurements. A Bendix datagrid digitizer was utilized during the photograph selection and within the 1:2,000 plots ground crews used optical dendrometers to measure four to six boles per plot.[8]

Forest and Brush Fires

One major surprise from the Gemini-Apollo photographs was the prominence and evident size of forest fires on some of them, especially enormous, deliberately set bush fires in the tropical savanna regions of Australia and Africa.

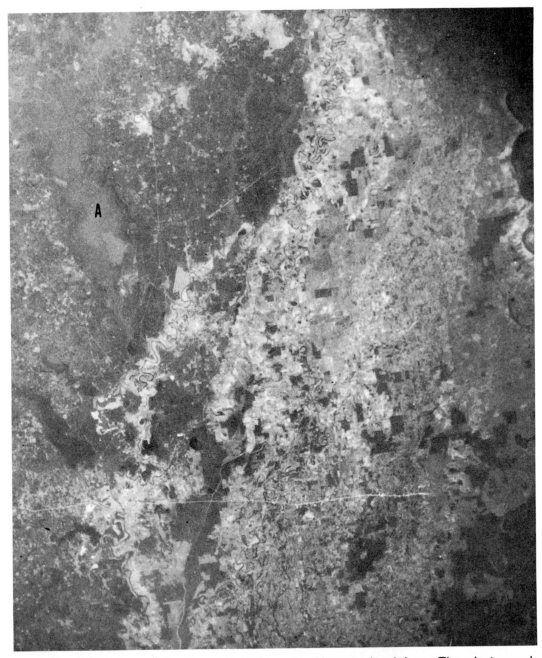

Figure 7.6. Flooding of the Ouachita River near Monroe, Louisiana. The photograph was taken by Apollo IX astronauts from an altitude of 169 kilometers (105 miles) on March 8, 1969. More than 203 mm (8 inches) of rain fell in some parts of the area in January and February. Area (A) in the upper left of the photograph is the major area of inundation. (NASA, supplied by Technology Application Center)

Gemini V passing over Northern Australia in the dry season of 1965 obtained a striking near-vertical scene of a group of huge forest fires along an irregular line about 50 kilometers (31 miles) in length. A dense smoke mass, driven north by the winter monsoon across the coastal plain and out over the Gulf of Carpenteria, covered more than 10,000 square kilometers (3,861 square miles). Investigation showed that cattlemen had set fire to large areas of savanna grass to open up new, and hopefully better, pastures after the monsoon rains of the following summer.[9] Another smoke plume from a large fire in the Cookstown area of Queensland, a more humid part of the continent, can be seen streaming out northwestward over the inland plateau for 160 kilometers (99 miles) on figure 7.7.

Slash-and-Burn Cultivation

In Africa, bush fires on many frames clearly indicate the role of the slash-and-burn agricultural system. In the Zambezi delta area of Mozambique, fifteen or more large bush fires of this type were photographed by Apollo VII astronauts in October, 1968. Similar clusters of fires are seen in frames of the southern Sudan, evidence of clearance in the bush fallowing system.

On Band C of the Apollo IX multispectral imagery fires showed strongly as bright spots or lines (infrared film responding to differences in heat values) and indicated the potential of the later multispectral scanner of ERTS-1 for the detection of forest fires.

Hydrology

Gemini and Apollo photography have aided hydrologists in their quest for evaluation of vital water resources. Aspects which have proved of particular utility have included snow packed in mountainous regions such as the Himalayas; soil moisture conditions in arid lands like central Australia; and hydrological status of reservoirs, playas, and irrigation systems.

A pronounced spiral turbidity vortex in the Salton Sea was identified as occurring whenever turbid irrigation water was released from the cropland of the Imperial Valley through the channels of the Alamo and New rivers.

Another feature of interest, which only small scale orbital photography probably could have shown clearly, is the sharply defined hydrologic-vegetational boundary between states or countries with different water management policies, such as between Texas and New Mexico, California and Mexico, and Israel and Egypt, to give three very striking and by now well-known examples.[10]

With population pressing on finite resources at local, regional, continental, and global scales, the cycle of drought and flood, too much and too little water, is everywhere being intensified. On many Gemini and Apollo frames, not originally taken with such information in mind, this process is apparent. Figure 7.8 is a dramatic view of Tung Ting Lake in the Yangtze floodplain in central China. It is easy to calculate how many square kilometers of the densely settled agricultural plain have disappeared completely beneath the flood waters, and how many more have emerged as silt banks and islands within the limits of figure 7.8.

Within the United States, the Apollo IX astronauts recorded for history the moment in March, 1969, when the Ouachita River was just past a record peak flood and was ponded back from Bastrop, Louisiana, into a vast 640 square kilometer (247 square mile) lake extending a hundred kilometers north into Arkansas (fig. 7.6).

Figure 7.7. Northern Australia photographed by Apollo VII astronauts in October, 1968, from an altitude of 241 kilometers (150 miles). Smoke plumes from forest fires (A and B) can be seen streaming northwestward for more than 160 kilometers (99 miles). (NASA, supplied by Technology Application Center)

Figure 7.8. Lake Tung Ting and the Yangtze River in the Human and Hupeh provinces of China were photographed by Gemini V astronauts on August 22, 1965. (NASA, supplied by Technology Application Center)

Pollution Studies

Soon after the first orbital pictures became available, environmentalists recognized that a new and valuable tool for monitoring pollution had come into their hands. Studies of air and water pollution multiplied, along with studies of more subtle varieties of ecological imbalance. Gemini and Apollo imagery in general was a potent factor in reinforcing the environmental movement and creating a new awareness of the problems and dangers facing man in the care and maintenance of a small planet.

Many visible examples of pollution have been noted by workers on frames which they were examining for other purposes. Among these examples are:

1. Noxious smelter fumes from the copper regions of Arizona
2. Large oil spills at sea
3. Petrochemical pollution and other smoke plumes in Texas and the Gulf Coast region, including a well-known instance of a carbon black plant near Odessa which really does show up black
4. Pipeline fires in the Middle East, including a giant-sized inferno in Iraq, believed to have been caused by sabotage, which was identified from a Gemini photograph before news of it was released publicly

Marine and Coastal Studies

Gemini and Apollo imagery has proved surprisingly useful for analysis of coastal and near-shore waters. Figure 7.9 shows Apollo frame AS9-3128 which has been analyzed in detail. Using color separations (red, green, blue) and applying theoretic concepts in respect to climatology, ocean currents, tides, surface temperatures, and other phenomena, a detailed panorama of the nearshore geomorphology and oceanography of a critical marine area emerged. The important quasi-permanent ocean temperature/salinity discontinuity off Cape Hatteras, North Carolina, and the Gulf Stream boundary are mappable directly from the image.

Many areas of interest to marine biology and marine resources have been identified through Gemini and Apollo photographs. R. E. Stephenson of the U.S. Bureau of Fisheries observed an eddy of turbulent waters of immense size in the Gulf of Aden on a Gemini XI frame of September, 1966. Subtle tonal variations in water color were delineated and, through noting the increase in red and infrared reflectance of these waters, were related to rich plankton blooms which provide the basic link in a food chain for the Aden fisheries resource. In the Gulf of Mexico, sediment streams and tongues of turbid water were identified as far as 180 kilometers (112 miles) off the Texas shore. These can be directly related to the abundance of shrimp in these waters.[11]

Both the scale advantage of the orbital view and the clarity and subtlety of tonal differentiation in the ocean make Gemini and Apollo color imagery particularly useful in marine studies. Upwelling zones such as those near the Canary Islands, off the Peruvian coast, and off Namibia are clearly identifiable. The "wakes" streaming behind islands, such as those of the Cape Verde group where currents are strong and persistent, were unsuspected before the Gemini photography revealed them. The subaqueous topography of shallow seas, particularly where floored by calcareous sediments such as the Bahama Banks, can be delineated effectively on several frames.

Figure 7.9. The coast of North Carolina, including capes Lookout and Hatteras, along the Albermarle and Pamlico sounds were photographed by Apollo IX astronauts in March, 1969, from an altitude of 187 kilometers (116 miles). (NASA, supplied by Technology Application Center)

Urban Studies

An analysis of urban areas is beset by problems of both definition and data because of their dynamic geographic nature. Reliable statistics on areas and populations are difficult to obtain, and even more difficult to compare on a world basis. Population totals are published; however, information for the major cities of the world is often outdated. Close scrutiny of urban data indicates that their reliability and international comparability varies enormously.

Gemini and Apollo photography for the first time made some reasonable global comparisons possible. With sample ground truth on urban population densities in various cultural regions, it is possible to make estimates from appropriate imagery of urban areas and populations that are far more meaningful and comparable than official data.

In 1970 Shanghai was classified as the world's second most populous city (after Tokyo), and in 1974 even as first, on the basis of an official figure for the city of over 12 million persons. Gemini frame VS-65-45768, however, did not support the notion that Shanghai rivals Tokyo or New York. Careful analysis of the frame indicated that in 1965 the population of Shanghai—that is, of the metropolitan area as generally understood—approximated 5.5 million. The 12 million figure is an administrative artifact, quite useless for comparative purposes since it includes 5,274 square kilometers (2,036 square miles) of rural Kiangsu Province within the *shih* (municipality) boundary.[12]

In addition to the problem of comparability of published statistics on urban areas and populations cross-culturally, there has been the difficulty in recent years of obtaining any reliable data at all, particularly up-to-date maps, from some countries. *Municipio* populations are available for Cuba, but Apollo photographs, and later Landsat, enabled estimates to be made of the true urban centers, which occupy only a small percentage of their *municipios*.

A well-known set of Apollo VI frames obtained from the hatch-mounted stereoscopic experiment is AS6-2-1462 to 1464. These show the Fort Worth-Dallas metropolitan region in vivid detail and have been examined optically and instrumentally. The drainage network of the region and the pattern of urban growth were particularly well displayed. Also assessed has been the role of space photography for analyzing the urban environment and related aspects such as transportation.

Figure 7.10, Gemini frame 5-1-44 taken on August 23, 1965, is another photograph which became well-known through frequent reproduction and publication. The frame shows an area of 15,000 square kilometers (5,792 square miles) around Cairo, the capital of Egypt. It became almost a model frame for analysis of urban and central place hierarchies. The sizes, hinterlands, and spatial relationships of urban central places and settlements in part of the Nile valley and delta are very clearly indicated and, with care, can be readily quantified.

Meteorology

The availability to meteorologists of Tiros and Nimbus satellite imagery on a regular and global basis has been responsible for a relative lack of interest in Gemini and Apollo photography by students of weather and climate. Nevertheless, some Gemini and Apollo frames stimulated comments and study when first released. Among these were a few instances where by combining verticals and obliques a three-dimensional view of tropical cumulo-nimbus thunder cells was obtained. The scale of the photography (much larger than Tiros and Nimbus)

Figure 7.10. Cairo and the Nile Delta, Egypt, were photographed by Gemini V astronauts on August 23, 1965. (NASA, supplied by Technology Application Center)

is ideal for examining meso-scale disturbances, particularly in the tropics. The astronauts took the opportunity to photograph a number of tropical storms and hurricanes: the orbital perspective and complete coverage these photographs offer have been useful in investigating the structure of these disturbances, using photodensimometer analysis and similar techniques.

A Unique Perspective

This chapter describes only a small proportion of the investigations which have used Gemini and Apollo photography. It can be stated, however, that there are few better ways to get the "feel" and an initial understanding of the broad scale of geography of any area than from this unique perspective. Although the Gemini and Apollo programs have been concluded, the photographs which were made by the astronauts on these missions remain important to scientists. The photographs are easy to obtain, and the altitudes from which they were made were not great enough to cause serious image degradation. Although the photographs were made from space altitudes, they display the surface of the earth in much the same way as aerial photographs.

NOTES

1. For consideration of technical matters, the reader is referred to J. H. Atkinson, Jr., "Atmosphere Limitations on Ground Resolution for Space Photography," *Society of Photo-Optical Instrumentation Engineers Journal* 1 (1963); and to S. K. Gosh, "Deformations of Space Photos," *Photogrammetric Engineering* 38 (1972):361-66.
2. R. N. Colwell, "Determining the usefulness of Space Photography for Natural Resource Inventory," *Proceedings of the Fifth Symposium on Remote Sensing of Environment.* (Ann Arbor: University of Michigan, 1968), pp. 249-89.
3. J. McKallor, "A Photomosaic of Western Peru from Gemini Photography," U.S. Geological Survey Technical Letter NASA-87, 1967.
4. E. M. Wilkins et al., "Spectral Variations of Cloud Reflectance Deduced from Apollo 9 SO-65 Photography," *Remote Sensing of Environment* 1 (1970):221-30.
5. R. J. P. Lyon, "The Multiband Approach to Geological Mapping from Orbiting Satellites: Is It Redundant or Vital?," *Remote Sensing of Environment* 1 (1970): 237-44.
6. M. Abdel-Gawad, *Geologic Exploration and Mapping from Space,* Science Center, North American Rockwell Corp., Calif., 1967; and Abdel-Gawad, "Geological Structures of the Red Sea Inferred from Satellite Pictures," *American Geophysical Union Transactions* 49 (1969):192.
7. L. Darden, *The Earth in the Looking Glass* (New York: Doubleday, 1974).
8. R. C. Aldrich, "Forest and Range Inventory and Mapping," *International Workshop on Earth Resources Survey Systems* 2 (1971):83-106.
9. J. Bodechtel and H. G. Gierloff-Emden, *The Earth from Space* (New York: Arco Publishing Co., 1974).
10. Apollo photographs AS9-3807A, AS9-26a-3799A, and AS7-6-1696, respectively.
11. S. Q. Duntley and A. R. Boileau, "Exploration of Marine Resources by Photographic Remote Sensing," *International Workshop on Earth Resources Survey Systems* 2 (1972):531-50.
12. L. A. Eyre, "Shanghai—World's Second City?," *Professional Geographer* 13 (1971):28-30.

8

Skylab Programs and Imagery

Noel Ring
Norwich University

>". . . a mere 15 years ago, when Alan Shepard spent 15
>minutes in space, we didn't even know if space
>photography was possible!"[1]

Introduction

To establish the feasibility of an orbiting space station, NASA launched three
Skylab missions during the years 1973 and 1974. Inhabited by three-man crews for 28,
59, and 84 days, Skylabs 2, 3, and 4, respectively, produced several hundred thousand
photographs which provide high resolution coverage, in many cases multispectral and
multiseasonal, of large segments of the surface of the earth. Although the Skylab missions involved concurrent activities of verifying long-term habitation in space and conducting astronomical observations, the operation of the Earth Resources Experiment
Program (EREP) sensor systems collected environmental data which firmly established
the value of manned space flight for terrestrial scientific research.

Based on their experimental role of verifying the viability of an orbital scientific
laboratory, the Skylab missions might have restricted the scope and variety of EREP
functions. By the mid-1970s, however, after increasingly sophisticated manned space
ventures, NASA could incorporate a myriad of sensor systems designed to collect data
about environmental conditions for several major fields of application: agriculture and
forestry; geology; continental, coastal, and marine water resources; and regional planning. In one sense, Skylab represented the culmination of decades of technological
progress and experience in remote sensing. In another sense, it served as the first stage
in the eventual establishment of permanent space stations to be routinely staffed via a
transit system to earth orbit, the Space Shuttle.

Employing a package of six major sensing systems during EREP passes, Skylab
astronauts in their platform about 435 kilometers (270 miles) above the earth obtained
data in a wide range of both visible and invisible wavelengths of the spectrum. The
heart of the EREP program consisted of two matching camera systems patterned after
low-altitude aerial photographic processes in use for some five decades. The six-lens
Itek multispectral camera and the Actron earth terrain camera augmented by auxiliary,
handheld Hasselblad and Nikon cameras, offered areal coverage up to 163-kilometers
square and resolution clarity down to 10 meters. EREP package sensors collecting data
in invisible bands of the electromagnetic spectrum included an infrared

spectrometer, a multispectral scanner, a microwave radiometer-scatterometer and altimeter, and an L-band radiometer. Figures 8.1, 8.2, and 8.3 show the vehicle and EREP components.

In preparation for assessing the scientific utility of Skylab's experimental programs, NASA administrators not only assigned principal investigators to analyze recorded data for a wide range of practical resource management applications, but also conducted a contest among elementary and secondary school students to design experiments for educational applications. Skylab astronauts, therefore, worked at the direction of a broad public in producing information of long-term value for planning future space laboratory missions. The success of their endeavors has sustained support for the forthcoming Space Shuttle program, which portends a continuance and enhancement of the types of EREP activities they conducted.

Sensor Systems

Careful planning to match sensors to survey functions was a hallmark of Skylab's EREP package, as illustrated in table 8.1. Similarly, a long tradition of applying low altitude aerial photography for soils, forestry, and engineering investigations, together with more recent high altitude military reconnaissance, strengthened the camera systems which were designed for this orbiting platform. Since the 1930s, much of North America and portions of Europe have been routinely recorded by standard aerial photography which involved gyroscopically-stabilized

Figure 8.1. The Skylab vehicle as photographed with handheld Hasselblad camera by departing astronauts from the Skylab 4 Command and Service Module. (Photograph No. SL4-143-4706, courtesy of NASA)

Figure 8.2. Space stations and portions of the Earth Resources Experiment Program (EREP) system as photographed with a Hasselblad camera from their Command and Service Module by arriving Skylab 3 astronauts. (Photograph No. 3-114-1660, courtesy of NASA)

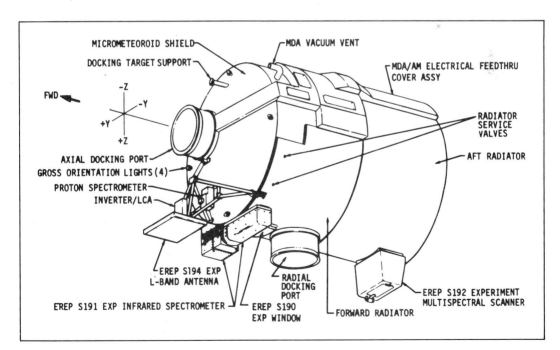

Figure 8.3. EREP sensor system and other salient Skylab features. (Courtesy of F. I. Tallentire, Martin Marietta Corp.)

TABLE 8.1
Skylab Sensors and Functional Applications

Sensor	Biogeography	Earth Science	Photography
S190 Multispectral Photographic Facility		Topographical mapping Agricultural stresses Pollution Population Land use	Films emulsions Multispectral lenses Filters
S191 Infrared Spectrometer	Absorption characteristics of chlorophyll	Atmosphere Radiation	Filters Cameras Cassegrain optics
S192 Multispectral Scanner	Absorption characteristics of chlorophyll	Climatology Agriculture Forestry Land use Geology Oceans Atmosphere	
S193 Microwave Radiometer/ Scatterometer and Altimeter	Distribution of biomes	Agriculture Oceans Atmosphere Sea States Biomes Climatology	
S194 L-Band Microwave Radiometer	Pollution effects	Radiation Sea States Brightness Temperature Atmosphere Faults	

Note: Modified from NASA, *Skylab Experiments,* vol. 2: *Remote Sensing of Earth Resources* (Washington, D.C.: Government Printing Office, 1973), p. vi.

cameras. High resolution vertical photographs were obtained along overlapping flight lines allowing precise scientific measurement and mapping from stereoscopic views of earth terrain. Early space flight planners, however, were so preoccupied by vehicular design and operational problems, as well as an obvious concern for human survival, that they relegated remote sensing activity to a peripheral if not negligible status.

Camera-conscious astronauts can be credited with awakening public interest in space photography. The evident potentials for applied research then incited scientists to prompt

NASA's expansion of the survey capabilities of manned space photography. Skylab represents the apex of nonclassified endeavors, having a 60 percent overlap and high resolution capacity.

Perhaps foremost among remote sensing developments in recent years has been a realization of the values of a multispectral analysis of environmental conditions. Use of film-filter combinations, concurrent coverage, and color-enhancement processes together prompted incorporation of a multispectral camera in the EREP package. The salient features of this system and its products are summarized in tables 8.2 and 8.3. The Itek multispectral unit which was used on Skylab missions consisted of six high precision cameras having matched and boresighted optical systems, each with an f/2.8 lens. The 70 millimeter film width provides sufficient size for tentative analysis and ease of enlargement of the two black-and-white, two black-and-white near infrared, and two color image renditions, one of the last being in color infrared.

Very high resolution is a prerequisite for accurate feature detection, just as multispectral coverage aids in this and in areal pattern mapping. The Actron earth terrain camera system which was used on Skylab missions involved a single f/4 lens assembly having a focal length of 457 mm. Three film types—color, color infrared, and black-and-white—in a 127 mm image format were variously used during EREP passes, as summarized in tables 8.4 and 8.5. Very notable among the achievements of this system is the discovery that land use patterns can be discerned almost as clearly from a 435 kilometer (270 miles) distance as from the vaunted "beer-can-in-the-parking-lot" type of coverage associated with the U-2 and RB57F high altitude aircraft camera systems. The earth terrain camera's obvious advantage is that its field of view far exceeds the much lower altitude aircraft reconnaissance systems. For a regional view from Skylab, see figure 8.4, Boston and its environs.

For purposes of land resource management, a combination of EREP sensor data products would be advisable, since each system and station has inherent advantages and limitations. With respect to the six stations of the multispectral camera system, each provides spectral and spatial differentiation capacities suited to particular applications. With some notable exceptions, the three infrared stations are said to be the best delineators of physical phenomena, whereas the three visible bands define most clearly the cultural landscape.

All of the infrared stations penetrate haze for effective discrimination of land-water boundaries, such as surface water, soil moisture, and drainage patterns. Referring to table 8.2, station 3, in color infrared, adds special utility in assessing vegetation vigor and areas of recent environmental disturbance by extraction, exploitation, or construction. In contrast, stations 4, 5, and 6 trade away the spectral discriminating advantages of infrared channels for greater spatial resolving powers in visible wavelengths. They are most appropriate for land use database mapping. Each has additionally unique capabilities, such as station 4, in high-resolution color for water penetration; station 5, which has the best image definition and resolution, for geologic mapping; and station 6 for a versatile array of soil, vegetation, and snow cover mapping. A definitive study of regional environmental resources would thus favor integrated data from several stations, e.g., stations 1, 2, and 5 or any other useful combination.

Although certain standardized, scientifically-directed photo coverage of earth resources occurred during the Apollo mission series, the sporadic, manual operation of modified Hasselblad 70 mm cameras continued as characteristic of space photography until the advent of Skylab's more sophisticated systems. However, Skylab also carried handheld cameras: a Hasselblad and a Nikon 35 mm single lens reflex used primarily for pictures of spacecraft operational procedures. These provided variety and special coverage of unusual, incidental con-

TABLE 8.2

Multispectral Camera Station Characteristics and Mission Film Rolls

Sta-tion	Filter	Filter Bandpass Micrometer	Film Type: Eastman Kodak	Estimated Ground Resolution Meters (feet)	Mission & Roll No.		
					SL-2	SL-3	SL-4
1	CC	0.7-0.8	EK 2424 (B & W infrared)	73-79 (240-260)	01,07,13	19,25,31,37,43	49,55,61,67, 73,A1,1B
2	DD	0.8-0.9	EK 2424 (B & W infrared)	73-79 (240-260)	02,08,14	20,26,32,38,44	50,56,62,68, 74,A2,2B
3	EE	0.5-0.88	EK 2443 (color infrared)	73-79 (240-260)	03,09,15	21,27,33,39,45	51,57,63,69, 75,A3,3B
4	FF	0.4-0.7	SO-356 (hi-resolution color)	40-46 (130-150)	04,10,16	22,28,34,40,46	52,58,64,70, 76,A4,4B
5	BB	0.6-0.7	SO-022 (Panatomic-X B & W)	30-38 (100-125)	05,11,17	23,29,35,41,47	53,59,65,71, 77,A5,5B
6	AA	0.5-0.6	SO-022 (Panatomic-X B & W)	40-46 (130-150)	06,12,18	24,30,36,42,48	54,60,66,72, 78,A6,6B

Note: After NASA-Johnson Space Center, *Skylab Earth Resources Data Catalog*, p. 7.

TABLE 8.3
Earth Terrain Camera Film Characteristics and Mission Rolls

Film Type: Eastman Kodak	Wratten Filter	Filter Bandpass, Micrometer	Estimated Ground Resolution Meters (feet)	Mission & Roll No.		
				SL-2	SL-3	SL-4
SO-242 (hi-resolution color)	None	0.4-0.7	21 (70)	81	83,84,86,88	90,91,92,94
EK3414 (hi-definition B & W)	12	0.5-0.7	17 (55)	82	85	89
EK 3443 (infrared color)	12	0.5-0.88	30 (100)	––	87	––
SO-131 (hi-resolution infrared color)	12	0.5-0.88	23 (75)	––	––	93

Note: After NASA-Johnson Space Center, *Skylab Earth Resources Data Catalog,* p. 11.

TABLE 8.4
Multispectral Camera Data Products

Scale Nominal	Image Size Inch (cm)	Enlargement	Products			
			Black and White		Color	
			Transparency	Print	Transparency	Print
1:2,850,000	2.25 × 2.25 (5.72 × 5.72)	1.00×	Positive Negative	None	Positive	None
1:1,000,000	6.41 × 6.41 (16.29 × 16.29)	2.85×	Positive Negative	Positive	Positive	Positive
1: 500,000	12.83 × 12.83 (32.50 × 32.50)	5.70×	Positive	Positive	None	Positive
1: 250,000	25.65 × 25.65 (65.15 × 65.15)	11.40×	Positive	Positive	None	Positive

Note: After NASA-Johnson Space Center, *Skylab Earth Resources Data Catalog,* p. 8.

TABLE 8.5
Earth Terrain Camera Data Products

Scale Nominal	Image Size Inch (cm)	Enlarge-ment	Products			
			Black and White		Color	
			Trans-parency	Print	Trans-parency	Print
1:950,000	4.50 X 4.50 (11.43 X 11.43)	1.0X	Positive Negative	None	Positive	Positive
1:500,000	8.55 X 8.55 (21.72 X 21.72)	1.9X	None	Positive	None	Positive
1:250,000	17.10 X 17.10 (43.43 X 43.43)	3.8X	None	Positive	None	Positive
1:125,000	34.20 X 34.20 (86.87 X 86.87)	7.6X	None	Positive	None	Positive

Note: After NASA-Johnson Space Center, *Skylab Earth Resources Data Catalog*, p. 11.

Figure 8.4. A city and its hinterland: Boston and environs as viewed by the Skylab 190A camera. (Photograph No. SL3-47-306, courtesy of NASA)

ditions on earth, but, more importantly, augmented the major systems by supplying oblique views of areas along the periphery of EREP passes. In fact, the Hasselblad and Nikon photography affords the only available coverage of some rather sensitive regions, such as the Soviet Union and China, thus, certain of Skylab's handheld products are of value as applied to a few regions and topics of concern (fig. 8.5).

Skylab's four nonphotographic sensor systems supplied a great variety of earth resource data, much of which is stored on magnetic tapes. The infrared spectrometer registered spectral radiance in the visible through near-infrared and also thermal infrared bands when sporadically targetted by the crew. Applications were particularly oriented to mineralogical conditions and to mapping extreme ranges in surface cover, such as ice fields and hot springs. The multispectral scanner traced a 74-kilometer-wide swath with a set of 13 detectors covering a total range of wavelengths from 0.41 to 12.50 micrometers (410 to 1,250 nm). The data, which is suited to a very wide range of applications, can be computer-processed in much the same manner as Landsat data-handling procedures.[2]

The microwave radiometer/scatterometer and altimeter was complemented by an L-band radiometer which operated at a different wavelength. While both of these systems were oriented to measuring brightness temperature of targets, a major application involved an altimeter-measured survey of sea levels and ocean floor depths.[3]

The six major EREP sensing systems aboard Skylab offered a multispectral coverage ranging from the camera systems at 0.4 micrometers (400 nm) through the L-band microwave radiometer at 1.42 gigahertz. The products of the two standardized camera systems are of principal applicability to most general concerns in remote sensing because they are most readily available for greatest areal coverage in renditions most easily interpreted. Of serious concern are the photo products, in that film positives, offering highest resolution, are rather small for ease of handling and require considerable magnification to benefit from the resolution scales

Figure 8.5. Southeastern New England taken by handheld Hasselblad camera at the same instance as figure 8.4 from Skylab 3. (Photograph No. SL3-122-2597, courtesy of NASA)

produced. In print format, the standard products can be more easily adapted to the mapping of spatial patterns, although loss of resolution is a problem. One solution to this dilemma is the purchase of both formats. Enlargements from negatives or rephotographing of portions of film positives or prints will produce prints of enlarged sections of particular interest, as illustrated in figure 8.6, 8.6A, and 8.6B. For general instructional purposes, slides are useful and can be obtained from such sources as are listed in Appendix A.

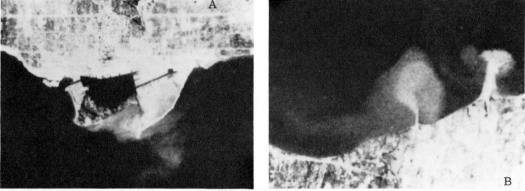

Figure 8.6. & 8.6A & 8.6B. Western Lake Ontario from Skylab 3. A and B are rephotographed enlargements of specific features of the image. (Photograph No. SL3-41-016, courtesy of NASA and the author)

EREP Passes

After the initial launching into orbit of the space laboratory, a procedure designated as Skylab 1, the space laboratory soon treated the public, perhaps jaded by spectacular lunar ventures, to another drama as Skylab 2 astronauts confronted a crippled craft. By a clever repair of the damaged power panel, the Skylab 2 crew amended what might have been an aborted space project and only slightly reduced the scheduled activities of the Earth Resources Experiment Program. During 11 EREP passes in 404 orbits from May 25 to June 22, 1973, they produced 5,275 frames of imagery and 13,716 meters of magnetic tape data. The Skylab 3 crew conducted 44 EREP passes during 858 orbits from July 28 to September 25 to obtain 13,429 frames and 28,529 meters of tape. Finally, between November 16, 1973, and February 8, 1974, from some 50 EREP passes during 1,214 Skylab 4 orbits, the crew collected 17,000 frames and 30,480 meters of taped data.

Although all of the more than 35,000 frames and 72,725 meters of taped data gathered from the over 100 EREP passes do not provide sequential or equally cloud-free coverage of particular earth locations, the seasonal variation and spatial distribution of the imagery allows an impressive number of practical applications. Indeed, the astronauts had been assigned some 172 EREP tasks involving nine major disciplines and 46 subdisciplines thereof. These applications included:

1. *Agriculture/Range/Forestry:* crop inventory, insect infestation, soil type, soil moisture, range inventory, forest inventory, and forest insect damage
2. *Geological applications:* mapping, metals exploration, hydrocarbon exploration, rock types, volcanoes, and earth movements
3. *Continental water resources:* ground water, snow mapping, drainage basins, and water quality
4. *Ocean investigations:* sea state, sea/lake ice, currents, temperature, geodesy, and living marine resources
5. *Atmospheric investigations*: storms, fronts and clouds, radiant energy balance, air quality, and atmospheric effects
6. *Coastal zones, shoals, and bays:* circulation and pollution in bays, underwater topography and sedimentations, bathymetry, coastal circulation, and wetlands ecology
7. *Regional planning and development:* land use classification techniques, environmental impacts, state and foreign resources, urban applications, coastal/plains and mountain/desert applications
8. *Remote sensing techniques development:* pattern recognition, microwave signatures, data processing, and sensor performance evaluation
9. *Cartography:* photomapping, map revision, map accuracy, and thematic mapping [4]

For guidance as to applications of EREP data to such concerns as listed above, the *Skylab Earth Resources Data Catalog* is especially well illustrated. Regrettably, however, no complete catalog of Skylab imagery now available correlates photographs either with these applications or by regional groupings. The Skylab 2, 3, and 4 catalogs produced by the Technology Applications Center (TAC) at the University of New Mexico list photographs in sequential order by camera system and provide excellent data as to percentages of cloud and snow cover, locational descriptions per country or state/province with indication of major settlements, and unusual environmental conditions, e.g., air and water pollution, active volcanoes, and strip mining.

A cursory perusal of the TAC catalogs will quickly reveal the severe imballance in the regional coverage of the world by EREP passes. As might be expected, the greatest proportion of cloud-free imagery is of arid areas, with the southwestern portion of the United States being especially well recorded. A special earth terrain camera swath from Louisiana to Maine has been compiled as a mosaic series of fifteen posters. The series—"Gulf States to New England"—is available from the Johnson Space Center, Houston.

Among other notable coverage by EREP passes is the bulk of major metropolitan areas in North America. Although few Canadian cities were clearly photographed by Skylab cameras, the Hamilton-Toronto-Peterborough complex in Ontario and several centers in the Maritime Provinces were adequately recorded. Urban area color photography includes the United States locations listed in Appendix A.

For other parts of the world, considerable EREP imagery exists covering the Western Hemisphere countries of Mexico, Brazil, and Argentina. Western Europe is poorly represented aside from a few photographs of the Mediterranean region, such as areas of Spain, Portugal, northern Italy, and Yugoslavia. North Africa is well photographed. Asian coverage is sporadic at best. In summary, Skylab's EREP pass production had some severe limitations in regional coverage, due partly to its equatorial orbit, but it did provide sufficient samples for very useful analyses of many environmental concerns.

Retrospective and Perspective

Unlike the remote sensing endeavors of earlier manned spacecraft missions, Skylab incorporated a highly sophisticated package of sensors to scientifically assess earth resources. The collection of a massive amount of environmental data across a large portion of the earth's surface during four seasons as recorded by standardized, multispectral, high resolution camera systems provided a baseline for evaluation of both past and future efforts. The limitations of the Skylab EREP pass program, as compared, for example, to Landsat coverage of earth resources, need not depreciate the impressive results obtained (fig. 8.7).

For earth resource management, Skylab offers a high quality static data base for regional development planning as a vital component in a coordinated remote sensing program. It can enhance the repetitive registry of Landsats 1 and 2, and so help in selecting high-priority areas of concern; it can broaden the areal scope of high resolution reconnaissance of high and low altitude underflight; and it can guide the efforts of future space laboratory ventures to provide needed coverage for augmenting present and planned earth resource surveys at various scales and locations. The enthusiasm for NASA's Space Shuttle program is in part to be attributed to the potentials revealed by Skylab, which can be carried as a payload on the Space Shuttle (fig. 8.8).

In many respects, the superb photographic array produced by the EREP passes is more easily handled by the untrained person than, for example, are the products of electro-optical scanning systems transmitted by Landsat. This assessment is based upon experience with children's reception of remote sensing materials in an environmental education project, "Landscapes of Vermont," funded by the U.S. Office of Education. An essential element of the program was the belief that people regard "the" environment as their own, thus prompting an application of locally-oriented materials to the understanding of land use change. The fairly comprehensive coverage of North America by Skylab imagery allows extension of local en-

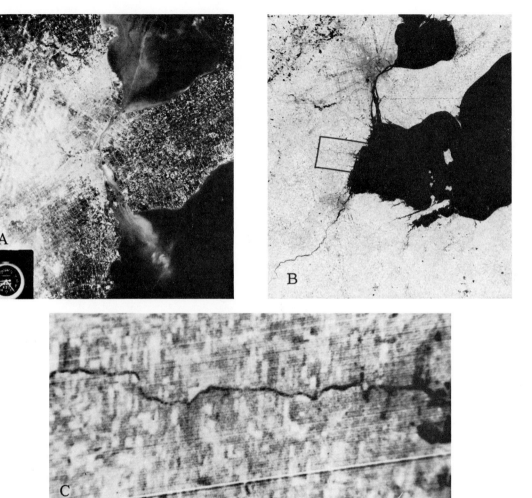

Figure 8.7. Contrasting coverage of Detroit-Windsor and environs: (A) fron Skylab in August 1973; (B) from Landsat 2 in August 1975; (C) enlargement of indicated portion of B. (Skylab Photograph No. SL3-83-152, Landsat Image No. S111-15220, both courtesy of NASA)

vironment interests to regional scales of concern appropriate to the regional character of many earth resource problems.

For years to come, Skylab will stand as a benchmark and model for manned space laboratory endeavors in remote sensing. The mission crews, designers, and data handlers associated with this project effectively launched a new era in space photography. Perhaps the greatest opportunity now confronting educators is the challenge of applying Skylab's products to their curriculum.

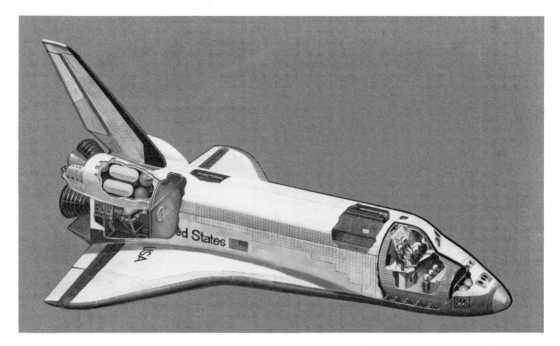

Figure 8.8. For future remote sensing missions, the Space Shuttle orbiter will carry Skylab-type space stations and other payloads in its freight compartment. (Photograph/Diagram No. 76-H-779, courtesy of NASA)

NOTES

1. R. Underwood, "Cameras in Space," *Industrial Photography* 25 (July 1976):22. An excellent summary of photographic progress in the United States space effort.
2. Of the nonphotographic sensor systems aboard Skylab, little imagery is available to or easily handled in remote sensing work at the introductory level. The nonvisible spectral sensing systems, as well as the two principal cameras, as well described in NASA-Johnson Space Center, *Skylab Earth Resources Data Catalog* (U.S. Government Printing Office Stock No. 3300-00586). This definitive book, compiled largely by Martin Marietta Corporation, should be included in all basic remote sensing reference libraries. A vivid example of Skylab's thermal infrared sensing capacities can be found in *Scientific American* 235 (September 1976): 74-75. This, combined with a sample of Skylab's infrared photographic coverage of the same area around Greenbay, Wisconsin (SL3-21-179), provides an excellent opportunity to mitigate frequent confusion as to the spectral registry of the two types of infrared sensing.
3. Skylab's contribution to oceanography is described and illustrated in the special NASA FACTS bulletin, NF-56/1-75, *Observing Earth from Skylab,* compiled by Frank Tallentire. An excellent summary of EREP operations, it is available for $1.00 (Stock No. 033-000-00627-6) from the U.S. Government Printing Office, Washington, D.C. 20402.
4. National Aeronautics and Space Administration, *Remote Sensing of Earth Resources,* in *Skylab Experiments,* vol. 2 (Washington, D.C.: U.S. Government Printing Office, 1973), p. 5.

9

Landsat Platforms, Systems, Images, and Image Interpretation

Benjamin F. Richason, Jr.
Carroll College, Wisconsin

Introduction

Deus ex Machina

The nomenclature of satellites in the United States space program reflects their mythological antecedents. Atlas, Mercury, Gemini, Apollo, and others made their appearances, in turn, as part of the space arsenal of engines and platforms which were conceived, launched, and monitored by NASA to provide a surveillance base for aiding in solutions to environmental problems. Even EROS (for Earth Resources Observation Systems) may have followed accidentally, if not erotically, in the tradition of heroic Greek mythology.

In the ancient Greek and Roman tragic plays, particularly those written by Euripides, the plot would become so involved, and the lives of mortal characters would become so complex, that only a god could provide solutions to terrestrial problems. The Greek and Roman audiences felt reassured when, in the epilog of many of the plays, a god was lowered from overhead by a rope onto the stage where he divinely ended the impasse in the plot which the characters had created. This device of introducing a god by means of rope, block, and tackle onto the stage for solving problems was known as *Deus ex Machina* (god from a machine).

More than 2,200 years later, an anology to *Deus ex Machina* is the remote sensing satellite which collects information about the environment and its resources, and sends this information to receiving stations on earth where modern-day scientists are using it to help solve the complex problems which have been created by characters on the earth stage. The term *lalia*—chatter—was used to describe the complex, verbose dialogue in Greek plays. Today, remote sensors "chatter" information at the speed of light concerning environmental problems and events, reassuring people everywhere that powerful tools are at work attempting to bring to an end the impasse between man and land and the types of economic and cultural imprints man has made on the land.

EROS Program

In an effort to collect and analyze information from synoptic imagery of the earth and its resources, the U.S. Department of the Interior established the EROS program which is administered by the U.S. Geological Survey. As part of this program the unmanned satellite ERTS-1 (Earth Resources Technology Satellite) was launched by NASA into a nearly polar, sun-synchronous earth orbit on July 23, 1972. ERTS-2 was launched into a similar orbit on January 22, 1975. On January 13, 1975, these satellites were designated as Landsat (for Land Satellite) 1 and 2, and the ERTS program became the *Landsat* program. On March 5, 1978, Landsat 3 was launched into an orbit similar to that of its predecessors. In addition to Landsat imagery and data, the EROS agency cooperates with and uses data from the Skylab Earth Resources Experiment Program (EREP) and from the Aircraft Program of NASA.

An integral part of these programs is the EROS Data Center which was established in 1971. Since 1973, this center has been located in the Karl E. Mundt Federal Building near Sioux Falls, South Dakota. This is the focal point for providing copies of remote sensor imagery and other space and high altitude data to the public. In addition to these activities, the EROS Data Center provides training courses and other professional services to those who use data derived from a variety of remote sensors.

The Landsat program has demonstrated the applications of remote sensing for inventorying, analyzing, managing, and monitoring the natural resources and environments of the world. Landsat imagery is used by geographers, geologists, foresters, hydrologists, biologists, and other scientists to investigate features and events at the surface of the earth.

Uses of Landsat data appear to be limited only by the imaginations of scientists and the technological state of their sciences. Cartographers have found that Landsat images can be used for preparing and up-dating maps. In geology, Landsat imagery has been employed in studying rift valleys, glaciated areas, glacier movements, soils, sand deserts, and the structural geology of mineral deposits. Volcanic activity, hydrothermal areas, and active faults are subjects which scientists have investigated with the aid of Landsat data. Landsat images have been used to monitor water reservoirs, lake fluctuations, and water turbidity, along with water impoundment inventories. Foresters and agronomists have used Landsat data to monitor forest fires, changes in vegetation, and livestock forage on western ranges. Land use planners have used Landsat imagery to monitor land cover and land use changes in urban, suburban, and rural areas. Applications of environmental monitoring by use of Landsat products are legion in studies of air pollution, wetlands, conservation, and oceanography. The chatter from the "god of a machine" is as beneficial on the earth stage today as it was on the Greek and Roman theatrical stages thousands of years ago.

Landsat Platforms

The Landsat Spacecrafts

The Landsat vehicles are panel-studded space observatories (fig. 9.1). The platforms of Landsat 1, 2, and 3 are essentially identical. Each weighs about 953 kilograms (2,100 pounds), has an overall height of about 3 meters (10 feet), and a width of about 4 meters (13 feet) when the solar cells, which provide power for the systems on board, are extended. The Landsat vehicles are mounted on a Nimbus weather satellite frame. The predicted lifespan had been a

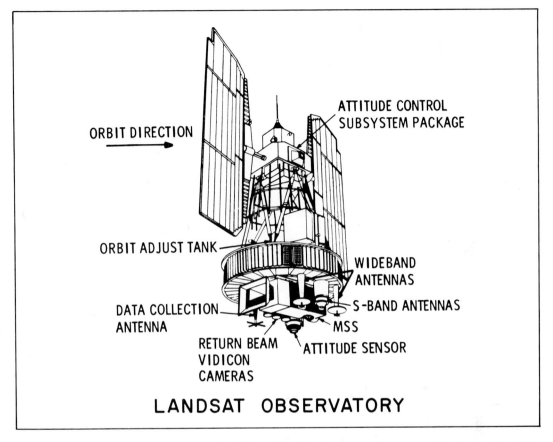

ORBIT DIRECTION

**ATTITUDE CONTROL
SUBSYSTEM PACKAGE**

ORBIT ADJUST TANK

**WIDEBAND
ANTENNAS**

S-BAND ANTENNAS

**DATA COLLECTION
ANTENNA**

MSS

**RETURN BEAM
VIDICON
CAMERAS**

ATTITUDE SENSOR

LANDSAT OBSERVATORY

Figure 9.1. The various payloads of the Landsat observatory are illustrated in this diagram. The positions of the Multispectral Scanner (MSS), Return Beam Vidicon Cameras, and the Data Collection System can be seen. (From EROS Reprint No. 167)

year, but it was not until almost five and a half years of operation that technical problems led to shutting down Landsat 1 on January 6, 1978.

While Landsat 1 was still in orbit, Landsat 2 was launched into earth orbit from Vandenburg Air Force Base in California. Although the launch date was scheduled for January 19, 1975, a voltage fluctuation in the first stage engine of the Delta launch vehicle delayed the event for three days. Once in orbit, the three-day delay resulted in a 135° difference in position between Landsat 1 and Landsat 2. An orbit adjustment brought the two satellites into the desired position of 180° apart. In its corrected orbit, Landsat 2 has an apogee of 918 kilometers (570.4 miles) and a perigee of 912 kilometers (566.7 miles), as compared to the circular orbit of 920 kilometers (571.7 miles) which was planned originally. Landsat 1, following several orbital corrections through February 4, 1974, had an apogee of 919 kilometers (571 miles) and a perigee of 896.3 kilometers (557 miles).[1] Landsat 3 was launched March 5, 1978. Ongoing adjustments of Landsat 2 and 3 will keep them in proper orbit. A fourth Landsat vehicle and payload system is under construction.

Landsat Orbits

Landsat platforms orbit the earth in nearly polar, sun-synchronous paths every 103.3 minutes, or 14 times each day (fig. 9.2). The paths are at azimuths of about 190° measured from true north. After 252 orbits, or every 18 days, Landsat passes over the same place on earth. Because Landsat 1 and 2 were 180° out of phase with each other, during the life of Landsat 1 every place on earth came under a Landsat vehicle every 9 days.

Because Landsat orbits the earth 14 times a day and because the earth is rotating on its axis from west to east underneath the tracks of the satellites, each successive north-to-south track of the Landsat vehicle is offset to the west by 2,875 kilometers (1,786 miles) at the equator (fig. 9.3). This westward displacement of successive orbit tracks is found by dividing the circumference of the earth at the equator (about 40,225 kilometers or 25,000 miles) by 14 revolutions by the satellite each day. These orbits result in a daily westward longitudinal shift of tracks at the equator of about 160 kilometers (99 miles). This means that orbit two on day one will lie 2,875 kilometers west of orbit one at the equator, and so forth. On day two, orbit one will lie 160 kilometers west of the path of orbit one on day one. Finally, on the eighteenth day, orbit one will be identical to orbit one on day one. On the nineteenth day, the first orbit will coincide with the second orbit on the second day, and the sequence will continue. In this manner the entire surface of the earth comes under a Landsat vehicle every eighteen days, and all areas can be

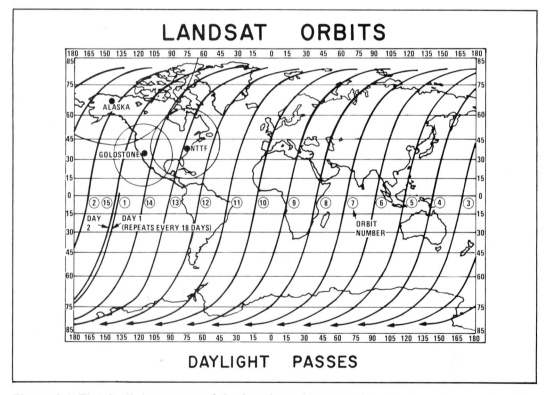

Figure 9.2. The daylight passes of the Landsat observatories are superimposed on the map in this figure. See text for explanation. (From EROS Reprint No. 167)

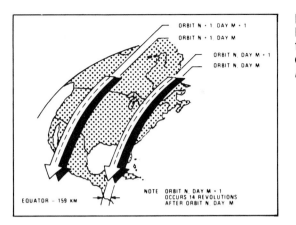

Figure 9.3. The ground traces of the Landsat satellites are illustrated in this figure for successive orbits on the same day, and for the following day. (From *Data Users Handbook*, NASA)

imaged by its sensors, notwithstanding the large gap of 2,875 kilometers between successive orbits.

The sensors image an area on the ground of about 185 kilometers (115 miles) on a side; however, there is about a 15 percent sidelap (26 kilometers) on adjacent orbits on successive days at the equator. The sidelap increases with increasing latitude, becoming 57 percent at 60°, and about 85 percent near the poles.[2]

The imaging passes of Landsat spacecrafts are from north to south on the lighted side of the earth. The satellites, traveling south on each imaging pass (fig. 9.2), cross the earth's equator at about 9:30 A.M. local sun time, and they pass over the same point on the earth at about the same local sun time every eighteen days. This does not mean, however, that local clock time will be the same for all Landsat crossings because designated time zones are used throughout the world.[3] On the other hand, it does mean that the altitude of the sun in the sky and the position of the sun along the horizon will be approximately the same within similar seasons at every crossing of the same parallel. Differences in position of the sun occur because of the apparent motion of the sun from season to season.

For example, Landsat crossed the parallel of 43°7′ N at 88°21′ W longitude at 3:34 P.M. Greenwich mean time (GMT) on January 21, 1976; it crossed 43°12′ N at 88°28′ W at 3:30 P.M. GMT on March 15, 1976; 43°8′ N at 88°38′ W at 3:23 P.M. GMT on July 1, 1976; and 43°13′ N at 88°31′ W at 3:17 P.M. GMT on September 11, 1976. This uniformity in the time each parallel is crossed by Landsat vehicles in an imaging mode provides essentially the same seasonal lighting and shadow effects for images of the same area. This is an important feature about Landsat images because it facilitates the identification and comparisons of earth features on successive image frames produced by the sensors on board. Table 9.1 shows the variations of sun altitude and azimuths for selected seasonal crossings at about 43° N latitude and 88° W longitude for Landsat 1 in 1976.

Imaging Systems

The sensors on board the Landsat spacecrafts include Return Beam Vidicon (RBV) television cameras (fig. 9.4), and Multispectral Scanning (MSS) systems (fig. 9.5). Because of some electrical failures on Landsat 1, the RBV system was turned off on August 5, 1972. A similar

TABLE 9.1

Landsat 1: Sun Altitude and Azimuths for Selected
Seasonal Crossings at About 43°N 88°W

Date	Sun Altitude	Sun Azimuth Along Horizon
January 21, 1976	18°	142°
March 15, 1976	33°	132°
July 1, 1976	52°	109°
September 11, 1976	38°	128°

malfunction caused the RBV sensors on Landsat 2 to be shut down shortly after its launch. Images continue to be made, however, by the MSS systems.

Multispectral Scanning System

The MSS system employs an oscillating mirror to scan the surface of the earth as it passes beneath the satellite. The MSS systems on Landsat 1 and 2 image scenes synchronously in the following wavelengths:

Band 4	500—600 nm[4]	Green
Band 5	600—700 nm	Red
Band 6	700—800 nm	Red to near infrared
Band 7	800—1,100 nm	Near infrared

The MSS systems collect sunlight reflected from objects on earth in proportion to the sensitivities of each MSS band. The multispectral scanners on the Landsat spacecrafts have capabilities of distinguishing different wavelengths and frequencies of energy which are reflected by different targets and scenes. Each target, because of the different materials of which it is composed, produces its own characteristic spectral signature. Green vegetation reflects more green light than red light, but the chlorophylls in vegetation reflect heavily in the near infrared portion of the spectrum.[5] Dry soils reflect more radiation in the red and near infrared wavelengths, while wet soils reflect less in all bands of the spectrum because of the light-absorbing properties of water in the soil.[6] Because water absorbs all wavelengths of light in the visible and near infrared portions of the spectrum, rivers, lakes, ponds, and coastal waters which are free of silt and other particulate materials will appear black on positive images which are generated from the MSS bands.

Although each MSS band produces images which are of value in themselves, all bands should be consulted for the optimum interpretation of earth features and events (figs. 4.5 and 9.6). Because of the different wavelength responses of the MSS sensors, some targets are not imaged in a particular band, while other objects may be degraded to the place where they are not usable for identification. The same target in the same scene, however, may be resolved with great fidelity in another MSS band. Notwithstanding the fact that certain spectral responses characterize each MSS band, attempts to generalize the capabilities of bands should be approached with caution because reflectance patterns may not be the same in all imaging situations.

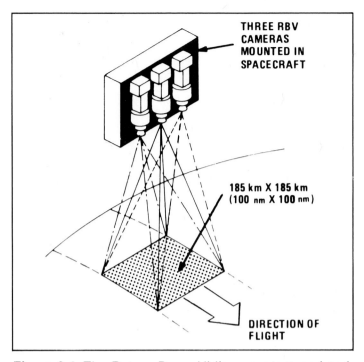

Figure 9.4. The Return Beam Vidicon system on Landsat. (From *Data Users Handbook*, NASA)

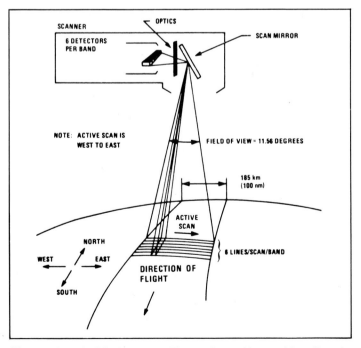

Figure 9.5. MSS System (From *Data Users Handbook*, NASA)

Band 4 (fig. 9.6, A) detects and registers blue-green wavelengths which are radiated from objects. This band is most useful for studying the characteristics of water bodies. In some instances, water depths and turbidity in standing water can be detected in scenes imaged in this band. Band 4 is sensitive to water pollution patterns, the presence and density of phytoplankton, and to such submarine features as shoals and reefs, as well as to atmospheric phenomena.[7] The disadvantages of imagery from Band 4 include the lack of discrimination of surface scenes because of Rayleigh scattering in the atmosphere, minimum penetration of clouds and haze, and the lack of gray-scale contrasts in topographic and cultural features.

Band 5 (fig. 9.6, B) is sensitive to radiation in the red wavelengths close to the outer margin of human vision. Generally, this band is best for producing images which distinguish topographic and man-made features. Imagery from Band 5 optimizes discrimination of urban and transportation infrastructures,[8] roads, railroads, and trails. Land use and land cover patterns are registered in greater detail on Band 5 than on Band 4. Band 5 appears to provide the best spectral region for imaging exposed bedrock and snow.[9] The EROS Data Center stated that if only one Landsat MSS image is to be acquired, Band 5 should be obtained because of its general imaging capabilities for a wide range of objects on the surface of the earth.[10]

Band 6 (fig. 9.6, C) registers targets whose reflectance is at a maximum in the red and near infrared spectral area beyond that of most human vision. This band can detect stress in vegetation. Also, cloud penetration is improved in this band as compared to Band 5; therefore, more surface detail is available when scenes are thinly overcast by clouds or haze. Water and land interfaces are discriminated in this band; however, Band 6 imagery can be mistakenly interpreted such that shallow water or heavy sediment loads appear to be land.[11]

Band 7 (fig. 9.6, D) is the second MSS near infrared band. It is best suited for distinguishing differences between land and water areas. Detailed observations of drainage systems, shorelines, and wetlands are maximized by use of imagery from this spectral band. On the other hand, Band 7 is least usable for studying the intrinsic characteristics of water bodies because water appears almost solidly black in the wavelengths to which this band is responsive. Maximum cloud and haze penetration is produced in this band; however, this has minimal applicability because, with the exception of hydrographic features, land use detail is very indistinct in Band 7. Urban patterns, on the other hand, may be pronounced on Band 7 imagery, particularly if there is a snow cover on the ground.

A fifth channel is included in the MSS package on Landsat 3. It is reported that this channel senses targets in the thermal infrared spectral region of the electromagnetic spectrum—specifically from 10,400 nm to 12,600 nm.[12] Radiation reaching the sensor in this spectral band is emitted from the ground surface as a function of surface temperature. Landsat 3, therefore, will have the capacity to image surface temperatures and to detect targets by contrasting the temperature of the target and its background.[13]

The intensities and wavelengths of radiant energy which are received by each multispectral scanner on Landsat spacecrafts are converted into electrical signals. These signals may be transmitted directly to earth receiving stations, or the electronic signals of objects may be stored on magnetic tape and telemetered to receiving stations upon command. The former type of transmission is a "real time" image, while the latter is a "near real time" image. Most imagery signals which are sensed over North America are transmitted in a "real time" mode. Signals that are collected from scenes in other parts of the world can be transmitted to a United States

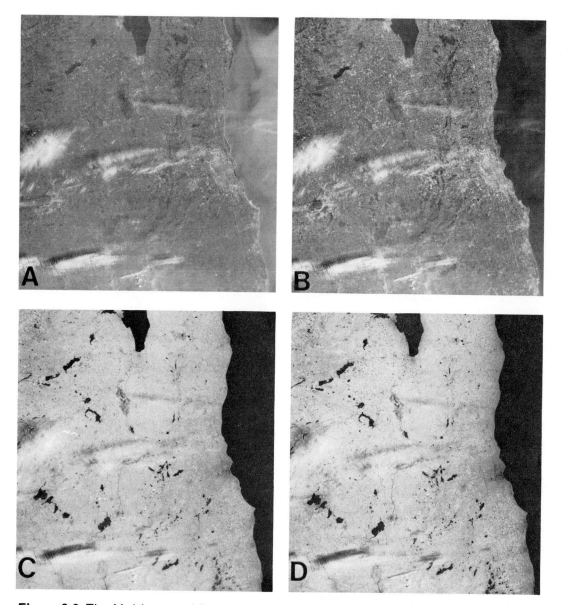

Figure 9.6. The Multispectral Scanner on Landsat senses and images earth scenes in four bands of the electromagnetic spectrum. A is a portion of a band 4 image, B of band 5, C of band 6, and D of band 7. See text for an explanation of the characteristics of images in each of the bands. (Images obtained from the EROS Data Center)

or Canadian receiving station from the stored magnetic tape signal only when the satellite is within range of a ground antenna.

Landsat transmissions are received at several United States stations, as well as at some foreign antennas. The antennas in the United States which receive Landsat signals are located at Goldstone, California, Fairbanks, Alaska, and Greenbelt, Maryland. The United States has assisted other countries in establishing receiving stations for Landsat imagery. Canada receives Landsat signals at its antenna in Prince Albert, Saskatchewan, and the data is sent to Ottawa for processing. Canada is also establishing a receiving station at St. Johns, Newfoundland. The Brazilian organization, Instituto de Pesquisas Espaciais, has a receiving station at Cuiabá, with processing facilities at San Jose dos Campos. Also, the Italian remote sensing organization, Telespagio, has established a receiving antenna and processing facility at Fucino near Rome. Chile has established a Landsat antenna at Santiago, and Zaïre, Iran, and Argentina are in the process of constructing receiving stations for Landsat imagery. The Landsat data which are received at the various United States antennas are processed into picturelike form and sent to the EROS Data Center. Here, the images are placed in the public domain and can be purchased by anyone.

By 1975, the MSS sensors on Landsat 1 were generating useful imagery over North America, ". . . but its capabilities to tape record data collected outside the range of recording stations had degenerated to the point where it was useful only over North America."[14] On March 3, 1977, it was reported that Band 4 of the MSS sensor on Landsat 1 was no longer operational.[15]

Return Beam Vidicon Sensor

Unlike the MSS systems which produce a continuous scan of the surface of the earth over which they are transported, the Return Beam Vidicon (RBV) systems shuttered individual scenes upon command when they were operational. During their short periods of operation, they sensed targets in the spectral range of 475 nm to 830 nm by three television cameras. On Landsat 3, the RBV system has been modified to include two panchromatic cameras which will generate side-by-side images rather than the superimposed images of the same target area which were characteristic of the systems on Landsats 1 and 2.[16]

When they were operative during the early orbits of the satellites, the RBV systems registered scenes in three bands of the spectrum. These RBV bands were designated as 1, 2, and 3 in order to differentiate the nomenclature of the RBV bands from those of the MSS bands. The wavelengths and color bands of the RBV systems are as follows:

Band 1	475—575 nm	Green
Band 2	580—680 nm	Red
Band 3	690—830 nm	Near infrared

The principal advantage of the RBV camera systems over that of the MSS systems is its geometric fidelity, or relative mapping accuracy. For this reason it was to have been used extensively for cartographic purposes. When the systems were shut off, cartographers decided to attempt to use the MSS images for mapping. The MSS imagery, when properly processed, yielded unexpectedly good imagery for mapping because of its global coverage, near real time transmission, orthogonality, relatively small spatial errors, and its ability to measure radiation beyond the capabilities of conventional film.[17] Furthermore, the MSS system possessed similar automation capabilities and wavelength-extensions into the infrared as that of the RBV systems.[18] For these reasons, many mapping projects which were originally planned for RBV imagery were completed satisfactorily by use of MSS data.

Data Collection System

The Data Collection Systems (DCS) on Landsat are designed to receive information about such environmental data as temperature, stream flow, soil moisture, and snow depth. Individual researchers provide the instrumentation at ground stations. As the satellites pass over a data collection station, environmental information is radioed to the satellite and recorded by the DCS at the same time as images of the scene are generated. Data from these platforms are made available to investigators within a day from the time data collection measurements are telemetered by the satellite.[19] In this way, ground truth data are supplied simultaneously with the imaging of the scene by sensors on board the satellite. Such ground truth data aid in better understanding and interpreting the Landsat imagery. Figure 9.7 illustrates the flow pattern of data received from the DCS, RBV, and MSS sensors.

Landsat Images

Image Format

The electronic images received from Landsat sensors are generated into photographlike prints by an electronic beam recorder. One of the first things a user notes about a Landsat MSS image is that it is not square or rectangular like conventional photographs. Figure 9.8 is a reproduction of one band of an MSS image. The base of the image is seen to be skewed to the left, producing a trapezoidal form.

The skewed peculiarity of Landsat MSS images is produced by the characteristics of the scanning sensor and the rotation of the earth. As the satellite travels from north to south over the lighted portion of the earth, the MSS scanner images the scene below by continuously sweeping from side to side at right angles to the track of the satellite. Six lines are scanned simultaneously in each of the four bands. Each sweep of the scanning mirror, therefore, produces the side-track, or easting and westing, dimensions of the image. The southward motion of the spacecraft provides the along-track progression of the image.[20] While these electromechanical processes are taking place, the earth is rotating the ground scene out from under the view angle of the sensors from west to east. Each scan line, therefore, is offset to the west, and the southern border of each Landsat frame will be offset to the left of the top, or northern, border of the print.

Annotation Block

An information, or annotation, block can be seen across the bottom of figure 9.8. The annotation block is located on that side of the print toward which the satellite was moving at the time the image was made. Reading from left to right, the annotation block provides a variety of information. The first seven places give the date on which the image was made (day, month, and year). In the case of figure 9.8, the date of the image is 13 November 1976.

The characters which follow the symbol C denote the format center of the image in degrees and minutes of latitude and longitude. On figure 9.8, the format center, which is the geometric extension of the yaw axis of the satellite to the surface of the earth, is N 43°4′; W 88°25′. The yaw of a spacecraft is a rotation in flight around its vertical axis.

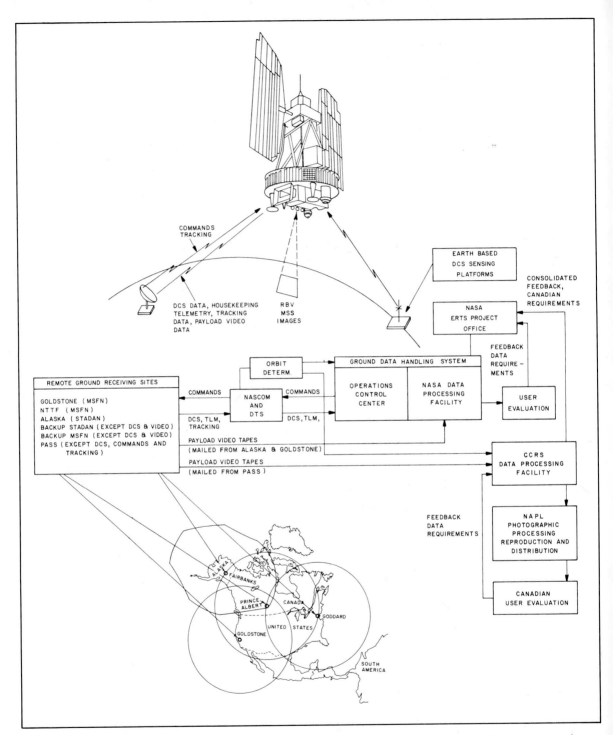

Figure 9.7. The diagram illustrates how Landsat spacecrafts receive reflectance and other signals at the sensors on board. In North America the electronic image signals are transmitted to both U.S. and Canadian receiving stations. (From *Earth Resources Technology Sateliite Data User's Handbook,* Canada Centre for Remote Sensing)

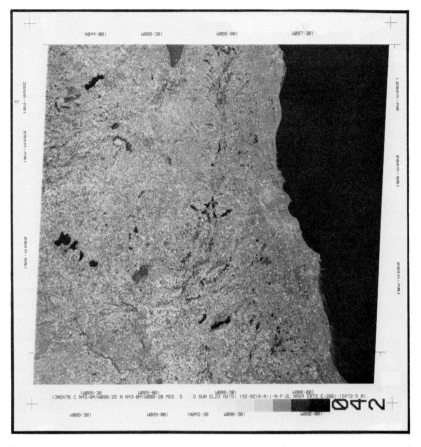

Figure 9.8. Multispectral Scanner Band 5 of a Landsat image of southeastern Wisconsin. See text for explanation. (EROS Data Center)

The characters which follow the symbol N denote the nadir of the image; that is, the point on the surface of the earth which is directly below the satellite. The line between the satellite and this point on the earth is at a right angle to the earth ellipsoid. The nadir of the image in figure 9.8 is N 43 °4 '; W 88 °20 '.

Next, information is given concerning the type of sensor which imaged the scene. MSS Band 5 means that this scene was imaged on Band 5 (600-700 nm) by the Multispectral Scanner. MSS 4, 6, or 7 would have indicated that the scene was sensed at wavelengths in Bands 4, 6, or 7, respectively. RBV 1, 2, or 3 in this annotation-block position would have indicated that the image had been sensed in one of the spectral bands of the Return Beam Vidicon camera.

Continuing to read to the right, D signifies that the scene was transmitted in "real time" to a receiving antenna. An R in this place would have denoted a recorded type of transmission, as when a scene was imaged in a part of the world out of reach of a receiving antenna.

Next, the positions of the sun at the time the scene was imaged appear in the annotation block. The scene illustrated in figure 9.8 was imaged at a time when the elevation of the sun above the horizon was 23°, and the position of the sun along the horizon was at an azimuth of 151° (or at a bearing of S 29° E).

The number 192, which appears next in the annotation block, indicates that the satellite was heading at an azimuth of 192°, as measured from true north. The next four digits are the orbit revolution number. Revolution 1 began with the first ascending node from south to north after the satellite was launched. The image in figure 9.8 was made on the 9,216 orbit.

To the right of the orbit number is a letter which denotes the ground antenna at which the telemetered image was received. N indicates that the signal was received at the Goddard Space Flight Center, Network Test and Training Facility, Greenbelt, Maryland. An A in this position would have meant that the image was received at Fairbanks, Alaska, and a G would have meant that Goldstone, California was the receiving station.

Following to the right, the 1 indicates that the image is 100 nautical miles (115 miles or 185 kilometers) on a side. A figure 2 in this position of the annotation block would have indicated a frame that is 50 nautical miles on a side, and a 3 would indicate that the frame is 25 nautical miles on a side.

The next four characters in the annotation block, N-P-2L, refer to image processing mechanics. They have little meaning for the reader, but are important in planimetric analysis of the imagery. In the case of the image in figure 9.8, N means normal processing procedures were used in generating the print; P means that the orbit ephemeris was predicted; and 2L means that the telemetered signal was compressed in low gain before transmission.

NASA-ERTS (National Aeronautics and Space Administration-Earth Resources Satellite) denotes the agency and platform which were responsible for the imagery.

E-2661-15472 is the frame identification number. This number is used when MSS images are ordered from the EROS Data Center. E denotes the project (ERTS), and 2 the Landsat vehicle. A 1 in this place would stand for Landsat 1. After 999 days the figures change to 5 for Landsat 1 and 6 for Landsat 2. The figure 661 indicates that the image was made 661 days after launch. The figures 15472 denote the Greenwich mean time in hours (of the 24-hour clock), minutes, and nearest tens of seconds at which the image was made. In the case of the image in figure 9.8, the time of imaging the scene was 15 hours, 47 minutes, 2 seconds GMT; or 3:47 and 2 seconds P.M., GMT. This would correspond to 9:47 and 2 seconds A.M. central standard time.

Next, the figure 5, again, denotes that the image was made in those wavelengths to which Band 5 of the MSS is sensitive. On a color composite of an MSS image (described below) the numerals 457 would appear in this position. The figure 01 is a regeneration number.

A gray scale appears below the annotation block on every MSS Landsat frame. This 15-step, black to white scale is produced by an electron beam recorder for each Landsat print, whether film negative, film positive, or paper positive. The scale is processed in the same manner as the photographlike print to which it is attached. The gray scale indicates the tone, or brightness level, of corresponding images on the print.[21] Individual blocks on the gray scale are related to electromagnetic energy which was received at the MSS sensor for each target within the scene.

In addition to the gray scale and the annotation block, the margins of the MSS image contain certain other useful information. Registration marks are placed in the four corners of the print. When the registration marks on images made in different spectral bands of the same scene by the same sensor are aligned and superimposed, all frames will be in registration, and all targets within the imaged scene will correspond exactly. The intersection of lines drawn from registration marks in opposite corners of the image will locate the format center. The format center of all four bands of the MSS, and the format centers of the RBV and MSS images, of the same target area will be identical.

The tic marks along the edges of the print locate the approximate latitude and longitude grid of the image. These are usually placed at intervals of 30 minutes of arc; however, at latitudes poleward from 60° the tic marks and grid numbers are placed at one-degree intervals to avoid crowding at the higher latitudes.

Scales of Imagery

Scenes are imaged by Landsat MSS sensors on 70 mm formats at scales of 1:3,369,000 (1 centimeter represents 33.69 kilometers, or 1 inch represents 53.17 miles). The EROS Data Center processes the electronic beam recorded images at the standard scales of 1:1,000,000; 1:500,000; and 1:250,000. Copies of Landsat images may be obtained at any one of these scales in the form of film negatives, film positives, or paper prints.

MSS images from Landsat are scanned in scenes which measure 34,253 square kilometers (or 13,225 square miles). People must adjust their thinking to this small scale when studying Landsat imagery. For example, the scene in figure 9.8 includes an area from Lake Michigan on the east to beyond Madison, Wisconsin, on the west; and from north of Fond du Lac, Wisconsin, on the north, almost to the Wisconsin-Illinois border on the south. This small scale, coupled with the manner in which scenes are formed, results in indistinct images with which most people are not familiar. Instead, their acquaintance is with photographs produced by aerial cameras at much lower altitudes, and, therefore, at much larger scales. For example, it would take at least 1,640 nonstereoscopic aerial photographs at scales of 1:20,000 (1 cm. = 0.2 kilometer or 1 in. = 0.316 mile) to cover the area imaged in figure 9.8.

Figure 9.9 illustrates the manner in which a ground scene is scanned by Landsat multispectral scanners. The MSS scans a swath 185 kilometers wide in six line elements, each of which is 79 meters (259 feet) square. The six-line scan represents an along-track distance of 474 meters (144.5 feet). The Landsat spacecraft must move along its path for a ground distance of 474 meters before the next six-line element is scanned. Each data element, or pixel (for picture element), along the direction of flight will measure 79 meters on the ground (474 meters ÷ 6). In the along-track dimension of the image, about 2,342 pixels, each 79 meters long, will be generated in a 185-kilometer range (185,000 meters ÷ 79). As the satellite proceeds along its path, the cross-track image velocity of the multispectral scanning mirror is about 5.6 meters per microsecond. After about 10 microseconds the pixel has moved a distance which corresponds to 56 meters (184 feet) on the ground. The image which is registered at that moment, therefore, includes 23 meters of previous data and 56 meters of new information.[22] The effective instantaneous field of view, therefore, will be 56 meters across-track. As the scanning mirror of the MSS sweeps across the 185-kilometer scene, it registers about 3,307 pixels in each of the four spectral bands.

The nominal pixel area—data area—sensed by Landsat multispectral scanners is 79 meters along the line of flight and 56 meters perpendicular to the flight path.[23] This produces approximately 7.8 million data elements, or pixels, within a frame which represents an imaged ground

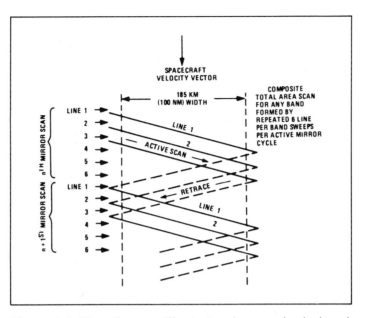

Figure 9.9. The diagram illustrates how a single band detector of the Multispectral Scanner scans the ground scene below the Landsat observatory. See text for explanation. (From *Data Users Handbook,* NASA)

area of 185 kilometers × 185 kilometers, or more than 31 million pixels within all four spectral bands.

At a scale of 1:1,000,000, each pixel measures 0.00079 centimeter (0.00095 inch) by 0.00056 centimeter (0.00067 inch). These small elements cannot be seen by the unaided eye, and even with optical magnification, they are extremely degraded. On the other hand, the pixels can be processed accurately by computers from the magnetic tapes on which they are recorded at ground receiving stations. If computer services are not available to a user, however, satisfactory pixel enhancements can be generated by combining optical and photo-chemical processes, to be discussed later in the section on Image Analysis and Interpretation.

Computer Compatible Tapes

In addition to the photographic products which are generated from Landsat signals of scenes, computer compatable tapes are available. The pixels of a 185-kilometer square area are stored on magnetic tapes. Four computer compatable tapes are required to produce the digital data which corresponds to images from the four spectral bands of the multispectral scanner.[24]

Probably, the imagery from Landsat sensors was meant to be processed by computers for analysis, plotting, and interpretation of objects. Many users of Landsat imagery, however, do not have access to the elaborate types of computers which are necessary to handle the millions of data from each image frame. For this reason, some investigators have successfully experimented with optical and photographic enhancements of Landsat imagery for interpretations of earth features which are produced at small scales and at gross resolutions by the MSS sensors. Enhancement by optical and photochemical processes will be discussed later.

Color Composites

MSS bands 4, 5, and 7 may be projected through proper filters and combined to produce a color composite of a Landsat scene (plate 6). The color composite is a false-color image which may be obtained as a film positive or paper positive product from the EROS Data Center. Many color composites have been produced for scenes of the United States and foreign areas; however, the EROS Data Center will generate a color composite image for any Landsat frame upon request, along with prepayment for the generation process and the print.

Color composite images resemble camouflage-detection, or color-infrared (CIR) photography. Color composites, therefore, do not portray the familiar colors of objects. For example, vegetation appears red on color composite. The chlorophylls in living vegetation reflect heavily in the near-infrared portion of the electromagnetic spectrum. Vegetation, therefore, will produce the "brightest" signature in those wavelengths ranging from .8 to 1.1 μm, or in the MSS Band 7. By exposing a sensitized color emulsion through a red filter to these "bright" signatures, vegetation signatures will be reproduced as red on the false-color print. Vegetation which is in a healthy, vigorous state of growth will appear redder than the signatures of vegetation which is in some form of stress because of drought, disease, insect infestation, or soil salinity. The deepening of red signatures for vegetation on sequential color composites can be detected as a growing season progresses from spring to summer, and the red hues are observed to decrease on summer to fall imagery. The recognition of vegetation differences in terms of varying shades of red provides the user with a powerful tool in interpreting Landsat color composites.

Persons using Landsat imagery should learn to recognize other features by the colors which are imaged on color composites. Clear water appears black because it absorbs all radiation from the sun. Water that carries silt, on the other hand, is imaged as light blue because of the reflectance from the particulate material in the water. Algae or duckweed blooms in water will register in various tones of red against a black or light blue background. Such surfaces as asphalt, cement roads, airport runways, parking lots, and rooftops appear in shades of blue-gray on Landsat color composites. Areas of new construction, quarries, land clearings, and fields which have been plowed, or from which crops have been harvested, appear as white, tan, or gray in color. Rocks and soils are imaged in colors ranging from blue, yellow, and brown in color composites.[25]

These colors of earth features on Landsat color composites are only approximations, and interpreters should build their own color keys for each Landsat frame studied. The colors on color composite frames vary with the angle of the sun from season to season, with seasonal vegetation differences, with weather and atmospheric conditions at the time the imagery was made, as well as with processing procedures.[26]

Landsat C will contain a thermal sensor in the MSS package. Eventually, images from this channel may be combined in color composites to produce improved color images by utilizing wavelengths that are entirely beyond those of human vision.

Computer-Enhanced Landsat Images[27]

The EROS Data Center has generated a new image product, called Computer-Enhanced Landsat Images. This type of product has been made possible by a laser beam filter recorder which has capabilities of enhancing signatures of Landsat imagery. The photographlike prints are produced from Landsat Computer Compatible Tapes. The black and white imagery can be registered and printed to produce color composites which are similar to those generated through filters from Bands 4, 5, and 7 of the Multispectral Scanner.

Computer-Enhanced Landsat Images provide several significant improvements over the conventional Landsat products. Most of the stripes, which result from fluctuations in the performances of the sensors, are eliminated. This improves an interpreter's ability to correctly identify a target. Contrast enhancement is increased by minimizing the effects of scattered wavelengths of radiation in the atmosphere. Edge enhancement of images is maximized by comparing the ''brightness value of each pixel with the average value of nearby pixels to determine edge presence, and (by increasing) the brightness difference of those pixels defining the edge (of an object imaged in the scene)''[28] In addition, distortions in the original image which are produced by variables in orbit, sensor operation, and platform altitudes are removed in the Computer-Enhanced Landsat Images.[29]

The Computer-Enhanced Images provide an increased value in optically viewing Landsat frames. These products, however, are relatively expensive, and users may find that they can produce some degree of density slicing and edge enhancement of pixels by other methods in their own laboratories without investments in costly equipment.

Image Analysis and Interpretation

In chapter 15, Guernsey and Mausel discuss computer applications in image analysis and interpretation of Landsat frames; therefore, this application will not be discussed here. Instead, this section will deal with other methods by which Landsat imagery can be processed for interpretation.

Some type of analysis and interpretation other than machine processing is necessary for the majority of Landsat investigators. Although direct optical magnification aids in the identification of gross land cover from Landsat imagery, it does not produce a permanent record for study and mapping. The usual scale of 1:1,000,000 for color composite Landsat products is adaptable for use with the Bausch and Lomb Zoom Transfer Scope if tone separation can be accomplished for accurate interpretation.

Photomicrography Processing

Both optical enlargement and image enhancement have been used effectively in analyzing Landsat imagery and object recognition. The essential hardware products for optical enlargement include an Aristo daylight-source easel, sturdy copy stand, and a 35 mm single-lens reflex camera equipped with a macro lens. Image enhancement can be performed effectively and inexpensively by producing a positive photographic transparency of a Landsat color composite transparency on high contrast color film. Kodak Photomicrography Color Film (2483) or Kodachrome Professional Film Type A are recommended for this type of image enhancement. Both film types are available in 35 mm cassettes.

Photomicrography Color Film is a high contrast film which has the ability to define semi-discrete tones. The film possesses high resolution, high definition, and fine grain.[30] A disadvantage to the use of Photomicrography film is its slow ASA 16 speed.

Without a brilliant transmitting light source, long exposures are necessary. Kodak warns, and experimentation verifies, that exposures of more than 1/10 second cause changes in the color characteristics of the emulsion. Color changes resulting from long exposures can be corrected, however, by the application of compensating filtration, such as Kodak filters CC1OY and CC2OY.[31] It is recommended that exposure tests be made using different filtrations when the enhancement is critical for interpretation and mapping.

Color changes in Photomicrography film which result from long exposures are not necessarily an impediment to analyzation of the final enhancement because tonal changes are relative to the problem of interpretation. The interpreter of original Landsat false color composites is required to make similar adjustments in observation. Once color tones have been normalized, identification of signatures may proceed in the same manner as when original false color composites are studied.

The Photomicrography slide-copy of a Landsat color composite frame, or portion of a frame, is a fine-grained, high-contrast, high-resolution product which may be used in several ways for target identification, measuring, and mapping. The photomicrography slide can be placed in a reflecting projector where the image is projected onto a first-surface mirror, which, in turn, reflects the image to a glass light-table surface. Any transparent sheet of tracing paper, such as K & E Albanene, which is placed on the light-table surface will render the image visible at a greatly enlarged scale. Detailed, large-scale mapping can be done directly on the tracing paper.

Eyton and Kuether[32] discuss the construction and use of a reflecting projector, which they call the "Throwback Projector." (See also chapter 18.) The reflecting projector which this writer constructed was modeled after the one Eyton designed. It consists of a 24 × 36-inch first surface mirror which is placed at a 45° angle in a "black box" at one end of a 20-foot guide rail. A 35 mm slide projector is mounted on a sliding support. This enables the projector to be located at various positions along the guide rail, providing a variety of magnifications on the light-table drafting surface.

The reflecting projector provides several research and laboratory applications in the use of enhanced Landsat imagery. The optically enlarged enhancement of a Landsat scene can be used for mapping land surface changes from repetitive coverage. Seasonal changes in vegetation cover and crop growth can be studied and mapped. Areas of variations in vegetation vigor can be observed and mapped on a repetitive schedule. Seasonal changes of water levels in rivers, lakes, marshes, and reservoirs can be determined and mapped. Construction sites and other changes in the urban scene can be discriminated by Photomicrography enhancement and optical enlargement.

Color Printing

A Photomicrography enhanced 35 mm slide of a Landsat image can be printed as a paper positive by one of several direct positive-to-positive photographic processes. A relatively new process of color printing has been developed by Ciba-Geigy Company, called Cibachrome.[33] This process can be incorporated into both research and laboratory uses for studying enlarged and image-enhanced Landsat scenes (plate 7).

The Cibachrome color print process does not require sophisticated equipment, the developing chemistry is simple, and the time to produce a high-quality positive paper print requires only 12 minutes. More important, the Cibachrome process does not require the production of an internegative; therefore, the resulting paper positive product is one of good resolution. The edge sharpness of the enhanced image detail is reproduced with exceptional sharpness and fidelity. The Cibachrome print provides, "high color saturation, excellent hue rendition, . . . and resistance to fading."[34]

To produce a Cibachrome paper positive of a Landsat Photomicrography enhanced slide transparency, the slide is placed in an enlarger and the image is projected onto an easel board in the usual manner of photographic enlargement. Individual targets may be isolated and magnified up to the capacity of the enlarger. Color variations may be introduced and tones further enhanced and separated by selecting various combinations of the secondary magenta, cyan, and yellow color filters. The filter combinations are inserted into the filter drawer of the enlarger between the light source and the condensing lens and slide transparency. Landsat images at an original scale of 1:1,000,000 may be enhanced, enlarged, and printed on Cibachrome direct positive paper at scales of 1:50,000 or larger. Discrete color separations can be identified of urban areas, forested tracts, crops, and field boundaries.

Posterization, Diazo, and Color Key Enhancement

Further tone separations of Landsat enhanced and enlarged images can be produced by posterization, Diazo, or Color Key processes. Individual tones appear as semidiscrete units on images produced by these processes.

Posterization is a process which provides high contrast film negatives or positives which can be printed, in turn, through a variety of primary and secondary color filter combinations on Cibachrome paper. In this process, a high contrast black and white film positive is made of each Landsat MSS band. Either film negatives or film positives of Landsat imagery can be used in this enhancement process.

In posterization each MSS spectral band is exposed individually on high contrast film, such as Kodak Orthochromatic Professional Type 3 film. The signatures of each spectral band appear on the orthochromatic film in proportion to their "brightness" on the Landsat image. If a negative film is used, it must be reversed into a positive by printing it again on Type 3 film. The resulting high contrast film positive is placed over a sheet of Cibachrome positive-to-positive paper which then is exposed through the desired filters. If precise registration is maintained throughout the exposure process, the developed Cibachrome print will appear similar to a Landsat color composite. Variation in filtration in the Landsat MSS bands will produce a variety of color combinations which will aid materially in the separation of tones, and thus aid in the identification of objects imaged on the false color prints.

Production of direct positive enhanced images of Landsat frames on Diazo or 3-M Color Key transparencies is less complex than posterization. Individual Landsat spectral frames can be printed on either Diazo or Color Key films, which are available in a multitude of colors. In the Diazo process, Landsat positive film transparencies are exposed on a selected color Diazo film by use of an ultraviolet light source. A commercial offset plate-making machine is useful for these exposures because the vacuum easel makes possible uniform contact between the Landsat image and the Diazo color sheet. In addition, a plate maker provides a constantly reliable source of light, and exposures can be controlled and monitored by the built-in timer. It is advisable to make experimental exposures for each band on the selected Diazo color film because of different densities of Landsat bands and the different responses to the light source by different Diazo color films. A step-wedge, or gray scale, should be exposed on the Diazo sheet at the same time to provide the user with a somewhat reliable measure of the degree to which each gray signal on the Landsat spectral band is imaged on each Diazo color film.

The Diazo color sheet is developed, or stabilized, in ammonia vapor. An ammonia vapor tube, or a simple pickle jar in which ammonia-saturated cotton has been placed, will produce excellent results. Color Key sheets and Diazo films are exposed in the same manner; however, Color Key sheets are developed and stabilized by a special chemical solution which is sold by the 3-M Company. Each Diazo or Color Key sheet displays only one color. If the blue Diazo or Color Key is exposed to Band 5, only the "brightest" signatures will be produced in blue; and similarly for color films used for imaging signatures from other Landsat MSS bands.

When the different desired monocolor transparencies have been produced for each Landsat spectral band, the transparent films can be superimposed and accurately registered to produce the color combinations and image enhancements which are necessary for interpretation. Colors can be changed by inserting or subtracting different color films. When the proper color combinations and edge enhancements have been determined, the series of Diazo or Color Key films can be fastened together to produce a color composite of the Landsat scene. The product can be examined on a light table, or it can be printed on positive-to-positive Cibachrome paper. Also, the registered Diazo or Color Key sheets can be sandwiched in the film transport of an enlarger and large-scale images of isolated scenes within the frame can be printed on Cibachrome paper in the same manner as color slides are printed (plate 8). Similar enhancement procedures can be applied to RB-57F high altitude infrared images, as well as to Skylab, Apollo, and Gemini photographs.

The identified tone-separates can be measured with a relatively high degree of accuracy by use of a polar planimeter on any of the optical-photochemical enhanced images. Areas of crops, forests, construction areas, mining pits, wetlands, and areas within cities can be identified and measured.

The production of Photomicrography enhancements, color positive prints, posterized tone-separates, and Diazo or Color Key enhancements provide interesting classroom and laboratory work in remote sensing. Furthermore, the ability to isolate and identify land cover by these methods is a valuable tool in the interpretation of Landsat images for research purposes in those institutions where the absence of elaborate and expensive digital processing equipment otherwise precludes the application of remote sensing techniques.

Interpretation of Landsat Imagery

The conventional parameters of image interpretation are employed in studying Landsat imagery; however, several others must be introduced, along with cautions in the use of the common principles in object recognition. In addition to the extremely small scale, low-level resolution, and false-color parameters of object recognition which have been discussed, such factors as atmospheric conditions, sun angle, and the applications of size, shape, shadows, pattern, texture, tone, and association need elaboration.

The Landsat spacecrafts operate far above the atmospheric zone of the weather systems of the world. Furthermore, the MSS sensors do not possess haze and cloud penetrating capabilities, although in some instances, the infrared sensor of Band 7 renders a lightly obscured scene somewhat more identifiable. That is because the infrared wavelengths are scattered less by the atmospheric particles than those wavelengths in the blue area of the spectrum. Figure 9.6 illustrates the problem which is produced by the atmosphere. Whereas Bands 4 and 5 show little or no ground detail in the clouded regions, Bands 6 and 7 have imaged some ground targets through the clouds and haze.

Although one or the other of the Landsat vehicles transport their sensors over all parts of the earth every nine days, the actual number of usable images of the ground is much less than this. The sensors are activated on all passes over the United States regardless of weather conditions and cloud cover. Out of a maximum of about 40 possible images of a single area during a year, 10 good ground-imaging passes are characteristic, although this varies from one part of the country to another. In the Midwest, Gulf and Atlantic coasts, and Pacific Northwest many more cloudy days occur in a year than in the Great Plains, Basin and Range, and Southwest. Alaska is particularly susceptible to few good imagery passes, not only because of frequent clouds, but also because of darkness during the low-sun or no-sun season. In southeastern Wisconsin in 1976, 16 passes, or 39.5 percent of the maximum possible passes, were made by Landsat spacecrafts when 10 percent or less of the sky was cloud covered.

Cloud cover masks earth scenes from the view of the satellite sensors. Although cloud imagery from Landsat may be of some value to meteorologists and climatologists, it is maddening to the investigator who is interested in surface features. Even scattered cumulus clouds can have an unfavorable effect on interpretation. If a small cumulus cloud is located over an area of primary interest at the time of imagery, the scene below it cannot be retrieved. Also, clouds cast shadows. The cloud shadows generally fall towards the northwest because of the altitude and position along the horizon of the sun at the time of imagery. Cloud shadows are usually dense enough to obliterate ground details. Care must be exercised not to interpret the black shadows of clouds as waterbodies or other features which register as black on the Landsat images. Reference to the sun angle and horizon position of the sun in the annotation block will tend to eliminate this type of identification error.

Changes in the elevation of the sun cause changes to occur in the reflectance of targets in the ground scene. The composition of the atmosphere may affect target reflectance. All of these factors bring about changes in the irradiance of a sensor. For example, it has been demonstrated that the radiation from sand, bare soil, and rock is very responsive to changes in the seasonal elevation of the sun, whereas most types of vegetation do not change their reflectance as much. Because of these reflectance differences, each Landsat scene must be evaluated individually.[35]

Certain ground scenes, such as snow, ice, bare limestone in quarries, and metal roofs, may reflect so brightly that the capacities of the sensors to electronically register the data are exceeded. In a way, the sensors are blinded temporarily by this type of excessive reflectance and cease to operate for a short period of time. This "blinding" of the sensors has the effect of either blotting out an image of the object or causing a streak to occur on the image until the electrical charge on the sensor is reduced to operating capacity.

Size, shape, shadows, pattern, texture, tone, and association are elements of object recognition which must be evaluated in the interpretation of images on aerial and space photographs. Some of the elements of object recognition, however, have little meaning when a person is interpreting the small-scale scene from the Landsat multispectral scanners.

The scale of Landsat images is very small, and features at the surface of the earth appear degraded on the photographlike print. Visual inspection provides only gross land cover and land use identification. Details of the surface, however, become meaningful when Landsat images are enlarged and enhanced. Compare plates 6 and 7.

Plate 6 is a color composite of Landsat imagery made on April 29, 1976. The scale before reduction as a plate was 1:1,000,000. The interface between land and water is very apparent, but other features are more difficult to discern. Major highways can be observed because of their brightness and linearity. Cities are evident because of their street patterns and bluish tones. Some landform features can be identified, such as the curving moraines in the center of the photograph, and the radiating drumlin field in the western part of the image.

Careful examination of plate 7 reveals field patterns and forested areas. The landforms are accentuated by shadows produced by the sun angle. The sizes of large features, such as cities and water bodies, along with the lengths of roads and streams, can be detected; however, sizes of individual fields, city blocks, and buildings are too small to be seen. Shapes of gross features on the land can be observed, but shape is not the important criterion for the identification of images in Landsat scenes as it is for images on suborbital aerial photographs. On the other hand, shape may be important in the identification of features which are too large to be seen on a single aerial photograph frame, such as meander scars, coastlines and bays, and large glaciers.

Pattern, texture, tone, and association appear to be the major criteria for object recognition on small scale Landsat imagery. The patterns of roads, drainage ways, swamps, marshes, and lakes are relatively easy to locate and identify. In large cities, such as Milwaukee on plate 6, the street pattern is evident. On the same plate, the patterns of landforms can be observed in the moraine and drumlin areas.

Variations in tone and color provide a primary basis for image identification on Landsat frames. Dark tones identify the central areas of cities and industrial sections. Light tones may indicate land that is less intensively used. Tone provides the interpreter with clues to the identification of forests, croplands, harvested fields, plowed fields, construction sites, and the location of stone quarries and gravel pits. Tone on the Landsat image is a function of the reflectance of sunlight from targets, and it must be evaluated carefully for each scene.

Texture and association are valuable criteria in the interpretation of Landsat images. Texture is the arrangement of patterns and associations of objects at scales where individual patterns and associations cannot be seen as entities in and of themselves. In other words, texture is the fabric of the scenes. A coarse texture may define an urban scene, while a fine texture may represent rangelands or fields cultivated to crops. Texture, therefore, may be used to distinguish urban patterns from rural field patterns, croplands from pastures, and rough topography from level areas. Interpreters can use the associations of features as a tool in object recognition once they have gained an intuitive feel for the area under investigation. For example, when the texture of the feature indicated a rural environment, and radiating roads, railroads, and street-pattern textures were absent, dark tonal areas would be interpreted as wet, depressions, or soils rather than as urban areas.

Although individual pixels appear minutely small on the 1:1,000,000 Landsat format, some features as small as 10 meters may be distinguished under some conditions. If an object is effectively distinguished from its background and from surrounding features, its signature on small scale imagery may be identifiable. Roads, because of their relatively high reflectance, are usually distinguishable. Bridges across bays or rivers may be identified. Cultivated fields surrounded by pasture or bare soil may produce identifiable signatures even at small scales.

Plate 7 illustrates the kinds of information which are available from a photomicrographically enhanced Landsat scene of the Milwaukee, Wisconsin, area. The street pattern, along with highway interchanges, in the city and its environs are evident. The central city can be distinguished from the suburbs. Parks, golf courses, conservancy areas, airports, and cemeteries can be recognized by their patterns, tone (colors), textures, and associations. The industrial area of the city can be seen, along with the harbor facilities. Individual structures, on the other hand, cannot be identified. In the harbor areas, however, piers, slips, and mooring basins can be identified because of their differentiated signatures. The photomicrography enhancement produces edge sharpness of tones which results in a coarse texture, thus allowing for increased ease of interpretation of individual targets.

Compare figures 9.10 and 9.11. These maps, which were prepared from Landsat imagery illustrate the value of the repetitive coverage by Landsat sensors. Figure 9.10 illustrates a large area of marsh in east-central Wisconsin on April 29, 1976. The map was prepared from enhanced Landsat imagery by use of the reflecting projector.[36] Figure 9.11 shows the same marsh area on September 29, 1976. Changes in the open water areas between April and September are evident. Field research revealed that water levels in the marsh dropped between April and September because of a severe drought in this area in 1976. In addition it was learned

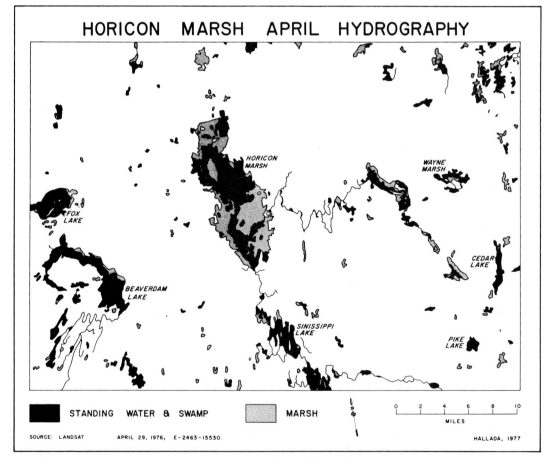

Figure 9.10. Map of the area around Horicon Marsh in east-central Wisconsin. The map was prepared from enlarged and enhanced imagery from Landsat. The scene was imaged on April 29, 1976. Photo Identification: E-2463-15530. (Prepared by Wayne A. Hallada)

that the wildlife managers opened the floodgates of the marsh in order to reduce the area of open water and thereby control the goose population of the marsh. The levels of most lakes in the imaged area fell during this period, and their margins were taken over by marsh grasses.

Field work can establish the types of land cover and, in some cases, the kinds of crops in fields. This ground truth information can be plotted on Landsat imagery. Where similar tones (colors), textures, and associations are identified, there is a maximum likelihood that the land cover is the same, or at least similar to, the areas which were positively identified in the field.

Seasonal variations in land cover, position of the sun, and object reflectance cause targets in the same area to appear differently, and in many cases more interpretable, on different images. A snow cover, for example, may have the effect of accentuating landforms, plowed streets in cities, and main highways in rural areas. The absence of foliage on deciduous trees in winter scenes makes it possible to distinguish broadleaf species from coniferous stands. Landforms appear more pronounced when foliage is absent. Early spring and autumn images make possible

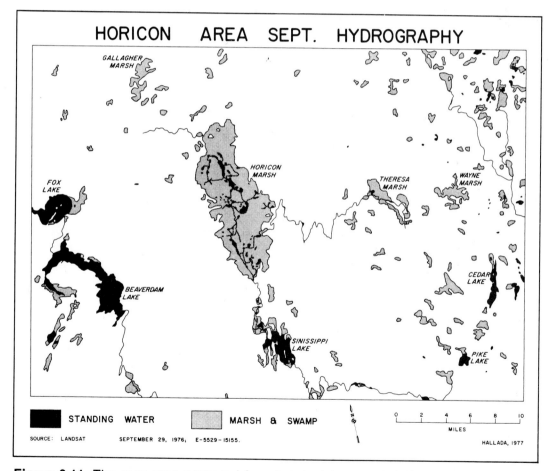

Figure 9.11. The map was prepared from Landsat imagery made on September 29, 1976. Photo Identification: E-5529-15155. (Prepared by Wayne A. Hallada)

the differentiation of field boundaries because these boundaries tend to be lost amid the general reflectance of all vegetation during the growing season.

Uses of Landsat Imagery

The many uses of Landsat imagery are discussed in the following chapters. These include the applications of remote sensing techniques to the analysis of landforms, agriculture, rural landscapes, urban areas, and industrial sites. Investigators in many sciences have used, and are using, Landsat images and products to study agriculture, forestry, rangelands, geology, water resources, marine resources, and the cultural landscape. Geographers and cartographers are using Landsat products for the production of maps and in land use studies.[37]

In agricultural sciences, Landsat imagery is being used to discriminate crops and range vegetation. Measurements of crops and timber acreage by species, the biomass of rangelands, vegetation vigor, soil associations and conditions, and the extent of irrigated fields are being studied with the aid of space imagery.

In geology, rock types and structures, volcanic deposits, landforms and physiographic regions, areas of potential mineralization, and faults are subjects of Landsat investigations. Hydrologists are applying Landsat imagery in the interpretation and mapping of water boundaries, floods and flood plains, sediment and turbidity patterns in water bodies, and lake inventories, as well as in the delineation of snow boundaries. Landsat imagery is used in the detection of living organisms in oceans, as well as in mapping circulation of currents, changes in shorelines, and the location of shoals and reefs.

Environmental applications of Landsat data are legion. Reclamation and strip mining operations are studied and monitored. The effects of air and water pollution, eutrophication, defoliation, and natural disasters are studied. Land use planners are making increasing use of Landsat imagery as they apply it to classifying uses of both urban and rural lands. Planners use Landsat data for categorizing land capabilities, for mapping transportation networks, and in assessing wetlands.

To obtain information on the depth and breadth of current investigations involving Landsat data, the reader is referred to the *Weekly Government Abstracts,* published by the National Technical Information Service. Included are weekly NASA Earth Resources Survey Program abstracts. Each report includes author, title of the investigation, and an abstract of the investigations which have applied remote sensing techniques. Information about these abstracts can be obtained from NTIS, U.S. Department of Commerce, Springfield, Virginia 22161.

NOTES

1. Stanley C. Freden, "The Landsat System," Appendix A in *Mission to Earth: Landsat Views of the World,* NASA-SP 360 (Washington, D.C.: U.S. Government Printing Office, 1976), p. 437.
2. Ibid. pp. 1 and 437.
3. NASA, *Data Users Handbook,* Document No. 71SD4249 (Greenbelt, Md.: Goddard Space Flight Center, 1972), p. I-5.
4. Very short wavelength energy of the electromagnetic spectrum is measured in small fractions of a meter. A nanometer (nm), formerly a "millimicron," equals 10^{-9} meter (or one billionth of a meter). Also, wavelengths in the electromagnetic spectrum may be designated in micrometers (μm), formerly called "microns." A micrometer equals 10^{-6} meter (one millionth of a meter). E. A. Mechtly, *The International Systems of Units,* NASA SP-7012 (Washington, D.C.: U.S. Government Printing Office, 1969), in *Manual of Remote Sensing,* ed.-in-chief, Robert G. Reeves (Falls Church, Va.: American Society of Photogrammetry, 1975), pp. 65-66.

5. Stanley C. Freden, "Survey of the Landsat Program," *Mission to Earth,* p. 2.
6. Ibid.
7. Canada Centre for Remote Sensing, *General Landsat Information Kit* (Ottawa, Ont.: Energy, Mines and Resources, March 1975).
8. Ibid.
9. Ibid.
10. *ERTS Data Fact Sheet* (Sioux Falls, S.D.: EROS Data Center, 1973).
11. *General Landsat Information Kit.* (Sioux Falls, S.D.: EROS Data Center, circa 1973).
12. NASA, *Landsat Data Users Handbook,* Document No. 76SD4258 (Greenbelt, Md.: Goddard Space Flight Center, September 2, 1976), p. 2-4.
13. Personal communication from Dr. Ray Lougeay, June 21, 1977.
14. Joseph Lintz, Jr., and David S. Simonett, eds. *Remote Sensing of the Environment* (Reading, Mass.: Addison-Wesley Publishing Co., 1976), p. 324.
15. Missions Utilization Office, Goddard Space Flight Center, *Landsat Newsletter,* no. 15, June 1, 1977.
16. NASA, *Landsat Data Users Handbook,* 1976, p. 2-3.
17. Alden P. Colvocoresses, "Applications to Cartography," *ERTS-1: A New Window on Our Planet,* U.S. Geological Survey, Professional Paper 929 (Washington, D.C.: U.S. Government Printing Office, 1976), pp. 12-17.
18. Ibid.
19. NASA, *Landsat Data Users Handbook,* 1976, p. 2-4.
20. Ibid., p. 2-3.
21. NASA *Data Users Handbook,* September 15, 1971, p. 3-5.
22. *Data Users Handbook,* 1976, p. C-4.
23. Ibid.
24. NASA, *Data Users Handbook,* 1971, pp. 3-11 to 3-12.
25. Freden, *Mission to Earth,* p. 2.
26. Ibid.
27. This section is based on information contained in, *New Special Order Product* (Sioux Falls, S.D.: EROS Data Center, December, 1976).
28. Ibid.
29. Ibid.
30. From Kodak Information Sheet supplied with Photomicrography Film.
31. Ibid.
32. J. Ronald Eyton and Richard P. Kuether, *Remote Sensing Photoguide* (Urbana: University of Illinois, Department of Geography, April 1975), p. 6-3 ff.
33. Ilford, Inc., *Color Print Manual* (Paramus, N.J.), p. 1.
34. Ibid.
35. NASA, *Landsat Data Users Handbook,* 1976, p. I-2.
36. The maps and image enhancements in this section were prepared by Wayne A. Hallada as part of a research project in remote sensing which was done under the author's supervision at Carroll College, Waukesha, Wis., Spring 1977.
37. Freden, *Mission to Earth,* p. 4.

10

High Altitude Color Infrared Photography

Benjamin F. Richason III
James Madison University, Virginia

Platforms and Products

There have been a number of developments and refinements in the use of aerial photography in the study of earth resources in the last decade. One of these has been high altitude color infrared photography. There is nothing new about exposing film from high altitudes or in the use of color infrared film. A recent development, however, is the nonmilitary utilization of color infrared film exposed from relatively high altitudes in the earth's atmosphere. To fully appreciate the applications of this type of photography, people must understand the meanings of ''high altitude'' and ''color infrared photography.''

High Altitude

The term high altitude is relative. What may be considered high altitude for the purposes of one form of research may be quite different for another. High altitude, therefore, is not defined in terms of any specific range in height. Instead, high altitude is determined by the normal cruising altitude of the two types of aircraft most commonly used in obtaining this type of photography. Thus, high altitude photography can be defined as film being exposed from heights ranging from around 18,300 meters (60,000 feet) to 21,300 meters (70,000 feet).

Color Infrared Film

Color infrared (CIR) refers to the type of film which is sensitive to both visible and reflected infrared radiation up to about 900 nm. This range is just beyond the visible portion of the spectrum. Like conventional color reversal film, CIR film has three emulsion layers. The emulsion layers of conventional color film are sensitive to blue, green, and red reflected light, while those of CIR film are sensitive to green, red, and reflected infrared. On CIR film all three emulsions are sensitive to blue light. This can be a distraction because blue light tends to be scattered in the atmosphere, thus "washing out" the colors in CIR film. For this reason, when CIR film is exposed, a yellow, or minus-blue, filter is used so that these blue wavelengths are not sensed by the film. When a minus-blue filter is used, CIR film possesses excellent haze penetration qualities and provides sharp images fron high altitudes. This is especially important in high altitude photography because the greater column of atmosphere results in more scattering of blue light.

Because of the shift in the sensitivity of the dye emulsion layers in CIR film, the color rendition is different from that found on conventional film. For example, healthy vegetation appears as green on conventional color film, while on CIR film it appears in magenta or red hues because of the high infrared reflectance of the chlorophylls in vegetation. This is why CIR film is sometimes called false color, or camouflage detection, film. Each of the three emulsion layers is coupled with a subtractive dye color and linked to either green, red, or infrared spectral wavelengths.[1] Green wavelengths are coupled with a yellow dye, red wavelengths with a magenta dye, and reflected infrared wavelengths with a cyan (blue) dye. The color rendition of CIR film is accomplished by a subtractive reversal color process where the dye responses are inversely proportional to the exposure of the respective emulsion layers when the film is processed.[2] For example, if a red object is being sensed, it will expose the magenta dye layer, leaving the cyan and yellow dye layers to form, subtractively, a green color on the photograph. Similarly, if an object, such as healthy vegetation, is highly reflective in the near infrared wavelengths, the cyan dye layer will be exposed, leaving the yellow and magenta layers to subtractively form a red color on the photograph.

Advantages and Disadvantages of High Altitude CIR Photography

There are several advantages to exposing film from high altitudes. Exposures from high altitudes provide a general view of landscape that is not obtained at lower altitudes. A relatively large area, therefore, can be studied in detail on a single photograph. In addition, because large areas can be covered, the total number of frames needed to cover a specific study area is less, which reduces cost. CIR film, in particular, provides the advantages of haze penetration from high altitudes and of being able to register information slightly beyond the visible range of the spectrum. Furthermore, images are displayed in color rather than black and white, making possible the discrimination of more targets.

There are disadvantages, however, to high altitude color infrared photography. The most obvious is that in high altitude exposures some ground detail is lost which might normally be imaged at lower altitudes. There is also the danger that clouds will block views of the ground. In addition, CIR photography does not effectively record subsurface water features. Also, false-color rendition might be considered a disadvantage because of its shift in the display of color.

Types of Platforms

Most high altitude CIR photography in the United States is obtained from research flights conducted by the National Aeronautics and Space Administration (NASA) Airborne Instrumentation Research Project (AIRP), formerly known as the Earth Resources Aircraft Project (ERAP). This program was designed to gather a wide range of remote sensing data in support of other satellite programs such as Landsat and Skylab. In addition to gathering support data, AIRP tests and evaluates various sensor data for future satellite systems. AIRP also collects data for atmospheric research, and demonstrates new techniques for the practical use of remote sensing technology.

Two types of aircraft are used in AIRP flights, a General Dynamics/Martin RB-57F (fig. 10.1) and a Lockheed U-2 (fig. 10.2). The RB-57F aircrafts operate out of Houston, Texas. They are primarily responsible for obtaining high altitude photography of the central area of the United States. The U-2 aircrafts are stationed at the NASA-Ames Research Center at Moffett Field, California. They obtain remote sensor coverage of the west and east coasts of the

Figure 10.1. General Dynamics/Martin RB-57F from which high altitude photography is made. (Courtesy of NASA)

Figure 10.2. The Lockheed U-2 is used by NASA for producing high altitude imagery. (Courtesy of NASA)

United States, as well as of Hawaii and Alaska. When flight missions over east coast areas are flown, the U-2s fly from NASA-Wallops Flight Center in Virginia.

Payloads of the RB-57F and U-2

The photographic missions by RB-57F aircraft normally are flown at 18,000 meters (60,000 feet). The sensor payload carried by this aircraft varies according to the needs of the specific research being conducted. Usually, an RB-57F mission carries a sensor package containing two Wild-Heerbrugg RC-8 metric cameras with a 152 mm (6 inch) focal length lens, as well as a Zeiss 305 mm (12 inch) focal length metric camera. The RC-8 cameras provide photographs at a scale of 1:120,000, while the Zeiss camera provides photograph scales of 1:60,000. In addition to these primary cameras, four to six Hasselblad cameras with either 40 mm (1.57 inch) or 80 mm (3.15 inch) focal length lenses. These cameras provide scales of 1:458,000 and 1:228,000, respectively.

Depending on the mission, various film/filter combinations are used. A typical mission usually will use one of the RC-8 cameras and a Zeiss camera containing Aerochrome Infrared 2443—a color infrared film. The other RC-8 camera will contain Ektachrome EF Aerographic SO-397, conventional color film. The Hasselblad cameras contain Plus-X black and white 2402 and Infrared Aerographic 2424 black and white film. A number of filter combinations will be attached to the cameras to supply information across a wide portion of the electromagnetic spectrum. In addition to these camera systems, thermal imagery is also sensed, sometimes being obtained from either an RS-7 or RS-18 infrared scanner.

The U-2 carries a variety of sensor packages. Because of the U-2's higher operating altitude, a small change in the scale of imagery is produced. On a typical mission, this aircraft will carry a Wild-Heerbrugg RC-10 metric camera with a 152 mm (6 inch) focal length lens; however, a 305 mm (12 inch) lens is used on some occasions. Depending on the lens used, this camera provides photography at scales of either 1:130,000 or 1:65,000. In addition to the RC-10 camera, an array of four Vinten cameras are used. The Vinten system, like the Hasselblad, provides multiband coverage of landscape scenes. This multiband coverage (475-575 nm, 580-680 nm, 690-760 nm, and 510-900 nm) is closely correlated with the spectral coverage of Landsat. The Vinten cameras have a 44 mm (1.75 inch) focal length lens which supplies photography at a scale of 1:445,000 at the normal flight altitudes of the U-2. Depending on the mission objectives, the film types are similar to those used on RB-57F.

Availability of High Altitude Imagery

The imagery obtained from the AIRP aircraft is available through the EROS Data Center in Sioux Falls, South Dakota. The format of the film varies from one type of camera to the other. The Zeiss, RC-8, and RC-10 metric camera systems furnish film formats of 229 mm × 229 mm (9 inch × 9 inch). The Hasselblad and Vinten camera systems supply film formats of 70 mm (2.76 inches).

High altitude photographic coverage is sporadic in the United States. In order to have an AIRP flight flown, it first must be requested by a research investigator or some other user agency; therefore, gaps in coverage occur in many places. Depending on the film source, there is a wide range of standard products available, including both black/white and color. These copies can be acquired as either paper prints or film positives. Color copies of a 70 mm film source can be obtained as 6 cm (2.2 inch) film positives or enlargements on 22.86 cm (9 inch) and 45.72 cm (18 inch) paper enlargements. Copies of Zeiss, RC-8, and RC-10 camera systems can be obtained as either 22.85 cm film positives or paper, as well as paper enlargements of 45.72, 68.58, and 91.44 centimeters (18, 27, and 36 inches).

When ordering photography from the EROS Data Center, the mission number, roll number, and frame number, as well as the general area of the state in which the photography was made must be provided. If this information is not available, the user can supply the EROS Data Center with the approximate longitude and latitude coordinates of a desired area. The center will do a geographic computer search and will provide the user with a computer print-out which lists all photography within a specified area. The print-out contains information about film source, photo-scene identification number, percent cloud cover, quality index, date, scale, altitude, percent overlap, sensor type, film/filter code, and format size. Such data are important in selecting the right frames for a particular study or classroom use. In addition, the longitude and latitude coordinates of the corner points of each photo strip are provided.

Most AIRP photography is available by a strip of photographic coverage of a single flight line segment. The corner point coordinates for each can be plotted on any map showing these coordinates, and the photo strip sketched in. Once this has been done, the center of the first frame in the strip can also be plotted, and, with the percent of overlap taken into consideration, each individual frame in the photo strip can be plotted. When an appropriate frame has been located in the area of interest, the photo-scene identification number can be used to order the photography.

AIRP high altitude color infrared photography possesses a number of characteristics which recommend it to both research and classroom acceptance.

Applications of High Altitude CIR Imagery

In comparing the conventional color and CIR photography taken with two RC-8 cameras, the most obvious property of the CIR photography is the color shift. This is apparent in vegetated areas—such as forests and croplands—because of the high infrared reflectance by healthy vegetation. This shift is evident in forested areas—A—and agricultural areas—B—on plates 9 and 10. Because of the different plant species in agricultural and forested areas, there is corresponding variation in the colors represented on the CIR photographs. While there is an obvious difference in the brightness of each color, the red hue is basically the same.

Another property of CIR photography is its ability to penetrate haze caused by blue-scattering. This fact is apparent in comparing plates 9 and 10. The conventional color photography displays the effects of blue-scattering, which results in the "wash-out" of color. This is the reason for the bluish appearance of the photograph. The CIR photograph has had this distracting blue light filtered out and displays crisp, clear images of ground detail and sharp color definitions.

Water

A further distinction between conventional color and CIR photography is the rendition of water. Conventional color film possesses some capabilities for penetrating water surfaces to reveal shallow features; however, it is not as useful as CIR film for differentiating between land and water. In part this is the result of the lack of color contrast between land and water, as well as the blue-scattering effect at high altitudes. Because water absorbs infrared radiation to a great degree, less of this radiation is reflected from water surfaces than from surrounding land areas. Water bodies, therefore, appear dark blue to black on CIR film. This causes land-water boundaries to be highlighted. The highlighting effect is aided by the color contrast between water and healthy vegetation where present. This factor is illustrated at points C on plates 9 and 10.

Soils

Another characteristic difference between conventional color and CIR films which are exposed from high altitudes appears in the characteristics of soils images. In most areas, soils are only exposed in plowed or newly seeded fields. Regular color film is useful in soil studies because it records the subtle variation in the color of the upper horizon, and from high altitudes it provides information about such variations on a regional scale. This color film, however, does not supply detailed information concerning soil moisture content. Because of the infrared absorption-effect by water, CIR film images the water content in soils much better than conventional color film. Most well-drained soils appear as white, tan, or bluish-white on CIR photography. Imperfectly drained portions of a field appear as dark gray to black, depending on the water content. The surface drainage pattern in individual fields, therefore, can be easily discerned and delineated (see area B and other similar areas, plate 10).

Mineral soils and organic soils can be differentiated on CIR photography. Because of the usually high moisture content of most organic soils, they appear as very dark grays or black. Care must be exercised in identifying these types of soils on CIR photography because of the hazard of mistaking organic soils for bodies of water (or cloud shadows) which might have the same dark color. An example of organic soils is found at point D, plate 10. The potential for misclassification can be seen in comparing the appearance of these soils to the water colors at point C. For the most part, however, there is no problem in differentiating the black color organic soils from the bluish white of most mineral soils. Because of the uniform reflectance of most mineral soils on CIR film, regardless of their specific mineral contents, all tend to appear bluish white in color. In classifying different soil types, especially where they grade into one another, a good color aerial photograph generally would be more useful.

There is a problem of classifying urban built-up areas and large exposures of bare soil on CIR film, particularly where the two are adjacent. Like the difficulty in distinguishing between organic soils and water bodies, urban areas and mineral soils tend to have similar reflectance properties, resulting in the same white to bluish-white colors. The similarities are due to the materials used in roads and structures in urban areas where these materials contain minerals— for example, concrete. This problem would be especially evident in many arid areas where the lack of vegetation and soil moisture would tend to make differentiation difficult.

Wetlands

CIR photography is better suited for imaging and identifying areas of wetlands than regular color film. Many times in gross land cover/land use classification schemes both wetland and brush categories are mapped. Because of their similar appearances on regular color film, as well as on black and white film, confusion sometimes arises concerning the classification of these two categories. With CIR the degree of confusion is reduced because of the ability of this film, even at high altitudes, to differentiate areas of varying soil moisture, as well as to distinguish among vegetative types.

In addition, land-water boundaries are easily delineated on CIR photography. This is important because in many cases a particular wetland area will be associated with an area of open water, with successions or patches of different wetland vegetation (swamps and marshes) located around it. Because of the amount of water present in wetland areas during most of the year, these areas are usually differentiated from adjacent agricultural areas by their grayish-green to dark magenta colors. Where confusion still exists, however, viewing the areas stereoscopically may be helpful in that most wetland areas occupy depressions or other low-lying sites. In many cases ground truthing may be necessary for positive identification. Wetland areas can be seen at points E in plates 9 and 10.

Croplands

High altitude CIR photography also differs from conventional color film in its assessment of agricultural scenes. As previously stated, different varieties of plant species are more easily recognized on CIR film than on conventional color because of the reflectance properties of vegetation in near infrared. This is especially true when planting and harvesting dates of crops in a particular area are known because as different crops begin to approach maturity, or otherwise change physiologically, the changes are highlighted on CIR film. For this reason, users may find special applications in agricultural studies by employing CIR high altitude imagery in their investigations.

CIR film exposed from high altitudes also brings out regional crop associations. Because of the general overview provided by CIR high altitude photography, especially the 1:120,000 scale imagery, regional groupings of crops can be ascertained and mapped, even with a limited number of photographs. The determination of crop groupings is important in studying the pattern of agricultural development in an area, as well as for gaining insights into regional agricultural practices. A regional assessment can be made concerning the correlation between different soil types and crop associations. This type of information is more accurately interpreted from CIR imagery than from conventional color or even panchromatic film exposed from the same altitude.

In addition to the regional assessment of crop associations, the acreage determination of these groups, or of individual fields, is also facilitated by high altitude CIR photography as opposed to regular color. The excellent haze penetration with high altitude CIR photography is an important factor in accurately delineating individual fields or groups of contiguous fields for making precise acreage determination. Accurate inventorying from time to time is extremely important for reliable economic forecasts of crop production. A greater degree of accuracy is normally obtained by using the 1:60,000 photography; however, with proper caution the 1:120,000 CIR photography can be utilized with enlargement or enhancement. Standard dot or grid cell transparencies can be utilized to inventory crop acreages on the 1:120,000 photography. Such inventorying is even more effective if acreage grids are used on CIR film positives because of the sharper image detail.

Another advantage of high altitude CIR photography in agricultural studies is the early detection of disease and insect infestation in crops and orchards. It is true that conventional color film exposed at similar altitudes can also detect plant stresses; however, with stronger color contrasts, CIR film is better suited for these types of agricultural investigations. CIR film has the capability to detect plant stress because of changes in infrared reflectance from plants which are affected by drought, salinity, insects, or disease. A particular crop may be diseased, but in the early stages of infection there may be no noticeable change in the appearance of the plant which can be discerned by the eye or conventional films. Slight alterations in the physiological structure of the plant, however, affect the near infrared reflectance because of diminished chlorophyll production. CIR film is sensitive to these subtle changes in the vitality of the plant. Even in some advanced stages of disease, signs of stress may be detected only on CIR photography.

The relative advantages of CIR photography in crop disease detection may be lessened when exposures are made from high altitudes. On small scale photography the differentiation between healthy and diseased vegetation may be diminished because of a decrease in resolution. This would be true where the areas of infection are limited to small sections of fields or to a few trees in a forest. In that case, more detail can be obtained by using the 1:60,000 photography, rather than the 1:120,000 scale. With proper magnification, however, some disease patterns can be detected on small scale photography, an example being diseased orchards. Where crop stress can be observed, high altitude CIR photography has the added advantage of showing the precise areal extent of crop damage. This is important in determining the economic impact of infection. On a single photograph, diseased sections over a large area can be identified and delineated without having to sift through a number of lower altitude photographs.

A further distinction in agricultural studies between CIR film and conventional color from high altitudes is the ability of the former to better identify and locate farmstead complexes, such as rural residences, barns, out-buildings, and confined feeding areas. On regular film exposed at high altitudes, because of blue scattering and the resulting "washed-out" color, identifying farmstead complexes can be difficult. This problem is reduced on CIR film because the color contrasts between vegetation and cultural features is apparent, making the interpretation of such features much easier and less time consuming. Areas of confined feeding are also easily identified on CIR photography by their location near farmsteads and the dark grayish soil color resulting from the increased organic content.

Urban Scenes

Urban areas offer another example of the advantages of high altitude CIR photography for landscape identification. Because of the strong color contrast between vegetated and nonvegetated areas, the delineation of the Central Business District (CBD) is greatly facilitated on CIR photography. This is seen in the downtown area of the city of Lansing, Michigan, located in the west-central portion of plate 10. The contrast between the area of most intense commercial activity—the CBD—and adjacent residential areas is apparent. Because of the lack of vegetation and the predominance of multistory concrete structures and parking lots, most CBD areas appear in white to bluish-white hues on CIR prints. This distinguishes them from surrounding residential areas with their tree-lined streets and grass lawns. In many cases the distinction is sharp because of zoning ordinances which define commercial and residential land use areas in a city. As plate 9 shows, the distinction between the CBD and surrounding residential areas is not as readily apparent on regular color film because of the lack of color contrast.

In addition to the CBD, there are strips of secondary commercial development on the periphery of the main urban area. Such strip development is found along the east-west street at point F in plate 10. In many cases the delineation of strip commercial development on CIR photography is easier than the delineation of a CBD area because of a stronger color contrast between vegetated and nonvegetated areas.

Another urban observation which is facilitated by CIR photography is that of identifying the relative ages of residential areas in order to determine the growth pattern of the city. CIR photography from high altitudes is very useful for this because two of the more important elements in making such distinctions are vegetation and street patterns. Vegetation is important in such a residential analysis because, in most cases, the older a residential area is, the more likely it is to contain streets lined with large trees. Somewhat newer residential areas may also have tree-lined streets, but these trees are usually not as large; consequently, such areas do not appear as red as do the older areas. With newly developed residential areas or subdivisions, the red to pink hues may be very subdued or even absent because of the lack of large trees and established lawns. With such an emphasis on vegetation in this delineation, the importance of CIR photography is obvious.

Street patterns may be an important element in identifying the age of residential areas. In many regions, older areas tend to have gridlike street patterns, while newer ones have curvilinear or wavy patterns, with cul-de-sacs in some places. Because of the sharp color distinction between vegetated and man-made features, such patterns show up very well on CIR photography.

Industrial Signatures

An important aspect of urban analysis is the interpretation of industrial areas within or adjacent to cities. CIR photography is especially useful for industrial interpretation because many heavy manufacturing plants give off at least some smoke and steam in their production processes, obscuring from above important features of the plant and surrounding areas. With its haze penetration capability, CIR photography reduces this atmospheric obstruction somewhat, making delineations easier. Because size and shape are important parameters in object recognition, the sharper details presented on CIR photographs make it possible to identify the sizes and shapes of the industrial structures. Some industrial structures can be identified on CIR photography because of the large tar roofs which distinguish them from surrounding commercial structures. These roofs appear in black tones on regular color film, but they are not as sharp or as dark as on CIR film. Conventional color film is probably better suited, however, for identifying nearby stockpiles of industrial raw materials, especially mineral ores.

Other Urban Images

Other important urban features which can be interpreted better on high altitude CIR photography include recreational areas, parks, and cemeteries. All of these facilities have one thing in common—large areal expanses of grass and ornamental vegetation which show up in light pink hues. This makes it easy to separate these urban features from the surrounding built-up areas. The golf course at point G on plate 10 is an example.

As in the case of rural studies, another important benefit of high altitude CIR photography in urban studies is the synoptic overview provided by high altitude exposure. The latter, in most cases, makes possible the study of a large urban area on a single photograph. The structure of large urban areas can be determined at a glance because of the distinct way color contrasts set them off from surrounding rural areas. The various components of urban structure can be understood when viewed on one image, rather than being required to construct a photo mosaic, or to view a number of overlapping prints. That includes the relationship between the CBD and surrounding residential areas, as well as the main industrial concentrations. Areas of recent growth and the direction of development are easily seen, especially in terms of age of residential areas and commercial strip development. Street patterns are better delineated on this kind of photography because of the ability to obtain a full view of overall patterns, as well as its sub-parts.

It should be noted that the previous discussion concerning the relative advantages of high altitude photography is not intended to rule out all other types of conventional high altitude aerial photography. Indeed in certain circumstances types of photography other than CIR film exposed from high altitudes should be used. However, because of its strong color contrasts, near infrared imaging, haze penetration, and synoptic overview, high altitude CIR photography does have a number of distinct advantages over other film types, with the result that it is best suited for a variety of different investigations.

Scale Factors

Scale is an important factor to be considered when any aerial photograph is studied, but it is especially important to investigations of high altitude color infrared imagery. Because the film in the Zeiss and RC-8 cameras is exposed at about the same height above ground datum, variation in scale results from the differences in the focal lengths of the lenses of the two cameras (plates 11 and 12). Not all AIRP missions are equipped with both camera systems; therefore, the variety of image scales may not be available for a particular area of interest. Where this is the case it is likely that the RC-8 will be the camera used, providing photography at a nominal scale of 1:120,000.

As plates 11 and 12 show, the differences in scale mean that a larger area of the surface of the earth is imaged with 1:120,000 photography than with the 1:60,000 photography. When 1:120,000 scale photography is used, more of a particular ground scene will be shown, and fewer photographs of that area will be needed. This is an important factor to consider in terms of cost per photograph and the inconvenience of working with a large number of photographs, either in the laboratory or in the field.

An additional disadvantage of large scale 1:60,000 photography is that it contains no sidelap between prints from adjacent flight lines. This produces gaps between flight-line coverage. These gaps with the Zeiss camera range between .8 to 5 kilometers (.5 to 3 miles) or more. Within a given study site, therefore, a part of the area will not be photographed. On the other hand, total photographic coverage is achieved on the 1:120,000 scale imagery.

Although there are scale disadvantages to the 1:60,000 photography, it has certain advantages over other types of CIR imagery. At the scale produced by the Zeiss cameras, scenes are imaged in more detail than that found on 1:120,000 RC-8 photographs. This reduces the amount of time needed to identify and delineate features and areas. If proper equipment is available, however, the 1:120,000 images can be enlarged to a point where they, too, can be used for making precise interpretations. This is especially true if film positives are used. Simple hand lenses often will be adequate for identifying most ground targets on small-scale imagery.

In terms of the detail of interpretation provided at each scale, there are certain advantages and deficiencies in both. With the Zeiss photography, for example, more information concerning surface soil drainage in individual fields is furnished by making possible more precise delineations between well-drained and poorly drained areas. Consequently, the boundaries between organic and mineral soils will also be better differentiated on 1:60,000 photography. This does not mean that such features cannot be distinguished on the 1:120,000, only that it becomes more time consuming and a little less precise to do so. In any case, when only making a differentiation between bare soil and vegetated areas, either scale photography will serve.

Specific agricultural studies are usually better facilitated on the 1:60,000 imagery. Different crop species are somewhat easier to identify at this scale because the presence of crops in rows can be identified as opposed to those crops that are not so planted. Spacing between rows, which sometimes can be an aid in interpretation, is not practical at either of the scales. The presence or absence of rows is an important factor to consider in trying to differentiate between general categories of cropland and pasture, especially permanent pasture. Also, it is of value to be able to distinguish between seeded pasture grass and varieties of grasses and scrubs found in brushland areas, an operation which is more feasible at a scale of 1:60,000 than at 1:120,000. The stereo model is important, too, because of the ability to differentiate the relative heights of scrubs and bushes as opposed to grass. In addition to crop studies, the Zeiss photography is preferable for the interpretation of rural farmsteads because individual structures are more easily identified, although a certain degree of magnification may be required.

In the interpretation of other rural scenes, especially where no specific feature is being identified, either scale will be sufficient. An example would be the delineation of forested areas and woodlots or deciduous and coniferous forests. If species associations are to be identified, however, the 1:60,000 imagery would be helpful. In delineating the stream drainage in an area, both sets of photography are useful, but the regional drainage pattern can be discerned better with the 1:120,000 photographs.

The 1:60,000 scale imagery is useful in the interpretation of a variety of urban features. Individual structures are identifiable at this scale. The differentiation between houses, stores, and industries takes less time and effort and is likely to be more accurate. Shapes and relative sizes are apparent at larger scales, especially where the stereo model is employed, and the boundaries of the CBD can be located. The limits of different aged residential areas are better delineated because of their characteristic sizes, shapes, and associations.

The Zeiss imagery is also somewhat better for urban transportation studies. Street patterns are more easily picked out, as well as railroad routes, associated rail yards, and loading sites. Parking areas also are more easily identified, including, with proper magnification, the density of vehicles in the lots.

Industry is another area where the 1:60,000 imagery is very useful. Although major industrial sites can be interpreted without much problem with the 1:120,000 scale photography, details of activity and possible stockpiling around a plant are normally better discerned at 1:60,000. The differentiation of various types of raw materials is a problem because of their similar colors. The amount and location of stockpiles can be identified at the scale of 1:60,000. Furthermore, with proper magnification, the finished product is probably also better seen at this scale.

For the most part, the Zeiss photography at a scale of 1:60,000 provides more detail of a number of different features. The 1:120,000 imagery has the advantage of providing more complete coverage of a particular area and with fewer photos needed. Depending on the type of study being done, each of these factors must be weighed in determining what scale high altitude CIR photography is appropriate.

Image identification can be improved by enlarging and enhancing high altitude CIR photographs. As explained in Chapter 9, CIR high altitude photograph prints or film positives may be enlarged, enhanced, and copied by a 35 mm camera equipped with a macro lens. The positive slide produced by this procedure can be further enlarged and enhanced on Cibachrome paper. Plate 14 illustrates this type of product.

Seasonality

High altitude CIR photography has been made for more than one season in some areas. Where multiseason photography is available, it was usually taken during the spring/early summer and again in summer/early fall. Unless the photographs were flown in support of some specific project requiring exposure at another time of year, the spring and late summer sequence provides maximum information concerning a number of features. The photographs in plates 12 and 13 were imaged in early June and mid-September, respectively. Depending on the type of information desired, each high altitude CIR seasonal photo has its own advantages and disadvantages, as a comparison of the two plates shows.

Late spring/early summer photography is best for any type of soils study. During the spring months the amount of bare soil on the photograph will be at a maximum because of field preparations and the general lack of crop growth, except for fall-sown crops. In late summer, crops are at maturity and ground cover is at a maximum.

Surface soil drainage patterns are easy to distinguish, especially at the scale of 1:60,000. Small streams and drainage ditches which are indistinct or are dry during late summer or fall are apparent in the spring when there is maximum flow. Also a lesser amount of foliage may be present in spring so that smaller streams are less likely to be obscured. The regional stream network is much easier to identify during this time of the year, and the 1:120,000 scale photography is very useful where an increased tonal signature is needed. With a high water table, the total extent of wetland areas is much easier to delineate in the spring than in the fall with its drier conditions.

Crop differentiation, on the other hand, is facilitated on high altitude CIR photography taken during the summer. During this time of year crops are at maturity and possess maximum tonal signatures. Spring photography is not as good because summer crops either have not broken the surface of the soil or are not high enough to be visible at these altitudes. Spring photography is better, however, for differentiating fall-sown crops, such as winter wheat, because at that time of the year these crops are practically the only ones covering cultivated fields. Such crops can be distinguished on summer photography, although not as well, because by midsummer most fall-sown crops will have been harvested, leaving only stubble in the fields.

While mature crops are not present on the spring imagery, it can be useful in differentiating between cultivated crops and pasture and other grasslands. Such differentiation is sometimes difficult on summer photography because of similarities in some crop signatures. Spring photography shows those fields which have been cultivated. If there is no evidence of cultivation in the spring, and the color hues and patterns are similar to those on the summer imagery, the chances are that the field is either pasture or currently not used for crops.

Farmsteads are much easier to identify on late summer photography than on spring photography, particularly in heavily cropped areas. Houses, barns, and other out buildings are difficult to interpret on spring imagery because they have reflectance characteristics similar to those of bare soil. Farmsteads are easily identified on summer photography where bare soil is reduced. In many cases the high reflectance of metal roofs aids in the identification.

Transportation studies, too, are more difficult in spring, especially delineating rural road patterns, because of the predominance of bare soil. With similar reflectance characteristics, roads tend to blend into the landscape and are hard to identify. Rural roads are best identified on late summer photography where differences between man-made structures and vegetation is at a maximum.

The seasonality factor often is not as important in urban areas as it is in rural landscapes because the reflectance characteristics are much more constant in urban areas, the result of reduced areas of vegetation, especially in and near the CBD. The season, nevertheless, can become a factor in urban analysis if the photography is taken early in spring prior to the leafing out of trees. This can be an important consideration when age of vegetation is being used as one of the indicators of the residential pattern of an urban area.

The identification and delineation of the exact boundaries of an urban area are difficult on high altitude CIR photography taken in the spring. Because of the large amount of land being imaged and the greater amount of bare soil reflectance in spring, color contrasts are not as strong in spring as they are in the late summer months. Identification as to where an urban area stops and rural cropland begins can be obtained from springtime photography, but it is a more difficult and time consuming operation than on summer photography. For the same reason, secondary commercial or strip developments are also more difficult to identify on spring photographs because of their location on the periphery of urban areas.

Educational Uses of High Altitude Color Infrared Photography

While high altitude CIR photography has a great many applications in research, it also has a wide range of uses in teaching in a number of fields. AIRP photography was originally intended for use in support of imagery obtained from orbital platforms, but more and more uses are being found for it in the classroom. In a real sense, like other types of remotely sensed imagery, the only limitation on its uses is the imagination of the instructor or user. Strictly speaking there is some limitation in the level of course being offered, the sophistication of the equipment on hand, and the degree of knowledge of the subject matter. Within a given classroom situation, however, there is a wide variety of possible uses and applications of this type of imagery.

One of the basic classroom uses of high altitude CIR photography is as visual aids in support of lecture materials. Paper prints can be projected onto a screen with an opaque projector, although some resolution is lost. Film positives are probably the best film product for lecture use because of the ability to project a particular scene as a transparency with an overhead projector. There is minor loss of detail, while a clear sharp image is retained. High altitude CIR photography of a number of different areas of the United States can be purchased for comparative study and used for lecture demonstrations.

The use of high altitude CIR photography is not confined to classes in aerial photography or remote sensing. It can be used in such courses as geomorphology, cultural geography, urban geography, or archeology. It is unnecessary for students in some of these courses to have an extensive background in aerial photography in order to appreciate what they are looking at because they are only searching for major patterns or gross distribution relationships. By utilizing high altitude color infrared photography, attention of students is focused on the screen because of the strong color contrasts. It also provides them with the large regional overview on a single photograph which is not afforded on other types of visual aids.

AIRP photography has a wide range of uses in laboratory situations. It can be used as laboratory support materials in a variety of courses dealing with environmental problems. Such laboratory materials are particularly useful if they cover the hometown or regional area of the classroom participants. The photography becomes much more interesting to students and helps them relate to it by identifying places they know. Unfortunately, not all areas of the country are covered by AIRP imagery, but if a particular area of interest is not covered, another one similar in appearance and content can be selected. Thus if students cannot view a familiar area, they might at least be able to view a similar one, such as the same physiographic region, and find it of interest.

Sophisticated equipment is not needed for basic exercises employing high altitude CIR photography. Either paper prints or film positives can be used, but paper prints can be obtained for about half the cost of film transparencies. On the other hand, in many cases film positives provide better resolution and color contrast. If elementary exercises are used, only a simple hand lens and a ruler with divisions of at least .01 of an inch are needed. It is a good procedure to put the film products between two pieces of clear acetate. Land cover details can be drawn with a pen or grease pencil on the acetate overlap without harming the photograph. When the project is completed, the marks can be wiped off and the acetate used again. When film transparencies are used, some type of light table is needed for illumination. In more advanced courses such equipment as pocket and mirror stereoscopes, scanning stereoscopes, and parallax bars can be utilized with the photographs.

In order to use high altitude CIR photography effectively, students should understand certain basic principles of the imagery. First, they must be aware that they are viewing reflected not emitted infrared radiation. Furthermore, they must become acquainted with the color shift in the scene and know why, for example, healthy vegetation appears in red hues on the image. Also, users should have some fundamental concept of scale so that linear measurements and area calculations can be performed. These types of understandings will give them an appreciation of high altitude exposures and the regional overview provided by this photography.

Users should be made familiar with the relative size and shape of various objects to be identified on the photography. This information, along with a basic knowledge of scales, will permit more accurate interpretations of the imagery. This is important because most persons probably have never been exposed to photography at scales of 1:60,000 or 1:120,000 and thus are likely to misinterpret objects as being either larger or smaller than they really are. Also, where applicable, a fundamental knowledge of stereoscopes is needed.

It is usually helpful in a classroom or laboratory situation to use high altitude CIR photography in conjunction with U.S. Geological Survey topographic maps. The position of individual photographs can be plotted on the topographic quadrangles, and a variety of information can be gathered from the maps for use in the interpretation of landscape features. By using such maps, students will gain an understanding of the nature of the photographic missions and the way they were flown. A variety of other supplemental information, such as soil survey reports, can be utilized with the AIRP photography.

A simple land cover/land use study is an example of the type of work that can be done with high altitude CIR photography. Such studies have wide applications in many areas of geography. The equipment for such a project includes a hand lens, several sheets of clear acetate, and three ink reservoir pens. When film positives are used, a light table is needed. If more detailed work is to be done, a pocket or mirror stereoscope would be necessary. For more sophisticated exercises, such as the preparation of a land cover/land use map of the area covered from the photography, a reflecting opaque projector would be required.

Before any interpretation can be done, a suitable land cover/land use classification scheme must be selected and adopted. Several schemes are in current use, but perhaps one of the most widely used is that formulated by the U.S. Geological Survey.[3] (A detailed classification scheme is not practical because of the scale of photography.) A classification scheme containing approximately twenty categories, however, can be used to identify and delineate such general features as cropland, pasture, wetlands, and forests. In addition to selecting the most suitable classification scheme, students should become familiar with the various types of urban and rural land cover/land use in the study area. Some background source material about the area should be provided so that students can make reasonable interpretations from the photography. This is especially true of rural areas where a knowledge of regional crop and livestock associations would be most helpful. This is the type of information many students do not have because of their urban backgrounds.

The three ink reservoir pens, with black, blue, and red ink, are used for delineating land use categories. The red pen is used to delineate the roads and railroads on the acetate overlays; the blue pen is used to outline the water features. When this has been done, the photograph scene will be subdivided, the road and water-feature lines providing boundaries. This will enable users to analyze systematically the scenes within each of the subdivisions.

Next, the individual classification categories are numerically coded for identification on the acetate overlay. For example, numbers and letters can be assigned to the categories, such as cropland, 1a; pasture, 1b; residential, 2; forestland, 3; and so on. Once this has been accomplished, the black pen can be used to delineate the different land cover/land use categories on the acetate overlay and then code them with the proper number. In this way the entire photographic scene can be systematically classified.

Examples of types of land cover/land use can be seen on plate 13 and compared with the corresponding ground photographs in figure 10.3. These examples include:

1. Downtown CBD area
2. Residential area
3. Shopping center
4. Industry
5. Sewage treatment facility
6. Coniferous trees
7. Golf course
8. Cultivated cropland—corn
9. Brushland
10. Wetlands, forested
11. Pastures
12. Bare soil, plowed field
13. Harvested crop—wheat
14. Deciduous trees

Additional work can be done by placing a dot grid over the acetate overlay and performing area calculations to establish the amount of land in each category. The detail from the acetate

Figure 10.3. Ground truth photographs for selected sites in plate 13: (1) downtown CBD area, (2) residential area, (3) shopping center, (4) industry, (5) sewage treatment facility, (6) coniferous trees, (7) golf course, (8) cultivated cropland—corn, (9) brushland, (10) forested wetland, (11) pasture, (12) plowed field, (13) harvested cropland—wheat, (14) deciduous trees. (Photographs by author)

sheet may be transferred to a base map for the purpose of drafting a land cover/land use map of the study area. To do this, a change in scale is usually needed. This requires the use of a reflecting opaque projector. Once the detail has been transferred to the base map, different categories can be color coded and shaded in. If a more sophisticated map is required, categories can be differentiated by the use of a variety of adhesive patterns, or screen tones. This type of map can be photographically reproduced for printing.

High altitude CIR photography can be used for more than a research tool. Although it has many uses in primary research, it can be equally important when used in classroom or laboratory work.

NOTES

1. *Remote Sensing with Special Reference to Agriculture and Forestry* (Washington, D.C.: National Academy of Sciences, Committee on Remote Sensing for Agricultural Purposes, 1971), p. 46.
2. Ibid.
3. J. R. Anderson, E. E. Hardy, and J. T. Roach, Jr., *A Land Use Classification System for Use with Remote Sensor Data*. U.S. Geological Survey Circular 671 (Washington, D.C.: U.S. Department of the Interior, 1972).

11

Application of Remote Sensing in Landform Analysis

Victor C. Miller

Indiana State University

Introduction

General Considerations

Landforms are four-dimensional configurations of the surface of the earth. The conventional dimensions used for identifying landforms are length, width, and height, along with time as a vital fourth.

By contrast, individual remotely sensed images are only two dimensional—length and width, whether obtained by cameras or by one of several scanning devices. The images may be large or small scale; they may show incredible detail or only general regional relationships; they may be black and white, color, false color, thermal, or radar; but they are all two dimensional. These two-dimensional images which are produced by aerial and space photographs and photographlike images impose limits on the interpretation of landforms.

Only the use of the stereoscope and stereoscopic pairs of aerial photographs brings into being the third dimension of landforms—height. The slope of an alluvial plain, the escarpment bordering a plateau, or the stair-step walls of the Grand Canyon appear as reality when viewed through a stereoscope. Place a pocket stereoscope over figure 11.1 and observe the height of landforms which is created in stereoscopic vision.

As to the fourth dimension—time, the slow rate of downcutting by a stream, the retreat of a scarp, the filling of an individual depression, or the upstream march of a nickpoint cannot be recorded by a once-every-eighteen days scanner from a height of some 900 kilometers (559 miles). The advantage of repeated, two-dimensional satellite imagery over stereoscopic aerial photographs is meaningless in the context of changing landforms.

This is not to say, however, that the only way to study landforms is on 1:20,000 stereoscopic pairs of aerial photographs. Persons who want to learn about landforms are going to learn very little, however, if they "train" on Landsat imagery while letting their stereoscopes collect dust.

In dealing with the application of remote sensing in landform analysis, this chapter is *not* going to do several things. It is not going to attempt to list all the ways in which geographers, geologists, and other earth scientists use remotely sensed images to identify, measure, and interpret the landforms of the surface of the earth. The inquisitive readers of this book want to know how they can become proficient in identifying, measuring, and interpreting landforms, not what others may have done.

This chapter is not going to attempt the impossible task of trying to teach the reader how to interpret *all* types of landforms, developed under *all* kinds of conditions, and in *all* kinds of earth materials throughout the world. Nor is it going to provide coverage of and instruction in all kinds of remote sensor imagery. Instead of surveying a few obvious and rather meaningless high spots in each type of imagery, the reader will be given an opportunity to learn something about landforms and how they are portrayed through a few selected examples of imagery.

In this chapter two types of imagery will be used in the analyzation of landforms: (1) the stereoscopically-viewed large-scale aerial photograph and (2) the two-dimensional, extremely small-scale satellite image (Landsat). These two imagery formats have several things in common. First, they are readily available to anyone who wants them; they do not have to be contracted for, but simply ordered from the federal government. Second, they can be obtained for virtually any area in the country, and they are not expensive.

The Fourth Landform Dimension—Time

William Morris Davis is famous for his structure-process-stage trilogy appropriate to landform development. Time is a fundamental factor in the *stage* of development of a stream valley, or in the *stage* of dissection in an area; however, there are other things to consider when the interpreter of landforms on remotely sensed imagery is dealing with the dimension of time.

Process is a factor which must be considered in the interpretation of a landscape. What has process—that is, the erosional and depositional work of wind, water, and ice—to do with time? Consider the fact that it rains on the Flint Hills of Kansas and it rains on the Coastal Plain of Mississippi. It also rains in southwest South Dakota. It has been raining, on and off, in all these areas for a long *time,* and in each one there are extensive tracts of terrain where dissection of the land has reached the mature stage. On the other hand, the topography of these areas—the landform assemblages—do not resemble each other.

Erosion by running water is a process common to all maturely dissected plateaus, plains, and coastal plain areas; however, landscapes are not products of running water alone. Even if they were, they would reflect the fact that in some areas there is more annual precipitation than in others, more frequent occurrance of heavy downpours, and greater variability of seasonal precipitation. Evaporation is significant in some areas but negligible in others. Lubricated gravity displacement producing creep and landslides is more common and important in some areas than in others. Vegetation and soils, along with chemical and mechanical weathering, are different in different places on the earth. It is not a matter of how *long* things have been happening, but also *what* has been happening.

Throughout long periods of time glaciers, waves, wind, running surface water, and subsurface water, along with tectonic processes, have been shaping the face of the land. Furthermore, structure must be considered in connection with the factor of time. The interpreter must think of structure not only in terms of composition, porosity and permeability, bedding characteristics, and the *age* of earth materials, but also of the structures impressed upon bedrock. The interpreter's "time" extends beyond *how long what* has been happening, to encompass the answer of the question: *to what?*

For example, the bedrock strata of the Badlands of South Dakota, Flint Hills of Kansas, and Coastal Plain of Mississippi are all virtually horizontal. But the bedrocks of these areas, individually and collectively, are not identical. Also, some of their strata are poorly cemented and some are extremely well cemented. Different landforms, therefore, developed through *time* in these three areas, despite the fact that they display several similar characteristics—horizontal sedimentary rocks, integrated drainage systems, and rock weathering. The landforms are different because the climate, vegetation, rock types, and other controlling factors are different among the three areas.

Photogeologists must not forget the fact that *all landforms are unique.* (All persons who interpret landforms on remotely sensed images are photogeologists.) Many landforms are similar, having similar geometric dimensions, characteristics, and similar origins, but each discrete landform must be identified as being what it is, not because it looks like some other landform of known structure or origin, but because all evidence demands that it be so identified. If this evidence is inconclusive or insufficient, and if the landform still resembles another of known identity to which the same evidence could point, then it may be *tentatively* identified, subject to field checking.

Everything discussed above applies to *eroded* topography—landforms produced by the destruction and removal of earth materials. The products of destruction and removal—erosion—are transported to other sites and *deposited*—by water, wind, or ice—to form a completely different family of landforms, including river deltas, glacial moraines, and sand dunes. These depositional landforms, too, may be attacked, eroded, and, perhaps, eventually obliterated. Yet these are all part of the fourth landform dimension—"time."

Tricks of the Trade

It is one thing to be handed a Landsat image or a stereoscopic pair of aerial photographs and be asked, "What are the landforms in this area and what are their origins?" It is quite another to be given the same imagery, asked the same question, and to be told, "The geographical coordinates of this area are . . . ." By knowing the location of the area imaged, an interpreter can turn to maps to learn the answers to three important questions—questions which should always be foremost in the interpreter's mind: What is the climate of the area? What is the regional geology of the area? What is the regional topography of the area?

Climate. The influence of climate on the soils, drainage (surface and subsurface), vegetation, weathering, gravity processes, and landforms of an area cannot be overemphasized. A striking example is the high rate of solubility of limestone under arid conditions.

Regional geology. There are some areas of the world for which reliable detailed geologic maps are not available; however, even for remote areas there is usually some geologic information. Geologic maps of the United States and Canada have been compiled and can be obtained from government agencies at reasonable prices. The geology of many states has been mapped, and each year the U.S. Geological Survey publishes many excellent geologic quadrangles.

From these maps, whatever their scale, the nature of the bedrock can be learned. Knowing that the area to be studied is in south-central Indiana, for example, immediately eliminates the necessity of considering the type of landforms which could not possibly occur there, such as cirques, folded ridges, and fault scarps.

There is a word of caution, however, which must be stated. Small-scale geologic maps of a state or of the United States usually depict the distribution of bedrock units. If interpreters are to study aerial photographs or other remote sensor images of such areas as northern Indiana, they will not see bedrock. Instead they will see landform assemblages developed on moraines, outwash plains, and lacustrine flats. Nevertheless, it is a good idea to check to see what types of bedrock, mantle resting on the bedrock, and soil blanket the area being studied. The geologic ages of these materials appear on geologic maps.

Regional topography. Knowing the climate and bedrock geology of an area, and whether or not the bedrock is hidden beneath unconsolidated overburden, such as glacial or stream deposits, the interpreter can then consult topographic maps of the region. It is a great help to know where the landforms of an area fit into the large geomorphic picture.

Case Studies

With this brief background and introduction to ways in which landforms come with time, case studies of specific landforms in Arizona, California, Kansas, Montana, Pennsylvania, and Virginia will be presented. Various climates, vegetation characteristics, rock types, rock structures, and landforms are included in these examples. These will be studied on different types of imagery. In these case studies, readers will be presented with a variety of problems in the interpretation of landforms on the different types of imagery.

The more that is known about landforms, the better they can be recognized, identified, and interpreted on aerial photographs and other types of imagery. Conversely, the more the users study remote sensing images, the more they will learn about landforms. In one way or another, landforms appear on all images of the land surface of the earth.

Case Study 1: Southeastern Kansas

Figure 11.1 shows a relatively large scale (1:20,000) stereoscopic pair of aerial photographs. The area may be located on the 1:250,000 Wichita, Kansas topographic map NJ 14-9 at latitude 37°23′ N, longitude 96°31′ W. Also, it can be located in the Cambridge NE Kansas Quadrangle at a scale of 1:24,000. North lies toward the *left* on the stereogram. Because it would be extremely difficult to think of the top of a stereogram as east, as in this case, and perhaps west in another figure, throughout this chapter the term "figure north" will be used. That is, as far as the case studies will be concerned, "north" will always be toward the *top* of the figure.

The most eye-catching features in the area are the contour-appearing light lines. When the photographs are viewed stereoscopically, these lines are seen to be the outcrops of a series of horizontal or nearly horizontal bedrocks. Most of the light-colored beds occur along the edges of well-defined topographic benches, or rock terraces. This indicates that these rocks are more resistant to weathering and erosion than the strata which separate them.

The area is located in the Flint Hills. The rocks are shown on the Geologic Map of the United States to be Permian in age (the last period of the Paleozoic era). They consist of shales and limestones, many of which contain chert. It is apparent in the photographs that this is not a particularly humid area because there are few trees and an abundance of bare rock outcrops. In this climatic setting, the limestone layers would be expected to resist erosion and to form a terraced landscape.

The drainage is well integrated, and the channels and valleys are well defined. Drainage density is not especially high. North of the east-west divide the topographic relief is several

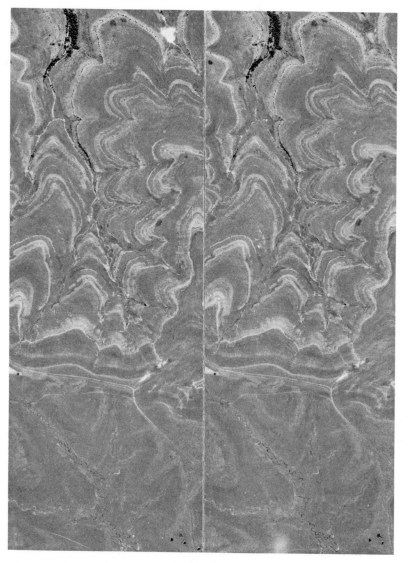

Figure 11.1. Stereogram of an area in the Flint Hills, Kansas. (U.S. Department of Agriculture)

times greater than it is to the south—actually 91 meters (300 feet) to the north and 18 meters (60 feet) to the south. Also the slopes in the north are steeper and the bench-forming units are expressed more sharply. Scarps, or ridges of outcrops, are found in the north and rounded slopes in the south. Topographic maps of the area confirm that the gradients of the north-flowing streams are greater. In about 8.5 kilometers (5 miles) the north flowing streams descend some 134 meters (440 feet), whereas the south flowing streams descend only 100 meters (328 feet). The positive correlation between gradient and slope is well illustrated.

The steeper gradient of the north-flowing streams imparts to them greater velocity and turbulence than is possible in the south (note the small meanders in the south). The rate of headward erosion by north-flowing streams, therefore, must be greater, the result of the southward migration of the divide between the north-flowing and south-flowing streams.

How much of the above landform/geology identifications, information, and interpretation is based on what can be *seen* in figure 11.1, and not on data derived from reference sources? A great deal, it would seem. Eyes peering through a stereoscope at the stereogram cannot ascertain such facts as topographic relief, actual stream gradient, rock age, and details of rock types. But they can tell you many things.

The stereogram *shows* that the relief, gradients, and slope magnitudes in the north are much greater than they are in the south; that the strata are nearly horizontal; and that there are light-colored resistant beds and darker nonresistant beds. The lack of vegetation gives a fair idea of the type of climate. Knowledge about the principles of differential erosion makes possible the assumption that the nonresistant units which underlie the smooth dark slopes are shales, and that the light-colored resistant beds are probably limestone (though they could well be a coarser rock, such as sandstone). It can be concluded with great accuracy that the south-flowing streams are downcutting more slowly. Their velocities, turbulence, and capacities are lower, and, therefore, their valley slopes undergo downwasting. In the north, the valleys are subject to undermining, due to rapid removal of shales. On the basis of what can be *seen*, the southward migration of the divide can be interpreted.

Case Study 2: Southwest Virginia

Figure 11.2 shows a stereogram of the valley of Big Moccasin Creek in southwestern Virginia at 36°41′ N, 82°32′ W. It can be located on the 1:24,000 Gate City Quadrangle and on the Geologic Map of Virginia. This area will be studied in a different way. First the topography will be observed, and as much as possible will be deduced from it. Then available references will be consulted to learn as much as possible about the geology and its impact on landforms. Finally, a question or two will be set forth to which the observations and reference data may or may not provide answers.

The outstanding feature in the area is the meandering Big Moccasin Creek (**bm**). It is flowing along a meandering course. It does not have a floodplain; therefore it cannot be termed a mature stream. To the northwest of the main road (**a**), the relief of the land is appreciable and the drainage density is quite low (coarse topographic texture). For the most part, the slopes are smooth and rounded. Southeast of the road, local relief is less and texture relatively fine; that is, the density of drainageways is high.

Observe that the fine-grained texture of the topography in the southeast is determined in part by undrained depressions (**b**), as well as by the basins of integrated drainage lines. Also, observe in this area the presence of numerous outcrop patterns (**c**) which define the fairly steep southeast dip of the rock strata. Several southeast-facing slopes (**d**), which are apparently nearly dip slopes, have been developed here as well. The long, high, uniform slope of the surface (**e**), which parallels the road to the north appears to be a dip slope, that is, a land area whose slope is controlled by the dip of the underlying bedrock. There are a few scattered light-toned features (**f**) in the uplands which appear to be outcrops, but they are not continuous enough to permit a determination of the underlying geologic structure.

Figure 11.2. Stereogram of the valley of Big Moccasin Creek, southwest Virginia. (Tennessee Valley Authority)

The valley-wall slopes—**g** and **h**—along Big Moccasin Creek are worthy of particular attention. They are quite high and, for the most part, fairly steep. They are not of uniform steepness; however, as they are followed along the creek, they are seen to display a back-and-forth, gentle-steep-gentle alternation which coincides with the meandering curvature of the stream. The slopes are steep on the outside (**g**) and gentler on the inside (**h**) of each major bend.

Interpretation. The rocks—at least those along and southeast of road **a**—are largely or entirely limestones. This area was not glaciated, and throughout this region the climate (note abundant vegetation) is undoubtedly sufficiently humid to permit intensive subsurface water solution of the limestone.

Although the structure of the higher central and northern area cannot be ascertained from the stereogram, it can be said that the limestones dip to the southeast and, presupposing no strike fault (horizontal movement) along the contact, overlie the rocks of the northwest. It would not be unreasonable to propose that the large dip slope **e** is developed on the top surface of the uppermost of a sequence of relatively resistant southeast-dipping strata which are older than the limestone.

Care must be taken in attempting to even tentatively identify the composition of the older rocks. For one thing, although the slopes developed on them may not be what would be called gentle, they are far less steep than they appear to be. The reason for this is that when photographs are viewed stereoscopically, the terrain model is considerably exaggerated vertically. Thus, all slopes appear steeper and all relief greater than they actually are. For example, the topographic map shows dip slope **e** to be inclined about twenty degrees. To the author's eye it seems to have a magnitude of at least forty degrees. This computes a vertical exaggeration factor of about 2.5 times.

If these were all dipping sandstones, they would be expected to form much higher terrain, and dip slopes would, in all probability, not be hard to find. If they were folded or faulted sandstones, their dissection would have produced ridges and ledges. Interbedded sandstones and shales could account for this relief, but a relatively higher drainage density would be expected. About the safest thing to conclude is that these rocks, whatever they are, are not soluble and do not seem to be coarse sedimentary rocks. They might, however, be massive crystalline rocks.

The geologic map shows that the area along and southeast of road **a** is underlain by limestones, and that in the entire area of greater relief, the bedrock consists dominantly of dolomite (carbonate rock). The low drainage density of the latter area can be attributed to high infiltration of water along joints. The relief and lack of undrained depressions may be explained by the fact that dolomite is far less susceptible to solution than is limestone.

The meandering course of Big Moccasin Creek and the absence of a floodplain along its valley suggest that at some former time and at a higher level, the creek flowed as a "mature" stream on a relatively wide floodplain, and that, for some reason, it underwent rejuvenation, and the meanders became incised (entrenched) in the bedrock. It might be unwise, however, to leap to this geomorphic interpretation. Consider the following questions.

Question 1. Is the sinuous valley of the Big Moccasin Creek the result of the downcutting of a meandering stream? There are two statements of fact that can be made: the stream follows a meandering course, and its valley has no floodplain.

Along most of its length, however, the valley displays a profile which is definitely not symmetrical. In addition, asymmetry alternates with symmetry, and the steeper slope (**g**) always is located around the outside of each bend. About the only places where there is symmetry of profile are where the stream is about to change from a northwest-concave to a southeast-concave bend and vice versa. Such regular shifting in asymmetry from one bend to the next can only have been brought about by the extention of meanders outward at the same time as the stream eroded downward. These are called *ingrown* meanders.

Mentally lift the stream halfway up from the level of its present channel to the tops of the valley slopes. Its sinuosity will then be considerably less than its present course. It will remain a meandering, but a much less pronouncedly meandering, stream. It is possible that if the channel is mentally high enough (that is, far enough back in the stream's history), it will become merely a *bending* stream.

The only problem is that figure 11.2 cannot answer the question with surety. It may have been a normally meandering stream which, for some reason, began and continued to cut down, while at the same time enlarging its meanders by simultaneously cutting laterally. On the other hand, it may have been a stream with an irregular course which, after beginning to cut into the rocks through which it now flows, developed the meanders as it continued to cut down. The answer is not to be found within the extremely limited area of this figure.

Question 2. Why does Big Moccasin Creek flow in a rock (dolomite) which is obviously so much more resistant to downwasting than limestone? After all, if Big Moccasin Creek were removed, the lowland would extend along the limestone belt (note the rise of the surface in the southeast corner of the figure). It would seem more logical for the major stream to follow the limestone and to avoid, by downdip shifting, the dolomite.

The course of a stream through or across resistant rocks when there are less resistant rocks is often attributed to superposition. For example, in this instance it might be postulated that Big Moccasin Creek was superimposed, through some kind of unconformity, across and into the dolomite. It should be noted, however, that the general trend of the course of the river is nearly parallel to the strike trend of the limestone. This relationship is exceptionally well shown on the topographic map.

Obviously, the study of such a limited area as that included in this figure cannot tell whether or not there ever was an unconformity through which the creek could have superposed, but the fact that the limestone dips to the southeast can be noted. If the limestone updip were to be projected into space and the stream were to be mentally raised upward, the stream would occupy the limestone outcrop belt.

The preceding suppositions suggest the possibility, and at least a reasonable probability that at one time the ancestor of the present Big Moccasin Creek developed as a subsequent stream (one whose valley was developed along weak rocks) along the limestone, and that as it cut down it somehow became "caught" in the dolomite. Since that time the limestone belt has continued to be lowered by rapid groundwater solution, shifting southeastward, while the river cut its valley and enlarged its meanders in the more slowly downwasting dolomite terrain. The question as to whether it was a meandering or a normally bending stream at the time it began to carve its present valley must remain unanswered.

Case Study 3: Southeast California

Figure 11.3 shows part of the northern fringe of the Avawatz Mountains (35°36′ N, 116°22′ W), which here constitute the southwestern edge of Death Valley. Refer to the 1:62,500 Avawatz Pass, California Quadrangle for the relation between the topography of this area and that of the surrounding area. The photograph scale is approximately 1:24,000 and north lies to the right of the stereogram. As previously noted, however, for this exercise "north" is considered to lie to the top of the stereogram.

This is an extremely deceptive area, for when first studied, its landforms are readily identified and some aspects of its history are as they seem. Upon closer scrutiny, however, the landforms and their relationships, the earth materials of which they are made, and the processes which produced them begin to elude the interpreter.

First, look at some of the easily recognized features. There are several well-defined alluvial

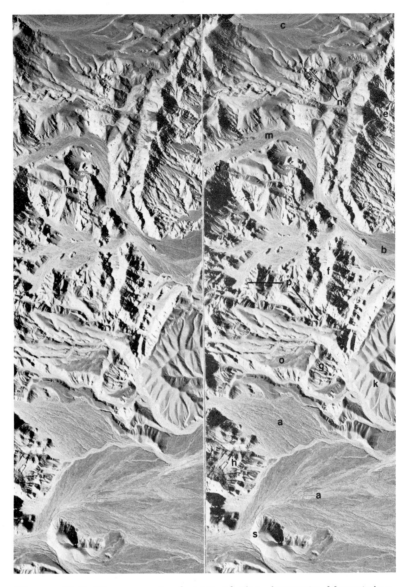

Figure 11.3. Stereogram of part of the Avawatz Mountains, southeast California. (U.S. Department of Agriculture)

surfaces, indicated as features **a, b,** and **c.** There are numerous intricately dissected bedrock areas (**d** through **h**). A relatively high area with a considerably coarser drainage texture is seen at **k.** Considering only these features and their relative positions, elevations, and the like, it might be assumed that this is an area where mountainous and hilly areas have been eroded, their debris carried by running water and deposited in adjacent lowlands and valley floors.

There are numerous conflicting observations to be made, however. It would be extremely difficult to list them all, but from north to south note the following:

1. Alluvial fan **c** lies topographically above valley floor **m**.
2. If surface **m** is the top of deposited material, it should follow that surface **c** represents deposition to a higher level.
3. Areas **n** are bedrock, and although they rise above **m**, they are generally lower than **c**.
4. These bedrocks seem to extend northward *beneath* **c**.
5. Surface **o**, apparently of alluvial material, although dissected, is higher than **m** and **a**.

Between **o** and **a** there is a remnant of an intermediate surface, which is just slightly above **a**. Bedrock occurs throughout area **p** and, as previously noted, at **g**. Perhaps the alluvium of areas **c** and **o**, and at other points which can be located, is not as thick as might at first be assumed. It may overlie a predeposition topography. It might also overlie, as a relatively thin blanket, a rather well planed-off erosional surface—a pediment. The dissected, fan-appearing slopes at **q** are immediately underlain by a sloping bedrock (pediment) surface.

Material **k** appears to be unlike either the definitely identified alluvium, or like the materials indicated to be bedrock elsewhere. It displays no complex internal structure, and its lower drainage density suggests high infiltration. It is of a uniform gray tone. The same material well might occupy the higher "alluvial" ground immediately northwest and southwest of the letter **m**, as well as hill **s**. It is suggested that this material is detrital fill derived from earlier erosion of the highlands to the west and accumulated throughout the lower area to much greater heights than present surfaces, **a**, **b**, **c**, and **o**. If this is correct, the volume of sedimentary material which had to be removed to permit the formation of the present sloping surfaces can well be imagined.

This interpreter does not know the area, and some of the above speculation, landform identification, and geomorphic interpretation may well be incorrect. That is not important; what is important is the fact, so well demonstrated in this area, that landforms not only exist to be identified, but to be *interpreted*—and sometimes interpreters are left with more questions than answers. But if they began with neither questions nor answers, there has been a gain.

Case Study 4: South-central Montana

The area in figure 11.4, situated at 46°17′ N, 109°45′ W, is not covered by large scale topographic maps. It is included on the 1:250,000 Roundup, Montana sheet NL 12-6. The contour interval of the Roundup map is 61 meters (200 feet); therefore, only a single contour line is mapped within the small area of the illustration. Merely locating the stereogram area on the map is a good exercise in map reading. The geological map shows Upper Cretaceous strata in this area.

In a way, studying figure 11.4 means peering at the birthplace of nonmilitary remote sensing. Prior to World War II, aerial photographs had been used by various individuals in several of the natural science fields. Then soon after the war, a small group of geologists, photo-intelligence specialists from the armed services, first undertook to map on stereoscopic photographs the geology of sizeable areas. They began in the folded belts of the Rockies, and the structure in figure 11.4 is an example of the type of structure and topography they encountered.

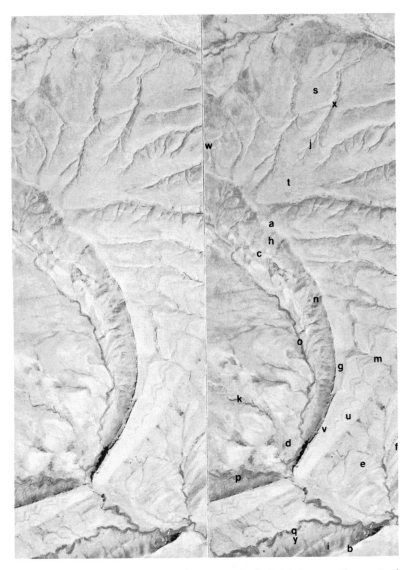

Figure 11.4. Stereogram of an eroded fold in south-central Montana (U.S. Department of Agriculture)

The interpreter should not have much trouble recognizing the structure in figure 11.4 as either the nose of a plunging anticline or part of an oval dome. The landforms are the products of differential erosion; the drainage is completely integrated and there is nothing "new" or "recent" in the appearance of the topography as a whole. The major, and many of the minor, landforms are determined by two geologic elements, structure and rock type.

Three stratigraphic units of these sedimentary rocks, stand out—literally—as having considerable resistance. These are the two ridge-formers, **a** and **b**, and the major bench-former, **c**.

Also, there are two thick less-resistant sequences, **d** and **e,** and some slightly more resistant units, such as **f,** within the nonresistant sections; just as there are thin nonresistant units, for example, **g,** associated with the ridgeformers. Most of the strata exposed in the scarp faces are nonresistant (e.g., units **h** and **i**).

Streams of several genetic types occur in this area. The most outstanding are resequent, obsequent, and subsequent. Streams flowing in the direction of the dip are resequent (**j, k,** and **m**). Those flowing in a direction opposite to the dip are obsequent (**n**). Those, both large and small, which have developed along lines or belts of nonresistant rock are subsequent; streams **o, p,** and **q** are examples. Slopes which are drained by resequent streams are resequent slopes; those drained by obsequent streams are obsequent slopes. Obviously, there can be no such thing as a subsequent slope.

Figure 11.4 shows an extremely interesting and, to the interpreter of landforms and structure, important geologic/topographic relationship. Note the gentle and long resequent slopes **s** and **t.** Across the divide, resistant unit **c** forms a prominent bench midway up the obsequent slope. The dip to the north and northeast is extremely gentle—remember that landforms and dips are appreciably exaggerated vertically. The ridge crest **a** in this sector is quite high.

In contrast, in the south the dip slopes, such as **u,** are steep; the ridge is narrow, its profile symmetrical, and its crest (**v**) low. In addition, unit **c** has no topographic expression south of the low-dip area. One must consider the possibility that the unit is lens-shaped and pinched-out, but it is not uncommon for bedrock of uniform thickness and type to be well expressed in one structural attitude and not in another.

There is a definite geologic/topographic correlation between dip magnitude and ridge height, width, and profile form in the case of differentially eroded dipping strata having different resistance. Low dip homoclinal ridges (rocks dipping in one direction) are wide, high, and assymetric; steep dip ridges are narrow, low, and symmetric.

The crest-forming bedrock at **a** is not the same one that occupies the crest along ridge sector **v.** In the latter area, strata **a** outcops along the obsequent slope. It "takes over" the crest near point **g,** where an intervening shale has permitted the downdip retreat of unit **v.** From this point north, unit **v** constitutes but one of several secondary resistant units on the resequent slope.

In an area of moderate dip, linearity immediately attracts the eye because under normal conditions moderately dipping strata are not eroded into linear forms and there is no reason for the streams to be particularly straight. For this reason, when there is a linear scarp or stream, it is a good idea to suspect control by some structural element which is steeply dipping or vertical. Such structures include faults and other fractures, although in some areas nonresistant, steeply-dipping dikes have the same topographic expression.

Study the several small linear tributaries which constitute line **w—x.** In places the bedrock appears to be offset vertically along this line, which means it could be mapped as at least a possible fault. The linear streams should be classified as subsequent. Other possible faults in this area may be noted.

Case Study 5: Central Pennsylvania

This case study deals with three Landsat images (figs. 11.5, 11.6, and 11.7) and a somewhat generalized geologic map of the same area (fig. 11.8). Labeled on figure 11.5 are parts of the

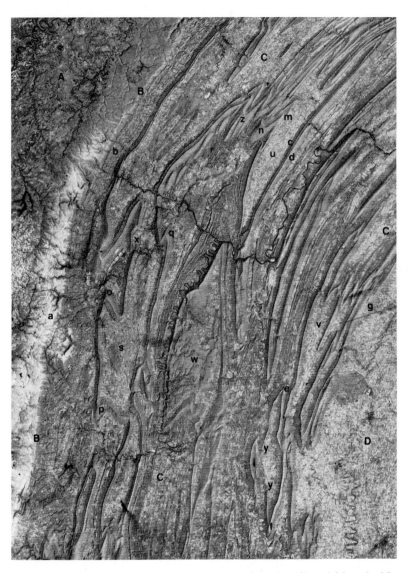

Figure 11.5. Infrared (band 7) Landsat image, dated March 23, 1973, of part of the Ridge and Valley and Allegheny Mountains geomorphic provinces of central Pennsylvania. The same area is shown in figures 11.6, 11.7, and 11.8. (EROS Data Center)

Allegheny Mountains (**A**), the Allegheny Front (**B**), and the folds (**C**) and Great Valley (**D**) of the Ridge and Valley Geomorphic Province. Figure 11.5 is an infrared (band 7) image dated March 23, 1973; figure 11.6 is infrared (band 7), May 16, 1973; and figure 11.7 is red (band 5), also dated May 16, 1973.

With the exception of highways, cities, and water bodies, virtually the entire area is covered by vegetation—some natural forest, some agricultural. Sparse woodlands will have a

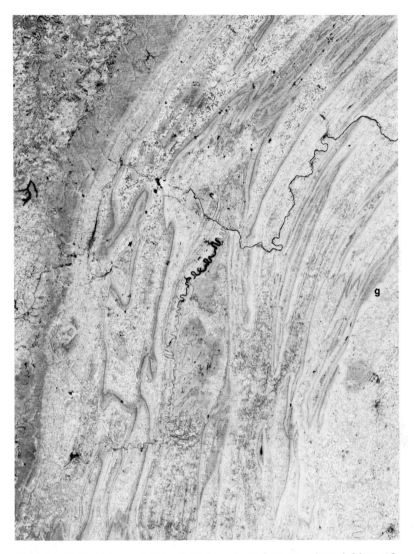

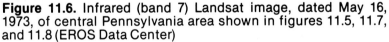

Figure 11.6. Infrared (band 7) Landsat image, dated May 16, 1973, of central Pennsylvania area shown in figures 11.5, 11.7, and 11.8 (EROS Data Center)

different reflectance than dense forests, just as pasture will not register the same as forest. Highly illuminated slopes contrast with slopes inclined away from the sun. The greatest effect of illumination contrasts occurs in the March image, figure 11.5. This is the time of the equinox, when the sun's altitude is much lower than in May. Areas of intricate dissection and appreciable, though not necessarily high, relief may not resemble areas of gentle slope and low relief, though under some conditions they might. In some seasons the effect of vegetation may obscure that of topographic difference, whereas in others the topography may dictate reflectance.

Figure 11.7. Red (band 5) Landsat image, dated May 16, 1973, of central Pennsylvania area shown in figures 11.5, 11.6, 11.8. (EROS Data Center)

The main streams are most clearly defined on figures 11.5 and 11.6, since infrared is almost completely absorbed by fresh water. Some drainage lines are discernable on band 5, figure 11.7, but it is likely that vegetation and shadows play important roles in making these "streams" visible. In figure 11.5, the snow (**a**) in the Allegheny uplands has "brought out" numerous drainage lines.

The low dip regional structures of the Alleghenies (**A**) is apparent from the slightly modified dendritic drainage pattern, a pattern in sharp contrast to the trellis pattern which,

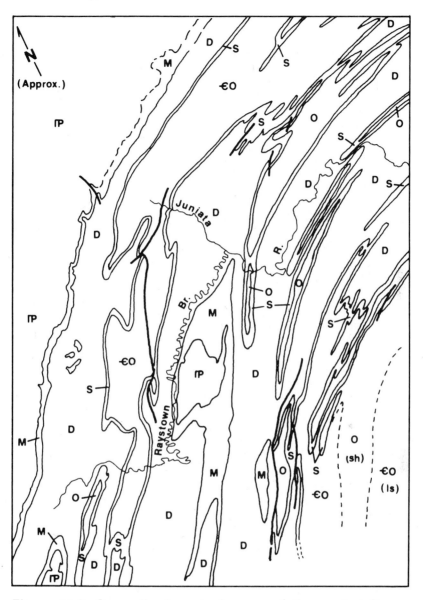

Figure 11.8. Generalized geologic map of the central Pennsylvania area shown on Landsat images, figures 11.5, 11.6, and 11.7.
CO—Cambrian-Ordovician strata
O—Ordovician strata
S—Silurian strata
D—Devonian strata
M—Mississippian strata
P—Pennsylvania strata
Faults shown by heavy lines. (After geologic map of Pennsylvania)

though its members cannot be seen, obviously must exist in the fold area (**C**). It is a fact well-known to geologists that the low dip present in the Alleghenies abruptly increases at the Allegheny Front (**B**), but, even without this knowledge, it is possible to interpret appreciable dip along **B** solely on the basis of the front's linearity. Along **b**, between the front and the first ridge, many strike lines indicate steep dip. These strike lines are most apparent on figure 11.5.

Without stereoscopic photographs, it is usually difficult to tell whether one area is appreciably higher or lower than another. The snow in the Alleghenies is a good indication of high ground, but snow cover is the type of criterion that can be utilized only if it happens to exist; it would make little sense to try to select imagery in hope of finding such an elusive aid. Of far greater value is the fact that the drainage divide between area **A** and fold belt **C** lies along or a short distance back from front **B**. This, of course, only indicates that the marginal part of **A** is topographically high; it could well be that the terrain descends toward the northwest, to a level comparable with that of the fold belt valleys and lowlands.

It is not difficult to determine which of the linear and zig-zag features in area **C** are ridges and which are valleys. For one thing, the larger valleys in belt **b** are under cultivation. For another, since the sun is late morning, and the ridge-shadow relation is known, it is, therefore, known that dark ribbon **c**, for example, is the northwest side and light ribbon **d** the southeast side of a ridge. Finally, a stream network which would form an elongate double-pointed pattern, such as is defined by sharp bends **e** and **f**, is inconceivable; therefore, the lines which form this design must be ridges.

What one might refer to as fringe belts (**g**)—the strips which extend along the lowlands immediately adjacent to the ridges—have a pronouncedly different appearance on each of the three images. Barely detectable on figure 11.5, they are very light toned on figure 11.6 (May, band 7), and dark on figure 11.7 (May, band 5). In March the natural forest is not in leaf; thus the high infrared reflectance of vegetation is not a factor. However, in mid-May, when the trees are in leaf, the contrast between the high reflectance of infrared by vegetation (figure 11.6) and its low reflectance in the red area (figure 11.7) readily explains the well-defined light- and dark-toned belts on these two images.

There is a particularly interesting study in tonal similarity versus tonal contrast in this area. Note that on the two infrared (band 7) images (figs. 11.5 and 11.6) the entire Great Valley (D) is almost uniformly light-toned. On the red (band 5) image (fig. 11.7), however, a slightly darker north-south band (bordered by lines **h-h** and **k-k**) can be seen. This suggests that there are at least two kinds of bedrock in the Great Valley, and perhaps, as a consequence, differences in topography, land use, and vegetation.

Having identified the narrow linear, curvilinear, and zigzag features as ridges, it remains to determine, if possible, their structure. Are these anticlinal, synclinal, or homoclinal ridges? At this scale, it would not be easy to distinguish among the three in the case of a linear ridge, such as at points **c** and **d**; a homoclinal ridge, a tight anticline, or a tight syncline *could* have nearly the same topographic expression. The answer lies at the points where the ridges bend, such as **e** and **f**. There are at least sixty such bends in this area. Many, such as **m**, are extremely sharp, whereas others, as, for example, **n**, are smoother and more rounded. The former appear to be synclinal noses, the latter anticlinal noses.

When all the anticlinal noses are studied carefully, it will be noted that more than half are pointed, indicating that only *some* of the anticlinal noses can be identified by form alone, and that all of the synclinal noses are suspect. Both noses **o** and **p**, however, appear to be anticlinal.

If this interpretation is correct, it follows that the dip along homoclinal ridge (hogback) **o-p** is to the west, and that the strata east of **o-p** are older and those to the west younger than the ridge-former. Nose **q** appears to be anticlinal. Therefore the dip along the ridge which extends south from **q** is to the east. North of **q** are two excellent outcrop V's (east dip). Lowland **s** is a structurally positive area. This generally anticlinal structure, complicated by secondary synclines, extends from the south border to the north border of the image area. Using the same technique as above, areas **u** and **v** are determined to be structurally positive and area **w** structurally negative.

There are several features and relationships which strongly suggest faults in this area. A homoclinal ridge is offset at point **x**. At points **y**, homoclinal ridges, instead of being connected, are terminated. The pointed fingerlike features **z**, on a large anticlinal nose, appear to be the topographic expression of fault displacements of a resistant bed.

There are many additional landforms in this area to which attention could be directed. Hopefully, those which have been discussed will prove sufficient to illustrate methods of identification and interpretation of topography and geology on extremely small-scale nonstereoscopic images. One final word: this particular area was selected because it reveals so much. It must be kept in mind that most areas are not as easy to interpret.

Figure 11.8 is a geologic map of the area. The scale is too small to show the structural and stratigraphic details; nevertheless, the resemblance between the map and the Landsat images is striking. Comparing the map with the images shows that all the ridges in the area **C** are developed on just two resistant units, Silurian and Mississippian sandstones. The lowland belts and the Great Valley are underlain by Paleozoic carbonates and shales. The Allegheny Front consists of Mississippian and basal Pennsylvanian strata; the latter also occupy the central part of large synclinal area **w**, which is bordered by inward-dipping Mississippian beds. The previously noted darker tonal belt in the Great Valley (fig. 11.7) is shown to be underlain by Ordovician shale; the surrounding lighter areas are Cambro-Ordovician limestones.

Feature by feature, compare the landforms and geology discussed above with the map. Space limitations preclude further discussion of this fascinating area.

Case Study 6: Central Pennsylvania

This case study deals with part of the area of Case Study 5 as viewed "stereoscopically" without a stereoscope. Stereoscopic aerial photograph coverage usually has about 60 percent forward overlap and 30 percent sidelap. As a result, an entire area may be viewed photographically in three dimensions. Landsat coverage, by contrast, usually overlaps and sidelaps but a few percent; therefore, most Landsat images must be studied separately as two-dimensional figures.

The normal stereoscopic model is achieved by simultaneously looking at two photographs, or images, of the same ground area, taken from two separate exposure stations. The topography of the area, vegetation, date of exposure, time of day, angle of sun, shape and position of the shadows, and all colors and tones—all are constant. The only variable is the location of the imaging device, and hence the "point of view."

A Landsat spacecraft passes over the same ground area once every eighteen days. The left hand part of figure 11.9 was imaged on March 23, 1973. Eighteen days later, April 10, the satellite was back in the same position and obtained another image. It returned to this position again on May 16, at which time the right-hand image of figure 11.9 was recorded. Both are band 5 images. During the thirty-six days between obtaining the left-hand and right-hand im-

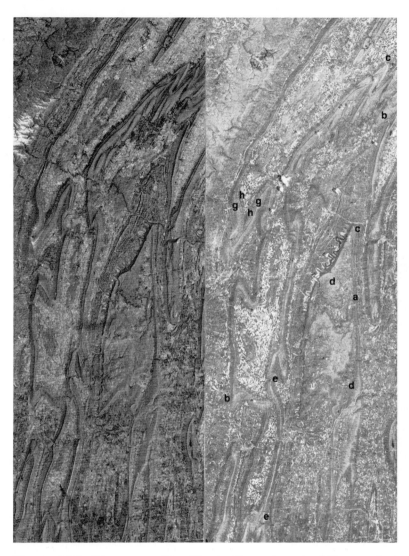

Figure 11.9. Stereogramlike illustration of part of the central Pennsylvania area shown in figures 11.5 through 11.8. Left half—March 23, 1973 Landsat band 5; right half— May 16, 1973 Landsat band 5. (EROS Data Center)

ages, the position of the apparent sun had moved appreciably, from about 0.5° N to 19° N. The altitude of the sun on the days of imaging was increased from about 43° to 59° (the latitude of this area is 40° N); the sun's azimuth along the horizon changed from 139° to 126°.

When figure 11.9 is examined, it appears to be a stereoscopic landform model. This is primarily explained by the fact that whereas in normal stereoscopic viewing the shapes and sizes of shadows, slopes, and other landform elements appear different because of different positions of the eyes "in space"; in this case, the shadows and tones actually have different shapes and sizes because they have been produced by different positions of the sun.

Vegetation changes appreciably between late March and mid-May. Unfortunately, whereas tonal values and contrast might be expected to vary consistently with variations in the sun's angle, those which reflect different conditions of vegetation do not; for example, the presence or absence of leaves depends on the season, not on slope orientation. Therefore, the three-dimensional "model" seen in figure 11.9 is not entirely trustworthy, since it is not produced solely by shadow differences. If used with caution, however, it can provide considerable information to the student of landforms.

Many of the homoclinal ridges (**a**) and anticlinal (**b**) and synclinal (**c**) noses "stand up" remarkably well on figure 11.9. When the figure is analyzed the dip direction at such points as **d** is apparent; the dip slopes actually can be seen. Also, there are several exceptionally well-defined V-shaped outcrops (**e**).

Included here is a good example of the type of deception which can be perpetuated by this kind of stereoscopic imagery. Remember that the Silurian sandstone ridges, which were identified in case study 5, are not maintained by a single resistant unit; there are two. One forms the actual crest of the ridge. Differential erosion has etched the second in such a way that it usually forms a shelf or bench extending along the obsequent ridge slope. In figure 11.9, however, there are places where resistant strata **h** appears to stand well above resistant strata **g**. Yet the topographic maps of this area show that the high ridge crest is actually developed on strata **g**, that the ridge height averages about 427 meters (1,400 feet), and that the elevation of the bench formed by strata **h** is some 152 meters (500 feet) below that of the ridge crest.

What is being seen, then, is not a stereoscopic model in the true sense. Instead it is a stereoscopiclike model which owes its existence, ultimately, to the fact that the earth's axis is tilted, and that from day to day there are changes in, among other things, shadows, tones, and vegetation. This might be termed "solar stereoscopy."

Despite its limitations, solar stereoscopy can be a great help in the study of landforms. Just being able to identify a single dip slope, anticlinal nose, or outcrop V, none of which is possible on a single Landsat image, is well worth the expenditure.

Case Study 7: Northwest Arizona

Figure 11.10, the infrared (band 7) Landsat image used in this case study, shows the border between the Basin and Range (**A**) and the Colorado Plateau (**B**) geomorphic provinces. Included is the lower (western) part of the Grand Canyon (**C**).

The objective of this case study is not to interpret the landforms of the area. Instead, it is to recognize the most important geologic features which have been mapped.

The recognition of a feature which is already known to exist in a particular place is just that—recognition; it is not interpretation. When tourists in Paris follow their city maps and catch sight of the Eifel Tower, they recognize it because they already know where it is and what it looks like. They assuredly do not interpret it! These tangential remarks are added because in the literature all too frequently the word interpretation is used when what is referred to is actually recognition.

In the study of aerial photographs and other remote sensing imagery, users may recognize a known feature, such as a previously mapped lava flow. They may identify a lava flow by its

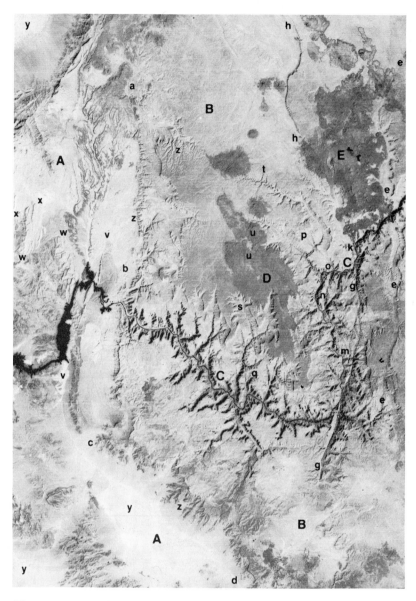

Figure 11.10. Infrared (band 7) Landsat image of the lower Grand Canyon area in northwest Arizona. (EROS Data Center)

shape, tone or color, and relationships to adjacent rock strata and structures. It is agreed that this latter type of identification does involve a rather elementary type of interpretation—an interpretation of the observable criteria which permits the recognition and identification of the landforms or the geologic feature in question.

The geology of the area in figure 11.10 is shown on the 1:2,500,000 Geologic Map of the United States (1974) and the 1:500,000 Geologic Map of Arizona (1969). The province border is Grand Wash Fault, upthrown on the east, which can be recognized extending north-south across the area through **a** and **b** to **c**, where it bends southeastward to **d**. The plateau area to the east of this line is characterized by gently dipping to horizontal upper Paleozoic strata, mostly limestones. These strata are cut by several long, steeply dipping faults, some of which are readily recognized. They generally trend north-south. Lying upon the plateau surfaces, and in places covering the fault-associated scarps, are several Tertiary and Quaternary basalts (**D** and **E**).

The easternmost fault is the Toroweap (points **e**), which is easily recognized—largely by the shadows along its west-facing scarp—extending southward from the Colorado River. Linear and curvilinear subsequent streams (**g**) clearly define the southern part of the next major fault, the Hurricane. To the north, part of Hurricane scarp (**h**) is marked by a shadow ribbon and a slight tonal break. However, that part of the fault which extends southeast from **h** to **k** is not well defined because the basalts of area **D** have flowed westward down the scarp face and have covered both the fault and the scarp. A branch of the Hurricane fault is mapped along line **m-n-o-p**. The fault line is followed by several subsequent stream segments. A third major fault can be recognized along the line **q-s-t**; it continues to the north, but cannot be identified beyond **t**. Note that at points **u** the fault appears to cut across the basalts.

In the Basin and Range segment, other faults can be recognized, one of which is marked **v-v**. Two others, **w-w** and **x-x**, striking more easterly, are north of the lake (Lake Mead). There are others.

Aside from contrasts in rock type—the Basin and Range province includes metamorphic and igneous as well as sedimentary rocks—areas **A** and **B** are also unlike as far as landform-producing processes go. In the plateau area, it can be seen that the principal activities are erosion by running water and weathering. In the Basin and Range area, deposition is widespread; large areas, such as areas **y**, have received thick deposits of detritus supplied by the dissection of the Grand Wash scarp and the several mountain and hill areas to the west.

A reminder is needed that in recognizing features on all satellite imagery such as this, scale is extremely deceptive. In many of the headwater areas and along dissected scarps (areas **z**), the stream systems have a delicate feathery appearance. The fact is that many of the tributaries are surprisingly large. For example, consider a tributary which has a total image length of 2.5 millimeters (0.1 inch). The original scale of the image is approximately 1:1,000,000; therefore the length of the tributary is about 2.57 kilometers (1.6 miles). If the channels of two adjacent tributaries are 1.27 millimeters (0.05 inch) apart on the image, their real separation on the ground is about 1.29 kilometers (0.8 mile). Although easily recognized as streams, they are not delicate feathers.

Editor's Conclusion

This chapter underscores the axiom stated in the foreword of the book that interpreters of remote sensor imagery need to be familiar with more than sensor systems and imagery. They must have in-depth training in the discipline which deals with the features studied on the surface of the earth.

Interpreters of remote sensor imagery are not going to be able to identify an esker, fault splinter, or cuspate spit on orbital or suborbital imagery unless they know what these features look like at ground level and what geomorphic agent or agents produced the landforms.

Images of landforms of one type or another are present on all aerial and space images. These features on the earth provide the stage upon which all human drama—economic, political, and social—has, does, and will continue to take place. Because of their universal presence, and their importance in agricultural, urban, industrial, and transportation site location, landforms are worth studying in detail.

It has been pointed out in this chapter that images of the earth's surface on aerial photographs and space images may be large or small scale; showing extreme detail on the one hand or general relationships on the other. Both scales have their advantages. Large scale aerial photographs make possible the study of landforms in extreme detail, as some of the case studies in this chapter illustrate. The landform characteristics and relationships of large areas, however, are best illustrated for study on space images which have been recorded by sensors carried hundreds of miles above the earth. Landform features such as shore lines, drumlin fields, karst plains, or mountain ranges, which may not be shown in their entirety on aerial photographs, are imaged in great fidelity by sensors aboard orbiting space vehicles.

Because landforms are three dimensional, they are studied best by means of stereoscopic viewing of overlapping pairs of aerial photographs. Until recently, the researcher and student were able to observe the three-dimensional image quality of the land only by means of a stereoscope and successive overlapping pairs of aerial photographs along a flight line or along adjacent flight lines. In this chapter a revolutionary new method of viewing and studying the land in the third dimension was introduced—stereoscopy from Landsat images. With this technique, extensive landform features in physiographic regions can be studied in the third dimension.

Currently, experimentation is underway concerning the combination of Landsat and radar images of the surface of the earth (plate 16). When this is done, the resulting image composite provides a unique emphasis of relief, along with an enhancement of some cultural features. In the years ahead landform analysis from sensors operating at different elevations and in different bands of the electromagnetic spectrum may make possible the interpretation of landform areas and features, as well as mantle, soil, groundwater, and mineral conditions, for solving many of the problems in agricultural and urban areas throughout the world.

12

Application of Remote Sensing in the Analysis of the Rural Cultural Landscape

James R. O'Malley
West Georgia College

Introduction

This chapter will introduce the reader to the application of remote sensing to the study of the rural cultural landscape. The rural cultural landscape is defined for the purposes of this discussion as encompassing the areas of the earth outside of urban settlements, not exclusive to, but including such aspects as settlement patterns, agricultural expressions, transportation routes, and historical remnants of culture.

Remote sensing as a tool for investigating the earth has been employed by scientists for almost two decades. This tool which evolved from aerial photograph interpretation has created an interdisciplinary link that has probed many areas of science. The rural cultural landscape, however, has escaped much of the energy and development involved in this "new science." The reader simply has to preview recent texts and symposium manuals to find a wealth of research concerning the physical landscape. With a few exceptions, such as Kohn, Steiner, Stokes, and Rehder,[1] the rural cultural landscape other than agriculture and land use has been neglected, or the works are buried in a myriad of government documents.

Such a situation is unfortunate because remote sensing brings as much potential enlightenment to the rural landscape as it has to urban and physical environments. It shares the common problem of scale with maps and field work which have been the research tools of cultural geography for decades.

The Rural Cultural Landscape

Alteration of the environment by people has been ongoing since the advent of culture. The degree and significance of this alteration is in many ways directly dependent upon the level of cultural development. The Bushman of the Kalahari make less of an impact on their environment than the mixed grain farmers of the United States Midwest. Cultural impressions nevertheless are present on the rural landscape in the form of land survey patterns, settlement patterns, field patterns, structure types and orientations, and a host of other micro cultural features.

Remote sensing provides a quick, repetitive collector which, if applied at the correct scale and sensor type, can produce valid and previously unmonitored data concerning the rural cultural environment.

The purpose of this chapter is to present the reader with an idea of the usefulness and problems of applying remote sensing techniques to the rural cultural landscape. Major themes that are included in this discussion are: (1) the advantages and limitations of remote sensing in acquiring rural cultural data at various scales, (2) sensor selection for rural cultural studies, and (3) examples of operational surveys.

Scale: Advantages and Limitations in Remote Sensing

The problem of proper scale selection has plagued users of maps and other abstractions of reality. The user of remote sensing as a tool in understanding the rural cultural landscape also finds scale an inherent problem. Platforms which vary from low altitude to high altitude aircraft and to satellites and spacecraft obviously produce images of different scale.

Scale, which is primarily a function of the altitude of a platform and of sensor configurations, influences the resolution of an image. Generally as image scale increases (large scale images are of small geographic areas) the ground resolution also increases. Conversely small scale images (large geographic areas) generally have poor ground resolution.

Small Scales

Small scale data images are produced by sensors from platforms such as Apollo, Gemini, Landsat, and others. These sensors produce scales of a macro or regional view of the rural cultural landscape. A perspective of several thousand square miles of the earth shows land survey systems and other macro cultural features which had been unobservable before remote sensing.

Figure 12.1 illustrates a cultural interface between Anglo and Mexican culture realms in the southwestern United States. Small scale images provide the users with a regional or "big picture" view which is needed to identify cultural regions and establish patterns on the landscape. Scales ranging to 1:1,000,000 and smaller are now capable of being produced by space imagery on a repetitive nature.

Figure 12.2 illustrates a small scale image of Southeast Asian agriculture which by western standards is very primitive. Repetitive imagery at this scale allows interpreters to monitor the frequency of shifting cultivation and the size of plots of a cultural activity which previously was inaccessible at such a scale.

Limitations of small scale imagery are generally manifested when an attempt to extract detailed information is made. Figure 12.2 reveals the frequency of shifting agricultural plots, as well as size of plots and density of plots; however, it cannot reveal type of crop, tools used, or number of plants being grown. These latter data units must be acquired from large scale imagery. Attempts to extract this type of data can only lead to frustration and erroneous data.

Other limitations of small scale imagery exist because, while many developed countries have access to satellite and other space imagery, many of the Third World areas which could make use of such studies do not. Such a situation is not the result of a few nations hoarding imagery, quite the contrary. It exists, instead, because of the lack of dissemination of interpreting techniques and hardware technology in many parts of the world.

Figure 12.1. Interface between the Anglo and Mexican culture realms near Imperial Valley, California. Note differences in field patterns and size, reflecting differences in the economic and agricultural systems of the respective cultures. Skylab 3, July 28, 1973; altitude 435 kilometers (270 miles). (From EROS Data Center)

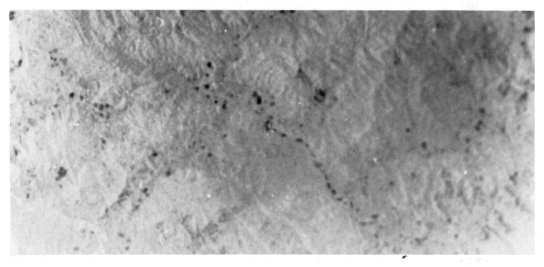

Figure 12.2. Dark signatures indicate shifting agricultural plots in Southeast Asia. Gemini IV. (From EROS Data Center)

Medium Scales

Since most photographic images produced from conventional aircraft vary in both altitude and camera focal length, medium scale images can vary in scale from 1:40,000 to 1:120,000. A perspective of several hundred square miles can be derived from remote sensor images at this scale. Images produced at a medium scale incorporate both advantages and limitations of small and large scale images. In many instances these medium scale images are a valid "trade-off" for the investigator wishing both large area coverage and high quality resolution.

Rural cultural landscapes are particularly suited to these medium level scales. Field patterns, transportation routes, and rural settlement patterns in general are depicted well at scales of 1:20,000 to 1:100,000. The investigator finds in medium scale imagery a scale that allows regional type studies with sufficient resolution for some site investigation. Figure 12.3 illustrates this point. Individual field size and orientation can be discerned while at the same time general relationships between wetland and agricultural environments can be described.

Figure 12.3. Medium scale image of agricultural and wetlands area. (From EROS Data Center)

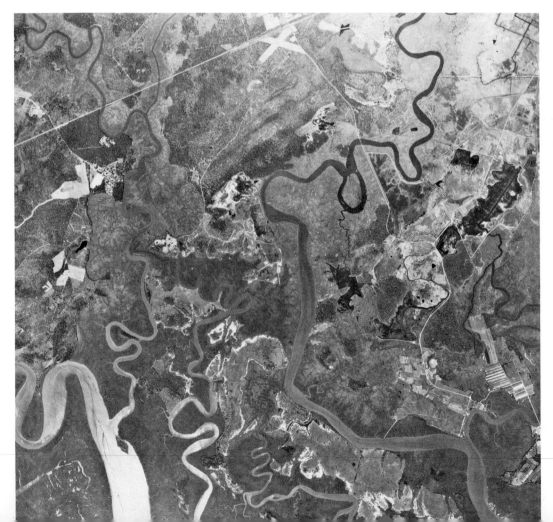

Limitations of medium scale imagery manifest themselves in two quite different ways. First, medium scale imagery may not provide the resolution for specific site examination that many rural landscapes require. In the second place, sufficient areal coverage may not be provided for complete regional examination. This latter shortcoming, however, can at least partially be overcome by forming a mosaic of several images.

Large Scales

Many aspects of the rural landscape are portrayed well by images taken from low altitudes. These images can range in scales from 1:20,000 to 1:2,000 and provide very good resolution with which to examine a site.

Rural landscape examination at these scales is usually site specific as a result of the small areal extent covered. Individual structures, their orientation, construction materials, and other specific data are only some examples of information which may be acquired from large scale imagery.

The obvious limitation of large scale remote sensing images of the rural landscape lies in the small geographic areas covered by single photographs. As with medium scale imagery, some of this problem may be overcome with mosaics of several images. Unless carefully assembled, however, these mosaics may present inherent problems of tonal and texture dissimilarities.

Scale selection is one of the most important steps that an investigator of the cultural landscape must make. Attempts to acquire specific site data from small scale imagery produces erroneous data just as would trying to determine the cultural interface shown on figure 12.1 from a large scale image.

The procedure of selecting only one image scale and then attempting to extract all the required information from it is all too often found in remote sensing interpretations. It is particularly critical when the rural landscape is involved. Rural landscape images exist in numerous scales, and only by evaluating an area at these many scales can accurate evaluation be made. Whenever possible, a multiscale sampling is a good idea, with specific goals then being achieved by utilizing the most appropriate scale for each goal. Various perspectives can be seen in figures 12.4 and 12.5.

Sensor Selection for Rural Landscapes

Selecting the type of sensor for rural landscape study is probably the most important decision an investigator has to make. Because of the great variation of energy range emitted by the sun and the numerous variations of sensors capable of capturing this energy, the correct selection is very difficult. Further discussion will be restricted to the three most common sensors: (1) photographic, (2) multispectral scanners, and (3) radar.

Photographic Sensors

By far the most common sensor type in terms of applications of studying the rural landscape is the image produced in the form of a photograph. These images are available in three basic types: (1) black and white, (2) color, (3) color infrared. These are the most frequently used and operationally sound photographic sensor systems. Some of the more outstanding advantages of photographic sensors are: (1) human familiarity with photographs which eases in-

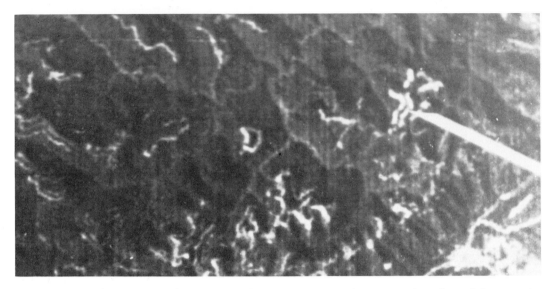

Figure 12.4. Strip mines in eastern Tennessee are shown on Landsat 1 image. Irregular lines indicated by arrow are visible from 910 kilometers (565 miles). (From EROS Data Center)

Figure 12.5. Strip mines on figure 12.4 are shown here at a much larger scale. (Courtesy of University of Tennessee, Geography Remote Sensing Project)

troduction to interpretation, (2) numerous film and filter combinations using the three basic film types, (3) relative ease of acquisition and low cost, (4) quality resolution, and (5) stereoscopic quality. Each of these advantages helps make photographic images the most commonly used by cultural researchers involved in remote sensing.

Black and White. These sensors are the oldest form of photographic image in remote sensing. Applications of these images lie primarily in inventorying rural cultural features. Structure counts, settlement patterns, and regional landscapes can be examined on images from these sensors depending upon scale. Variations in texture and tonal characteristics are of primary importance in interpretation using black and white imagery.

Color. The amount of potential cultural information available on photographs increases significantly when color is added. Additions are particularly in the areas of detecting changes in the rural landscape. In many cases these culturally induced changes will occur as color modifications before they actually have a dimensional change.

Color Infrared. These images represent a significant development in photographic sensors for rural cultural applications. Many aspects of the cultural landscape become evident when reflected infrared imagery is used. The primary advantage of near infrared imagery is the increased spectral range of this sensor. Its ability to "sense" through dust and haze and to separate numerous signatures in a particular scene enhances its uses for interpreting the cultural landscape. Thermal infrared sensors, which sense emitted energy rather than reflected wavelengths, allow additional information to be extracted from otherwise uniform landscape. The primary restrictions on color infrared and thermal infrared lie in the realm of increased cost, but the increased detection power of these sensors generally outweigh their disadvantages.

Multispectral Scanners

While not technically or physically limited to spacecraft or satellites most of the multispectral imagery that is available was derived from such platforms. The best known of the scanner systems producing imagery by this type of sensor are mounted on the Landsat platforms. These systems have the ability to record the same reflected energy levels as do black and white, color, and color infrared photographic images.

The primary advantage of space borne multispectral scanners lies in the repetitive large area coverage they provide. In most cases resolution is not of sufficient quality to conduct detailed visual investigations, but general rural cultural patterns are clearly defined.

Radar

The most significant difference between radar sensors and more conventional photographic and scanner sensors is their active energy nature. These sensors emit energy and do not depend upon reflective solar radiation. In addition, radar has considerable penetration power through both cloud and vegetative cover which adds considerable landscape detection capability. Disadvantages are manifested in two ways: (1) lack of familiar signature profiles as a result of being an active sensor and (2) generally inferior resolution.

Applications for interpreting the rural landscape with radar sensors are potentially numerous. The ability to penetrate vegetation and cloud cover could open areas of investigation which have been concealed from other sensors. Historical settlement patterns, agricultural activities, and other cultural features could be revealed if resolution limitations can be overcome.

With proper scale and sensor selection, remote sensing is and can be an important tool for rural landscape investigation. These detached data collectors cannot, however, replace conventional field work, which has been incorporated into remote sensing in the form of ground truth. Ground truth can establish consistent signatures for previously known phenonena and, in addition, reveal new information which might otherwise be missed in a review of the image.

Operational Applications of Remote Sensing
to the Rural Landscape

The rural cultural landscape is composed of phenomena of both a contemporary and historical nature. These range from remanent field patterns to recent vegetative alteration of forests on the grounds of nuclear power plants. Many of these cultural phenomena are easily observable from remote sensing platforms. In fact, many are being studied from a perspective never before possible. Figure 12.6, an early Landsat image of East Tennessee, portrays the linear cultural landscape of the area. Topographic and soil restrictions led people to occupy the valley floors. Settlement geographers have been aware of this orientation for decades; however, only after viewing these early Landsat images did the total settlement picture become apparent to many investigators of these rural landscapes.

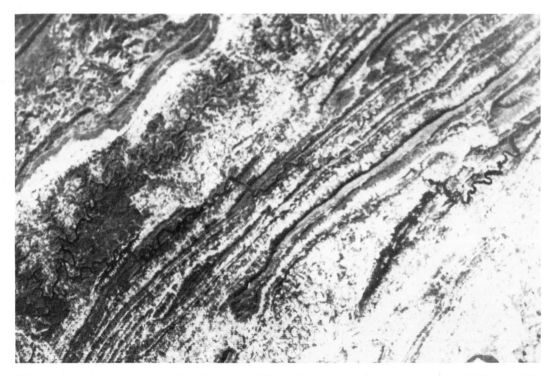

Figure 12.6. Linear cultural landscape of East Tennessee from Landsat 1. Altitude 910 kilometers (565 miles). (From EROS Data Center)

The following discussion is a series of applications of various remote sensors to specific rural landscapes. The examples should not be considered inclusive of all sensors or applications; instead, they are a representative sample of the information and data which can be extracted after minimal training in image interpretation.

Historical Application

Many historical features which represent previous cultures on the landscape have until the advent of aerial photography and remote sensing been investigated only by the most hearty fieldworkers. In many cases, surrounding vegetation and their earth-bound perspectives prevented these investigators from realizing the full significance of the landscape in question.

Early seventeenth and eighteenth century rice and indigo plantations along the southeastern Atlantic coast of the United States have been topics of much historical and cultural emphasis for years. Remote sensing allows investigators not only to locate these activities, but also to count and measure field sizes, measure irrigation systems, study associated settlement patterns and structure locations as well as to calculate distances from markets along riverways and overland transportation routes. Figure 12.7, an infrared image taken at high altitude (about 18,300 meters or 60,000 feet) of the Savannah, Georgia area depicts such information. To the north and west of the city of Savannah, remanent as well as active agricultural activity is found on an abandoned rice and indigo plantation. Numbers indicate individual phenomena as follows: (1) remanent fields, (2) active fields with grains planted as waterfowl feed, and (3) irrigation ditches of various widths and lengths.

Figure 12.8 of Darien County, Georgia, reveals still more information. A remanent rice and indigo plantation with many structures intact is located to the right of center (see circle) of the image. The physical interface between dry ground and the wetlands is worth noting on this image. These early agriculturalists took advantage of this area by locating structures on the dry areas and farming the wetter lands. The use of natural wetlands to enhance irrigation is quite evident on Figure 12.8. Under magnification the associated farm structures can be counted and identified.

From the images in figures 12.7 and 12.8 information that would take days of field study can be derived in hours. Relationships among rural cultural activities can also be seen on photographs or on other types of images. Investigators do not have to remain in the field for fear of later misinterpretation; today, remotely sensed images can provide them with continuous reinforcement of their on-site conclusions. The historical investigators have seldom had such a tool with which to work.

Survey Systems and Culture Areas

Survey and settlement patterns are two basic aspects of any rural landscape. Few, if any, methods of study can geographically portray survey systems and culture areas at more varied scales than remote sensing.

Examples of varied rural landscapes produced by different cultures can be seen on figures 12.9 and 12.10. These two Skylab 3 images illustrate the small scale perspective of regional landscapes through remote sensing. Figure 12.9 is an illustration of a rural landscape dominated by the French "long lot" system of land tenure. Survey patterns are long and narrow with frontage on waterways. This area of Louisiana, settled and surveyed initially by

Figure 12.7. Remanent plantation fields near Savannah, Georgia, shown in upper left of photograph. U-2; altitude 18,288 meters (60, 000 feet). (Courtesy of State of Georgia, Department of Natural Resources)

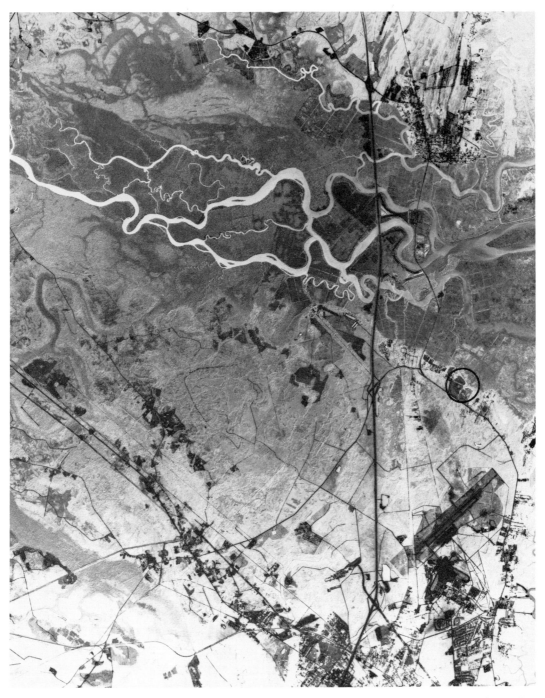

Figure 12.8. Remanent plantation fields and structures in Darien County, Georgia. (Courtesy of State of Georgia, Department of Natural Resources)

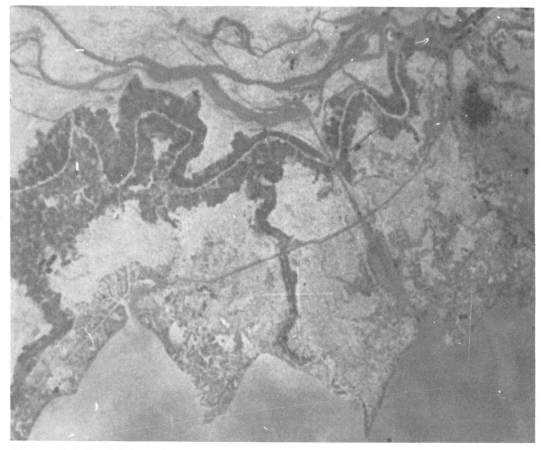

Figure 12.9. French long lot survey system borders Mississippi River near New Iberia, Louisiana. (From EROS Data Center)

French settlers, still portrays the survey system of three centuries ago. From this image information on the relationship between land holdings and waterways can be gathered.

Figure 12.10 illustrates the very different survey pattern of Iowa and Illinois. The area was surveyed and settled by the township and range system which produced the "checker-board" pattern shown. Figure 12.11 illustrates the same survey system at a much larger scale on which individual fields and structures are visible. Figure 12.12, of the same survey system, indicates how dominant a cultural survey system can be even on a heavily vegetated environment. Figure 12.6 of Eastern Tennessee shows a third rural landscape produced by a different land survey system. The metes and bounds system based on natural landscape features produces none of the geometric patterns produced by the other two. This Landsat 1 image taken at 910 kilometers (565 miles) also illustrates that broad patterns can be seen at small scales through remote sensing. This ability allows regionalization of large areas based on one of the most revealing of culture traits—survey systems.

Culture areas on the rural landscape can be formalized out of many different traits that are

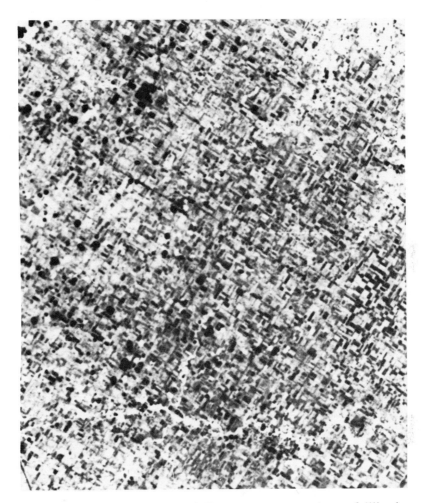

Figure 12.10. Township and Range survey system of Illinois and Iowa is depicted by checkerboard pattern even from an altitude of 435 kilometers (270 miles) from Skylab 3. (From EROS Data Center)

visible from the aerial perspective. Figures 12.1 and 12.13 illustrate the interfaces of two quite different cultures existing on the same physical environment. Figure 12.1, a Skylab image, depicts differences in field size and irrigation systems on either side of the United States and Mexican border near Imperial Valley, California. The application of remote sensing at this small scale cannot provide indepth structure counts, but it can allow the user a view of different economic and agricultural systems at work. Figure 12.13, another view from Skylab of the border between the United States and Mexico, also illustrates differences in broad cultural use of the rural landscape.

The use of remote sensors from space platforms has exposed rural landscapes in a regional perspective. While these sensors cannot give all of the data and information needed for total

Figure 12.11. Rural landscape patterns dominated by rectangular grid survey system. (Courtesy of State of Georgia, Department of Natural Resources)

Figure 12.12. Rectangular survey system patterns can be seen through the forest cover. (Courtesy of State of Georgia, Department of Natural Resources)

Figure 12.13. Culture regions similar to those in figure 12.1 can be seen on either side of the image. (From EROS Data Center)

understanding of the rural cultural landscape, they do provide a basis for more indepth study using other scales. With proper application certain cultural characteristics such as relative field size, geometric pattern, orientation, and generalized land use can be ascertained.

Sensor selection for image application in determining survey systems and defining culture areas is quite open; however, since field boundaries are unusually well defined when crop type is discernible, color infrared is perhaps the most desirable of images. Other sensors which produce conventional color do adequate jobs if field patterns and cultural interfaces are highly visible. Scale selection is the critical aspect of investigation. Too small a scale (1:1,000,000 plus) tends to obscure field patterns and orientation, while larger scales tend to reduce the area of geographic coverage too much. The general rule of thumb should then be to use as small a scale as will allow identification of the individual survey units. Of course this will vary with the topic that is under study.

Agricultural Activity

Since agricultural activity will be pursued in greater depth in another chapter this discussion will enumerate only some of the more anomalous activities depicted on the rural landscape.

Figure 12.14, a Landsat image of southwestern Georgia, is illustrative of one of the more unusual agricultural landscapes that can be detected by remote sensing. The circular signatures in the lower portion of the figure are irrigation areas in southwest Georgia. Such an occurrence would be expected on the rural landscape in a subhumid environment, but not within humid southeastern United States. The repetitive large area coverage of a region by remote sensors, therefore, can record changes in agricultural land use before they become "common knowledge." Such a regional discovery at a small scale then can be investigated by remote sensing at larger scales. Figure 12.15, a low altitude image of a portion of the same area in southwestern Georgia, reveals still additional information, such as field size. While fieldwork in the area or agricultural statistics of the county or state would indicate the presence of these irrigation activities, they would not reveal size, distribution, crop type, or other information as

Figure 12.14. Circular images are irrigation patterns seen from Landsat image at an altitude of 910 kilometers (565 miles). (From EROS Data Center)

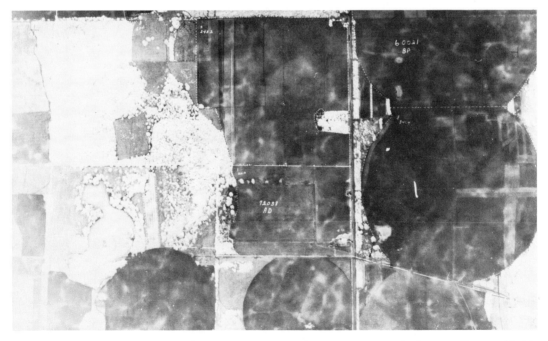

Figure 12.15. Low altitude image of circular field patterns similar to figure 12.14. (Courtesy of State of Georgia, Department of Natural Resources)

easily as the remotely sensed image. Inventorying and initial investigations of agricultural change is but one application of remote sensing to the rural landscape.

Other Rural Landscapes

Rural landscapes depict significantly different patterns when viewed through remotely sensed images. Figure 12.16, a Landsat multispectral scanner image of the coastal plain of Georgia and South Carolina, depicts such a landscape. The light tones that dominate the image are of cleared agricultural land. In contrast to these agricultural patterns, the dark circular pattern near the center of the image marks the United States' atomic energy facility south of Aiken, South Carolina. The forest vegetation which abounds within the government confines produces a different signature from that of the agricultural cultural landscape. The aerial magnitude of the facility and vegetation differences are visually apparent when viewed on this small scale image.

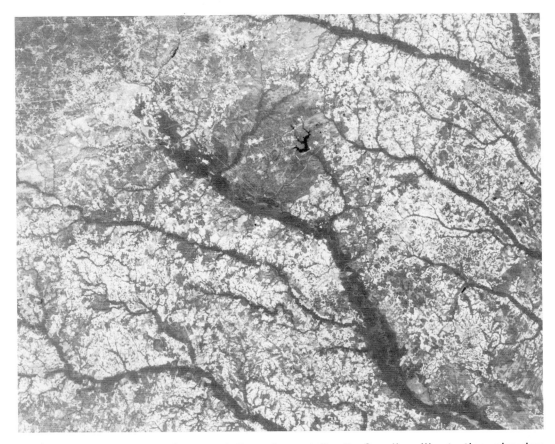

Figure 12.16. Landsat 1 image of Georgia and South Carolina illustrating circular vegetative pattern within a nuclear power facility. (Courtesy of State of Georgia, Department of Natural Resources)

Conclusion

Remote sensing, while not the total answer for investigating the rural cultural landscape, does provide an additional tool for the investigator. The two major considerations that must be taken into account in applying remote sensing are scale and sensor type. Incorrect scale selection usually leads to attempts to extract more data from an image than is available on the image. Correct sensor selection is equally important in that some components of the rural landscape are well enhanced on an image of one sensor type and not so well on another. When correct scale and sensor selection is made, most aspects of the rural landscape are captured. Consequently, researchers of the rural cultural environment, ranging from historical landscapes to contemporary landscape alterations, should make greater use of this new tool.

NOTES

1. C. F. Kohn, "The Use of Aerial Photographs in the Analysis of Rural Settlements," *Photogrammetric Engineering* 17 (1951):759-71; D. Steiner, "Use of Air Photographs for Interpreting and Mapping Rural Land Use in the United States," *Photogrammetrica* 20 (1965):65-80; G. A. Stokes, "The Aerial Photograph: A Key to the Cultural Landscape," *Journal of Geography* 49 (1950):32-40; J. B. Rehder, "Geographic Applications of ERTS-1 Data to Landscape Change," *Symposium on Significant Results Obtained from ERTS-1,* vol. 1, section A, 1973, pp. 955-65.

13

Application of Remote Sensing in Agricultural Analysis

Paul M. Seevers and Rex M. Peterson
University of Nebraska-Lincoln

Introduction

Agriculture is probably the most global of all human economic activity. Agriculture is practiced in every physical environment where it is possible to raise crops or domestic animals. Agriculture includes animal husbandry and raising of crops on subsistence plots, haciendas, plantations, state farms, agricultural communes, corporate farms, family farms, and other types of land divisions. Farms range from those that produce only one crop to those that produce a variety of crops and animal products. In some parts of the world land holdings are so small and application of labor so intensive that a farmer works less than one acre of land, in contrast to other areas where one man farms hundreds of acres. (Compare figure 13.1 in a tropical moist climate of Southeast Asia with figure 13.2 in Nebraska.) There are almost numberless variations in agriculture because of different blends of physical environment, cultures, economic systems, and individual practices in farming.

The diversity of agriculture results in a diversity of needs for information, some of which can be supplied by remote sensing. It is logical to look to remote sensing as a tool in agriculture because the technique has the capability of covering large areas rapidly, cheaply, and sequentially.

Remote sensing of agriculture means backing off and looking at fields from a remote perspective in contrast to close-up conventional viewing. All too often it is expected that many or all of the answers can be provided by study at a single scale, such as from Landsat, but this is normally not the case. There is information to be derived from each different scale of imagery, and each scale can be expected to provide data which will be of benefit to studies at other scales.

The following is a sampling of agricultural information which can be provided by remote sensing, or is expected to be provided in the near future:

Inventories of crop types and acreages in various crops
Crop yield estimates for major crops such as wheat
Soil associations
Inventories of land irrigated by center pivot systems

Figure 13.1. Infrared photograph of western Nebraska from 19,800 meters (65,000 feet). Black circles are 0.8 kilometer (0.5 mile) diameter fields irrigated by center pivot systems. Nonfield areas represent two types of grasslands: the upper are on sandy soils and the lower on more clayey soils dissected by erosion. (NASA photograph)

Figure 13.2. Medium altitude photograph of an area in Thailand. Note rows in fields, individual trees, and garden plots. (Department of Defense photograph)

Inventories of irrigated land

Presence, severity, and extent of certain diseases or insect infestation

Evenness of applications of fertilizer and irrigation water

The information needs for the various aspects of the agriculture industry are so numerous that they cannot all be discussed here. This chapter, therefore, will be directed toward generalized applications of remote sensing to agriculture. Emphasis will be on sensors that have proven merit, rather than those which are still experimental.

False Color Infrared Film

Remote sensing data is principally a remotely recorded measure of energy reflected from the surfaces of materials. The more complex the surfaces are, the more complex the recorded energy patterns. Certain portions of the electromagnetic (EM) spectrum are more valuable than others for obtaining data about the agricultural components of our environment. It appears that the most significant segment of the EM spectrum for agriculture is the near infrared portion from 0.8 to 1.1 micrometers.

Near infrared energy can be detected by scanning systems, as on Landsat (bands 6 and 7), or films can be sensitized to that wavelength. Although black and white infrared film is available, color infrared photography has the most uses in terms of inventory and analysis of agricultural lands; therefore, attention will be directed to color infrared film (also called false color infrared film).

Color infrared film utilizes a combination of an emulsion layer sensitive to visible green, a second layer sensitive to visible red and a third film layer sensitive to near infrared (in the 0.75 to 0.9 micrometer region). Because the near infrared portion of the spectrum is affected very little by atmospheric haze, the resulting photographs have a surprising clarity of detail compared with conventional photography. In addition to the quality of detail, the combination of visible and infrared sensitivity results in color differentiations which make some surface features much more apparent.

Physiological Basis of Infrared Reflection

It is important to understand that the relationship of vegetation to infrared wavelengths reflected from it has a physiological basis. Although the evidence substantiating the theory of infrared reflectance from plants is largely circumstantial, it is generally accepted that infrared reflectance relates to the number of interfaces between cell walls and intercellular spaces which exist in the leaves of plants. Most often this is related to the spongy mesophyll (soft tissue) layer of cells in leaves, where the irregular cell shapes usually result in the greatest number of the cell wall-intercellular space interfaces. Presence and maintenance of critical angle reflectance interfaces is considered to be the principal reason for infrared reflectance. If the critical angle of interfaces is changed, as by stress from moisture deficiency, for example, the amount of infrared reflection is changed.

Factors Affecting Crop Signature

It has been demonstrated that the intensity of near-infrared measured from a given area can be related to several aspects of the vegetation present, such as species, plant density, and stress. Many species can be differentiated by the color differences on the imagery, that is, they have a unique appearance or signature. Certain kinds of stresses, such as disease and moisture, can be detected by relative degree of loss of infrared reflectance—in this case the crop signature is changed by the stress in plants. Relative vegetative density (number of plants per unit area) also can be estimated. The ability to distinguish the above aspects of vegetation by detecting relative amounts of near-infrared radiation is a key factor in the application of remote sensing in agriculture. This means that films and scanners sensitive to near-infrared radiation are key remote sensors in agriculture.

An important factor in maximizing the capability of color infrared films to differentiate species is the fact that most cultivated crops are essentially "pure stands" of an individual species and variety. Depending on the scale and resolution of the film being used, it is possible in many cases to identify the crops present with a high degree of accuracy. Crop identification, however, is not always as easy as it might seem because plant density and health or vigor of the vegetation also affect the level of infrared reflectance recorded by film. Within the area shown on one high-altitude photograph (fig. 13.1), there are usually wide ranges in plant populations, nutrient applications, and moisture availability within any single crop. Management practices (meaning the way fields and crops are planted, fertilized, and tilled) are choices of individual farmers. Management practices on different farms, therefore, must be added to differences in the natural environment as variables that alter the appearance of a crop on color infrared photos.

Unirrigated crops, dependent as they are on natural moisture, are usually more dependent on water than on any other factor which affects the amount of reflected infrared. Plant density is governed by the anticipated availability of water. For example, many irrigated fields are planted with twice as many plants as in unirrigated fields. Compounds of nitrogen, phosphorous, and potassium, along with smaller amounts of calcium, magnesium, sulfur, iron, boron, chlorine, copper, zinc, molybdenum, and silicon are essential for the growth of plants. These compounds are called plant nutrients. The amount of nitrogen fertilizers applied by farmers follows the same pattern as plant density in that the greater the anticipated water supply, the higher the nitrogen application. Other plant nutrients are often applied along with nitrogen but are much less significant in determining infrared response. How much good is done by an application of fertilizer nutrients depends on moisture availability. If the combination of nutrient level and moisture availability is less than optimum, the plants come under stress conditions. This combination can be infinitely variable not only between fields of the same crop but within the same field of a specific crop. If the variability within the one crop in one field becomes great enough, it cannot be recognized as that crop by an infrared signature alone.

Moisture is not the only form of stress which can affect a crop signature. Insects and disease also produce stresses which affect the physical structure or metabolic processes of the plant and will determine to a large extent the degree to which the crop signature changes. For example, insects which devour parts of plants but do not greatly affect the plant's metabolism have to severely infest a field before damage is discerned by the appropriate level of remote sensing. On the other hand, a lesser infestation by an insect or organism which affects a high percentage of the total plant's metabolism could well change the crop signature drastically.

Stage of growth of crops is another variable to be noted. Planting dates of a specific crop will vary as much as two weeks or more within a specific area. This results in variable ground cover in the early stages of growth and differing maturity dates as harvest time approaches. In crops where the flowering stage results in a significant signature change, the variation in flowering dates adds to signature variation in the crop.

The greater the degree of uniformity with which a cropped area is managed, the higher the level of accuracy with which the area can be inventoried or monitored by remote sensing. Corporate holdings, for example, are much more uniformly managed than a series of privately owned properties. Irrigation projects have relatively uniform water availability resulting in more uniform appearance of crops on remotely sensed images. Areas of the United States which have annual precipitation rates that normally provide adequate moisture for crops will show fairly uniform individual crop signatures. An implication of this is that a familiarity with the area being dealt with will greatly improve the probability of success of inventorying or monitoring at the level of accuracy desired.

External Influences on Crop Signatures

Still other factors influence the appearance of a field on color infrared film, but these variables are external. They include the time of day when the photograph was taken (which affects sun angle and amount of light), the angle between the camera lens and the crop, and exposure and filters. Variations caused by these factors can be minimized by planning successive flights for the same time of day along the same flight lines, by using the same camera filters, and by keeping exposure as close to constant as light conditions permit.

There have been investigations into the use of different filters and combinations of filters in the use of color infrared photography for agricultural interpretations. The general results have been that certain color correction filters added to the basic minus-blue filter result in enhancement of specific color tones, but there has been no indication that these enhancements aid in the agricultural interpretations. All of the possibilities, however, have not been explored, indicating a need for further evaluation in the area of filter enhancement. For those wishing to experiment with color infrared film in their 35 mm cameras, it is essential to use filters that exclude blue light.

Crop Monitoring with 35 mm Camera

There has been a recent and rapidly growing interest in the use of color infrared film in 35 mm—and 70 mm—hand-held cameras on low altitude flights to monitor cropping problems. These problems include stresses from less than optimum moisture, various insect infestations, diseases, nutrient deficiencies and other cropping variabilities. Color infrared photography appears to be the most sensitive means for monitoring cropland problems; however, little has been done to evaluate the many specific problems encountered. It has been demonstrated that many of the problems can be recorded by color infrared film, but there are no definitive studies regarding timing of flights, severity of problems detectable, and if they can be detected prior to becoming visible to the human eye. At the present time, it can only be said that the technique is able to detect problem areas in fields, but will not specifically identify the problem. It is mandatory that the problem area be investigated on the ground to determine exactly what the problem is.

There are a growing number of farmers who photograph their fields from an altitude of about 150 to 300 meters (500 to 1,000 feet) with color infrared film in 35 mm cameras, then examine the slides for problem areas. Often the problem area can be related to something correctable, such as deficient sprinkler heads on an irrigation system, or a portion of a field that was not fertilized. Low-level color infrared photography appears to be a monitoring technique which has considerable promise.

Summary of Color Infrared

From the above discussion of color infrared film, it can be seen that many factors influence the appearance of a crop on the film. These factors are: (1) moisture; (2) the amount of nitrogen and other plant nutrients, the effectiveness of which depends on moisture availability; (3) plant density; (4) stage of growth; (5) infestation by insects or disease; (6) pure or variable stands (is the crop a pure stand or a mixture of crop, weeds and "volunteer" plants of other crops?); (7) evenness of application of moisture, fertilizer, insecticides, and herbicides; and (8) management practices, which result in many combinations of the first seven factors, depending on how each farmer plants and nurses his crops, waters them, feeds them, and protects them from weeds, insects, and disease.

Landsat Data

With the launch of Landsat 1, scanner-derived data became available to the general remote sensing user audience. Since the synoptic coverage provided by scanners on artificial satellite platforms is particularly well suited to the large area coverage required to deal with agricultural lands, much emphasis has been placed on evaluation of satellite data for agriculture. Landsat data has a close relationship to color infrared photography because it deals with nearly the same spectral bands. The difference is that in Landsat the data is electronically derived by a scanner and is recorded in separate bands. This allows for convenient examination of the bands individually as well as in combination. A combination can be generated which results in a false color image analogous to false color infrared photography. Picturelike images derived from Landsat can be interpreted in much the same manner as color infrared photography, except that the resolution of Landsat is at least one order of magnitude less. Compare figure 13.1 (a photograph from 18,288 meters, or 60,000 feet) with figure 13.3 (a Landsat image from 917 kilometers, or 570 miles). Figure 13.2 is typical of resolution attained by an aerial camera several thousand feet high. Note the individual trees and rows in fields.

The decreased resolution of satellite data means that the amount of detail which can be interpreted is much less. Even though the individual data element (picture element or pixel) represents approximately 1.1 acres of ground area, the minimum area resolved is usually 5 to 10 acres. This means that in an area such as Southeast Asia (fig. 13.2) smaller fields would not appear as separate areas on Landsat imagery.

The ability to resolve a ground feature on Landsat imagery is very much dependent on the contrast between the feature itself and its surroundings—in other words, its signature. Nominal resolution for Landsat data is considered to be approximately 76 meters (250 feet); however, many gravel roads, approximately 15 meters (50 feet) in width can be clearly seen on Landsat images because of the extreme tonal contrast (signature) between the road and the vegetation adjacent to it. In many cases, therefore, contrast is a more limiting factor in resolution than picture element size.

Figure 13.3. Central Nebraska as seen in near infrared by Landsat. Elongate dark area in center of image is a river valley with subirrigated croplands. Circles are center pivot irrigation systems 0.8 kilometer (0.5 mile) in diameter. Areas without field patterns are grazing lands. (Landsat image no. 1726-16420-5, 19 July 1974. EROS Data Center)

Use of Landsat in Agriculture

Use of Landsat data for agricultural interpretation centers primarily around the visible red band (band 5), the red and very near infrared band (band 6) and the near infrared band (band 7). The significance of band 5 is that much of the visible red energy is absorbed by plants rather than being reflected. This results in areas of actively photosynthesizing vegetation reflecting very little visible red energy, making these areas appear dark on images. Once the vegetative activity shows zero reflectance, increases in plant density or activity cannot be measured. This results in an inability to measure plant density above a certain level, as the scanning instrument is incapable of additional response. Bands 6 and 7 show the typical infrared response, in that bright areas have more healthy and active vegetation and a greater density of plants. The result is a brighter appearance on black and white infrared images in those areas where more infrared is reflected from plants.

All of the factors which complicate the evaluation of color infrared photography can also complicate the interpretation of the bands of the Landsat data. Environmental and management variations will affect the radiance levels of the cropland vegetation. Stresses, nutrient or moisture deficiencies, and relative decreases in numbers of plants per acre will cause increased reflectance in band 5. The same factors will cause decreased reflectance in bands 6 and 7.

At stages of growth where a significant amount of soil shows through the plants, the moisture contents and physical makeup of the soil also will affect the radiance values measured by the scanner aboard Landsat. The same soil with greater surface moisture will usually cause a decrease in reflectance for all bands. Soils with greater organic matter content also will show a decrease in reflectance for all bands. Dry sands, and those with essentially no organic matter content, will show maximum reflectance in all bands of Landsat imagery.

Because a single Landsat image covers large areas, (approximately 33,670 square kilometers or 13,000 square miles) the view will include ranges of soil and moisture variability. Changes in planting dates with change in latitude will appear, even on one Landsat frame. General management changes due to variability in other environmental factors will be expressed over the larger areas. What all of this usually means is that the signature for a specific crop does not remain the same over the entire area of a single Landsat image. The signature of a specific crop, therefore, must be determined for an area with similar environmental and management characteristics. That signature is then used to inventory or monitor only for that area. This area is usually only a fraction of one Landsat frame.

Landsat data is acquired on a routine basis and reasonably frequently for any given area. This provides the advantage of having multitemporal (several calendar dates) data available to take advantage of the crop calendar in evaluating agricultural areas. Difficulty arises, however, in superimposing images acquired at different times and in being able to determine the exact land areas to superimpose for comparison. Once this problem has been resolved, the multitemporal approach would seem to hold the most promise for inventorying agricultural lands.

Landsat Digital Tapes

Landsat data are available in two forms, a photographlike product and digital tapes. For a quick broad overview of agricultural lands, the image product is usually adequate. Quantitative evaluation of photographlike products by densitometer or density slicing usually has not been satisfactory. Comparisons within a single frame can often be made, but comparison between frames usually results in confusing data. Digital tape data are available, but require some form of computer to manipulate the data. Cost of acquiring the data for an image frame is increased by a factor of five over the photographlike product.

Analysis of digital tape data usually requires a rather sophisticated knowledge of the data gathering system and the relationship of the data to the ground scenes imaged. Recent development of interactive computer display systems, where the operator can interact with or modify the computer display of the data on a video output, has resulted in more rapid and complex analysis of Landsat data. The fact that digital data can be analyzed on a picture element basis results in a much more detailed examination of the data. The confounding factors mentioned previously are still present, however. The interactive systems can make some attempts to compensate for them, provided the operator has a high level of familiarity with the general conditions of the scene he is analyzing. Individual field signatures (training sets), which are used to program the computer to recognize crops, apply only over the area where the crop signature is relatively constant. This area is generally a small portion of a total Landsat frame, thus making it difficult to automatically analyze total frame areas at a high level of accuracy.

Prediction of Crop Yields from Landsat Data

Use of Landsat imagery for prediction of crop yield was evaluated in a study of wheat production in several counties of southwestern Kansas. Actual measured yield was compared to predicted yields of the Agricultural Reporting Service and to yield estimates utilizing Landsat data. The Landsat derived estimate was closer to the actual measured yield. This was due primarily to the fact that acreage estimates could be more accurately made from the satellite data than from a statistical sample of landowner response to a questionnaire about crop yields. Data on current weather and average crop yields for the area, however, were obtained by ground truth.

The success of the initial project led to an attempt to expand the technique to large areas (LACIE-Large Area Crop Inventory Experiment). Preliminary indications are that success will not be achieved as rapidly and as easily as in the pilot project.

Crop Calendars

A device for maximizing the effectiveness of color infrared photography for inventorying and monitoring croplands in areas of diverse cropping is the crop calendar. The crop calendar is a listing, usually by areas of significant difference in latitude, longitude, rainfall, average temperature, and other environmental factors, which indicates the range of dates for planting and harvesting of the crops which are grown in that area. Table 13.1 is a simplified crop calendar for eastern Nebraska. By using the information in a crop calendar, an overflight can be scheduled to maximize the information gained from the aerial photography. For example, selecting a flight time when fields of small grains (wheat and oats) are the only fields which do not have growing vegetation would provide photos that show small grains with a rather unique signature if more than one flight can be obtained relative to the crop calendar. Remote sensing can provide data where high levels of positive identification can be made. Obviously the cost relative to accuracy of data provided will be a judgement made in the number and timing of data acquisitions.

In eastern Nebraska two overflights of an area are necessary to distinguish the crops listed in table 13.1. Optimum time for overflights are late May and late August in this area. Iden-

TABLE 13.1

Crop Calendar for Eastern Nebraska

Crop	Approximate Planting Time	Approximate Harvesting Time
Corn	May 3-May 22	October 8-November 3
Winter wheat	September 12-September 27	July 2-July 13
Oats	March 31-April 15	July 12-July 24
Soybeans	May 11-May 29	October 1-October 18
Alfalfa	Spring: Fall:	First cutting: May 30-June 11
	April 10-May 1 August 21-September 8	Second cutting: July 5-18
		Third cutting: August 12-23
Pasture	Continuous coverage of vegetation	
Annual forage	April 18-May 9	August 20-September 9

Source: "Agroclimatic Calendar for Nebraska," University of Nebraska College of Agriculture and Home Economics, ed. R.E. Neil, D. SB 498, 1968.

tifications are made by a combination of color hues and the crop calendar. For example, a field that has a certain magenta color in May but is blue-gray in August is oats or winter wheat. In May the lush green vegetation produces a magenta color on the color infrared film—the bare plowed ground in the same field in August produces a blue-gray color.

Conclusion

As with the interpretation of remote sensing data for any other field of interest, the monitoring or inventorying of agricultural lands requires selection of a scale suited to resolving the features of interest. In the case of agriculture, however, it is sometimes necessary to sacrifice detail for the ability to monitor large areas quickly. Therefore, careful consideration must be given to the cost, time frame, and accuracy level which is acceptable.

For vegetative interpretation, the infrared aspect of plants is a highly significant factor. It would appear that infrared reflectance is the most sensitive means of monitoring what is happening in vegetation. The level of infrared which is measured from plants is directly related to species, vigor of growth, and plants per unit area. Theoretically these infrared relationships allow for identification of crops, evaluation of their relative state of growth, and the level of production which may be present on a given area. These relationships will also tend to confuse one another in the analysis procedure. Therefore one must utilize as much experience and secondary information as possible in the interpretation process. The fact that cropland usually represents pure stands of vegetation, which, in the case of irrigation, are relatively uniformly treated with regard to nutrient applications and water, makes infrared particularly suitable to dealing with agricultural lands.

Color infrared film provides a means of recording infrared levels from vegetation (in combination with green and red of the visible portion of the spectrum) for almost any scale of photography. Utilization of multiple-date flights in conjunction with the crop calendar allows for relatively accurate inventory of croplands.

The Landsat satellites have provided a new dimension to remote sensing data for agriculture. Synoptic coverage allows for viewing large areas and to make some qualitative judgements about the state of agriculture of the area. Individual bands of the spectrum provided by the scanner-derived data permit examinations of these bands for specific contributions to interpretive techniques. Satellite data suffer from an inability to resolve some of the ground features necessary for interpretation. Combining bands of satellite data in the proper manner results in false color images which can be interpreted in much the same manner as color infrared photographs.

Interpretation of satellite data suffer from the same confounding variables as color infrared film. Variability due to management, moisture availability, and various stresses result in variability of a crop signature in any one frame. This prevents simple training set programming for automated interpretation of entire frames.

Repetitive coverage of specific areas by the satellites allows use of multitemporal interpretation techniques and the crop calendar. Interactive computer display systems are the most promising manipulative tool for computerized analysis of satellite imagery.

In using remote sensing for agricultural analysis it is necessary to understand the limitations of the available data, recognize the problems posed by environment and people on the data, and not expect too much data. Remote sensing is a tool which can and will continue to provide agricultural data.

14

Application of Remote Sensing in Forestry

William H. Hilborn
University of New Brunswick

Introduction

The forest resources of the world are widely distributed and diversified in response to variations in climate and soil. Forests comprise a major component of the vast biological system which is characteristic of much of the land surface of the earth and, as a consequence, they influence not only that system in a major way, but they affect climate, soil, and water which comprise the nonliving components of the total environment.

The term forest resources is used deliberately to imply something more than just groups of trees. Early history indicates that man was most often interested in the forest as a home for game animals and hunting. Modern forestry is deeply concerned with the forest because of its influence on water supply and local climate, its recreational value, its habitat for wildlife, as well as its value as a renewable resource for building materials, pulp and paper, fuel, and other forest products. Moreover, the term "forest" in some parts of the world includes plant communities other than trees as we consider them in North America. For example, such things as bamboos and palms, which are botanically quite different from our familiar conifer and broadleaf trees, are classified as forests in some areas.

Management of the forest resources of the world varies widely in degree of sophistication throughout the world. Also, use or exploitation of the forest resource is quite variable depending upon the need for or profitability of forest products in local or world commerce. World demand for forest resources is such that areas of abundant, cheap timber, however remote, are being examined, and traditional, high yield areas closer to densely populated markets are coming under intensive management for multiple use.

Under such circumstances, forestry workers for many years have experimented with and adapted a variety of remote sensing techniques for inventorying and monitoring forests and the condition of forest stands. A number of foresters have specialized in the development of remote sensing to the extent that they have made significant contributions to its technology entirely beyond the needs of forestry. Certainly the broad nature of forestry as an environmental science has led foresters into experiments with a wide variety of remote sensing equipment and into the development of such equipment which has been designed specifically for forestry purposes.

271

Inventory—The First Task

As with any resource, one of the first tasks, prior to management and use, is an inventory of the resource to provide information about such things as size, distribution, quantity, and quality. If the resource is changing over time because of growth, depletion, quality variation, or in some other respect, the inventory may become not only a first task, but a periodic task. Inventories add nothing to the value of the resource, and the cost of inventories, which may be considerable, must be regarded as a prepayment for efficiency in management or development for the future. Such inventories, therefore, must be efficient, and the data collected must be comprehensive and up-to-date. Remote sensing was recognized as an important technique in forest inventory almost as soon as the technology was developed in the early 1960s.

Before remote sensing was developed, methods of forest inventory involved measurement and mapping of the forest along narrow strips spaced at uniform intervals by parties of two or more men on foot. The width of the strip might be "half-a-chain" (about 10 meters) and the spacing ten chains (about 200 meters) giving a 5 percent sample on small areas of one or two square kilometers. On larger areas, the proportion of a sample was reduced, often to a fraction of 1 percent. The forest was mapped only along the strips, the areas between were later filled in by interpolation on the map manuscript. This method was perhaps adequate during the initial exploitation of the extensive virgin forests of North America, but foresters were quick to see the advantages of the overall view provided by aerial photography.

War has always been a great spur to technology, and World War I provided, for the first time, all the resources necessary for remote sensing of the forest in the form of photographic surveys from heavier-than-air craft. In 1919, there were surplus military aerial cameras, aircraft, and demobilized pilots available. R. Thelen, of the U.S. Forest Service, published an article in 1919 outlining the use of aerial mapping for resource studies, and in that year the first known forest survey using aerial cameras was carried out. The project was organized and led by Captain David Owen, a pilot from Anapolis Royal, Nova Scotia, to survey the timber holdings of the Southern Labrador Pulp and Paper Company of Boston, Massachusetts. The expedition sailed for Labrador from Nova Scotia on July 18, 1919, returning the end of August with 13,000 photographs taken at altitudes from 610 to 2,743 meters (2,000 to 9,000 feet) using two Curtiss JN4 aircraft.

Within the next few years, airphoto forest surveys were undertaken in the United States, Canada, Burma, and elsewhere. By 1927, the first of three basic forest inventory techniques using remote sensing in the form of air photos had been introduced.

Inventory Techniques with Air Photos

Forest inventory techniques using air photographs fall into three basic classes: combined air and ground surveys, aerial surveys using stand volume tables, and aerial surveys using tree volume tables. Direct, subjective estimations of stand volumes might be considered as a fourth class, but this technique requires such an intimate knowledge of ground conditions gained by long experience in a limited area and is so subjective that it is not a technique with general applicability.

Forest inventories involve the determination of the wood volume on an area basis—cubic meters per hectare of forest, for example. (One hectare equals 10,000 square meters or 2.471 acres.) Such quantitative information cannot be measured directly. In ground surveys, it must be compiled from individual tree measurements, usually the diameter at breast height (dbh) of the trunk or bole and the total height of each tree. Breast height is standardized at 140 centimeters (4.5 feet) to reduce errors resulting from the taper of the tree bole. The wood volume of individual specimens of the same species with the same height and dbh may vary considerably because of variations in taper, and usually tables giving mean volumes for various dbh's and heights are used. Within localized areas, dbh alone may be used in a volume table because height is often correlated to dbh where the trees are growing under similar conditions.

Combined Air and Ground Surveys

The kind of measurements mentioned above cannot be made on aerial photographs because, although tree heights can usually be measured efficiently on aerial photos,[2] dbh cannot. As a result, the earliest forest inventory methods using air photographs did not use photographic measurements directly in the determination of volume. Rather, the photographs were used to subdivide and classify the forest into more or less homogeneous stands as a basis for subsequent ground sampling and for area determination of the forest classes recognized. This kind of classification or stratification can improve the efficiency of ground sampling enough to pay for the cost of the aerial photography and interpretation, especially where large areas and difficult ground access are involved. The method is still in general use because of the detailed information that can be obtained.

Qualitative and quantitative forest classes can be recognized on air photographs, and usually a combination is used, including descriptive classes of cover type as well as size and density classes. The number of classes in each category and the make-up of the categories may be quite varied depending upon the purpose of the survey, the allowable cost per unit area, the scale, quality, and type of photographs, the nature of the forest, and the skill of the interpreter. Great care is taken in the planning stages to match the level of sophistication of the various elements to provide reliable information suitable for the purpose of the inventory. A very different classification is used for a large reconnaissance survey of tropical forests than for a management survey in the relatively simple boreal forest.

Cover type. Cover type is usually based on species and species groups that provide useful descriptive information for a particular inventory. It could be a simple classification by proportion of conifers (softwoods) compared to broadleaved deciduous species (hardwoods), such as the following:

Symbol	*Composition*
S	More than 70 percent softwood by volume
M	Mixed—30 percent to 70 percent softwood by volume
H	Less than 30 percent softwood by volume

More elaborate cover type classifications based on detailed species analysis or forest site quality may also be used.[3] As an alternative, or in addition to species and species groups, the concept of utilization may be introduced into cover type classification. For example, stands of a northern commercial forest might be classified as sawtimber, poles, saplings, or reproduction.

Forest type differences can be seen on nearly every aerial photograph of a forested area. Figure 14.1 illustrates several different forest cover types based on species composition, mean

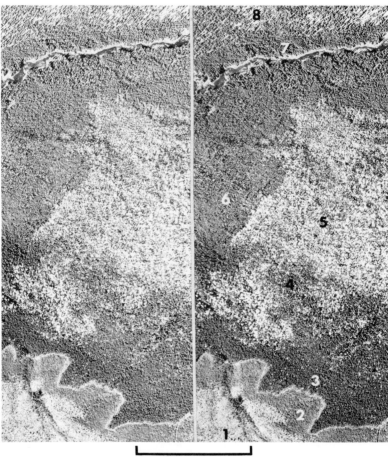

500 meters

Figure 14.1. Eight cover types are distinguishable on this stereo pair: (1) seeded in after fire, mostly deciduous brush; (2) seeded in from no. 3 after same fire, mostly conifers; (3) dense, medium size conifers; (4) mixed, about 80 percent conifers; (5) large scattered deciduous with understory of conifers; (6) large scattered conifers with dense understory of conifers; (7) large medium-dense conifers on stream bank and terrace; (8) very open conifers on steep, south-facing valley wall. (Province of New Brunswick, Department of Natural Resources)

height, and crown closure. Some of the stands shown result from fire, others from topographic position, two common causes of stand differentiation. Patterns of forest cover variation can be detected on aerial photographs from very large scale (1:1,200 for example) to very small scale, even on some images taken from manned satellites. Larger scales show size and density variations best; smaller scales are not so useful for this purpose. The smallest scales mainly show differences related to variations of spectral reflectance resulting from huge forest fires or cut-over areas and broad species changes. Variations in photographic quality and type affect the interpretation of different cover type characteristics. Good photography of appropriate scale and type is essential for accurate interpretation.

Species identification is a complex problem of interpretation. H. C. Ryker, in one of the earliest studies on species identification, incorporated five of the eight image elements used in species identification—size, shape, tone, texture and site.[4] Shadow, pattern, and association were not mentioned. Ryker used shape and tone as his main criteria for separating four species of western United States and described their photographic appearance in detail. Throughout the literature, a number of articles give some image characteristics for the identification of various species or species groups in many forest regions of the world. Usually the descriptions are based on the image elements mentioned above, but there are few comprehensive, illustrated manuals available, such as the one by V. G. Zsilinszky for species of Ontario.[5] Identification of species on air photos must be learned through local experience and study.

Table 14.1 is a dichotomous key for the identification of the northern conifers of New England and the Maritime Provinces of Canada. To be most effective, such a key should include photographic illustrations and line drawings. The key is intended for use with panchromatic photography at scales from about 1:6,000 to 1:16,000. To use with color photography or for other scales the key would require considerable revision to make use of their differing characteristics.

Area determination. Cover type areas can be determined by outlining each stand and preparing a forest cover map on which stands can be located and their areas measured and compiled. Alternatively, the total area of each cover type can be estimated by classifying cover type at a number of points located on the photographs. The number of points required for reliable area estimates can be determined by statistical methods. If one point per photograph is sufficient, the center of the photograph is used. If more than one point per photograph is required, they may be selected systematically or at random on the central portion of each photograph.

The proportion of points classified as a particular cover type represents the proportion of the total area occupied by that cover type. The total of all individual cover type areas equals the area of the whole. If the point classification method is used in areas of substantial topographic relief, cover types associated with higher ground elevations will be imaged at larger scale, and the estimate will be biased in their favor except where only the photograph centers are used.

Aerial Stand Volume Tables

It was early recognized that if reliable stand volume tables could be prepared using measurements obtainable from photographs alone, forest surveys could be undertaken with little or no expensive ground work. It was assumed that stand size and density were related to the stand volume. Two measures of size can be estimated on aerial photos, average stand height and average crown width. Density is nearly always expressed in terms of crown closure, that is the proportion of the stand area under tree crowns (always less than 100 percent, for even in the densest stands, there are many spaces between the crowns).

In practice, a variable proportion of trees, particularly the smaller ones, are obscured on virtually all medium and small scale aerial photographs, so that estimates of average height tend to vary according to the density of the stand, the elevation of the sun, the species, and other factors. Determination of average crown width suffers from the same problem. Consistent estimates of crown closure, using diagrammatic aids, have been made by experienced inter-

TABLE 14.1
Air-Photograph Interpretation Key to Northern Conifers

1a. Crowns proportionally large, more or less irregular especially in mature specimens, spreading. Branches often protruding, tops not narrowly conical, usually rounded or flattened 2

1b. Crowns smaller, regular, conical to narrowly conical or almost columnar. Tops conical, more pointed, well defined 6

    2a. Light-toned, irregular crowns, prominent branches or ragged and open crowns. . . . 3

    2b. Darker toned more irregular shaped crowns 4

3a. Prominent branches, "star" shaped in plan. Solitary or in small groups in mixed stands. Pure stands often on sand plains or well-drained sites . . . . *white pine*

3b. Ragged, open crowns. Branching sparse, not prominent. In mixed stands solitary or scattered, often taller than other species. Sites well drained . . . . *overmature white spruce*

    4a. On sand plains or dry sites. Broad, rounded, rough, open crowns . . . . . *red pine*

    4b. On moist or ill-drained sites, rarely in pure stands 5

5a. In swamps. Crowns irregular, open or ragged, if conical, blunt-topped. Some specimens may lean precariously (see also 10) . . . . *white cedar*

5b. On moist sites. Dense, broadly conical to irregular crowns. Crowns appear fuzzy. Very dark shadows . . . . *eastern hemlock*

    6a. Mostly in uniform, pure stands on sand plains or excessively dry sites. Crowns small, irregular, pointed. Tones medium . . . . *jack pine*

    6b. In pure or mixed stands, rarely on excessively dry sites. Crowns conical to columnar 7

7a. On moist to boggy sites. Crowns narrowly conical or columnar 8

7b. On moist to well-drained sites. Crowns conical, rounded, or pointed-topped 9

    8a. Crowns deciduous, light-toned especially in autumn, open. In mixed stands often taller than other species . . . . *tamarack*

    8b. Crowns darker, more dense, narrowly conical to columnar. Stands may be irregular in height . . . . *swamp type black spruce*

9a. Crowns very symmetrical, smooth. Heights irregular. Tops pointed, rounded, or blunt . . 10

9b. Crowns narrowly conical, not so symmetrical, not noticeably rounded or pointed . . . 11

    10a. Crowns symmetrical, round-topped, dense, light-toned. Mostly in pure stands and clumps on upland, limestone sites . . . . *white cedar*

    10b. Crowns symmetrical, pointed in front lighting (blunt in back lighting). Dark-toned shadows . . . . *balsam fir*

11a. Crowns narrow conical to columnar, rougher branching than balsam fir. In stands heights may be irregular . . . . *black or white spruce*

11b. Crowns more broadly conical, larger. Branching tends to be more prominent . . . *red spruce*

Note: Red spruce has the broadest, most luxuriant crown, while black spruce has the narrowest. Black spruce will occupy drier and poorer sites. Red and black spruce hybridize making distinction of the eastern spruces very difficult.

preters for many years, although a suitable device for direct measurement has not yet been developed. Scanning densitometers may prove satisfactory with larger scale photography.

    Stand volume tables have been compiled for a variety of species and species groups, nearly always for a specific region, giving average volumes for stands, usually in 10-foot height classes, 10 percent crown-closure classes, and 5-foot crown width classes. In many cases, statistical analysis showed that little volume variation was associated with variation in crown width and this variable is frequently omitted, volume being classified by stand height and crown closure alone.[6]

Aerial stand volume tables have by no means supplanted the combined air and ground inventory methods. Sometimes they are used to improve the initial aerial photograph classification of such surveys, but the increasing sophistication of inventory data often necessitates ground examination for all but reconnaissance surveys. Furthermore, aerial techniques cannot be used where a substantial component of the forest is hidden by taller trees or obscured by other factors, such as excessive shadows. Finally, because of the variable nature of the data which must sometimes be used to prepare the tables, their standard errors of estimate may be excessively large.

Aerial Tree Volume Tables

Another way of reducing the need for ground measurement in forest inventory surveys is through the development of aerial tree volume tables. This method corresponds to ground methods where all trees within a small sample plot are measured and their individual volumes determined from tables and totalled for the plot. The mean volume of a number of such sample plots located at random over a large area gives an estimate of the mean volume of the whole area. The results can be analyzed statistically to determine the probability of error as a result of sampling.

The aerial photograph measurements used to determine individual tree volumes are tree height and crown width. With favorable aerial photography, the height of most trees can be measured accurately except where the base or top is hidden. Similarly, crown width can be measured, but irregular branching may pose a problem in determining the average width. Because of vertical exaggeration in stereoscopic viewing and the unexcelled overview provided by aerial photographs, both height and crown width can be measured more efficiently, if not more accurately, on medium and large scale aerial photographs than on the ground, provided the base and top of the tree are not obscured.

As mentioned earlier, dbh and height are well correlated with volume of individual trees. Crown width as a substitute for dbh for use with aerial photos was early recognized, but the development of aerial tree volume tables was not pressed, probably because of difficulties in obtaining suitably large-scale photographs and in making the necessary photographic measurements. Technology to improve these deficiencies has evolved considerably, but equally important has been the development of statistical methods and the use of computers in objective evaluation of the techniques.

It has been shown by various workers that, although not as accurate as ground measurements, reasonable estimates of volume may be determined from large-scale aerial photographs by measuring the crowns and heights of individual trees. A number of technical problems have been associated with the development of operational procedures; some of these remain only partially solved.

Problems of Large-Scale Aerial Photography

Image movement during the fraction of a second required to expose a negative may be significant. For example, at a scale of 1:1,500, image motion (I) resulting from forward motion of the aircraft is 1/1,500 of the aircraft movement during exposure. If the aircraft is flying 250 kilometers (155 miles) per hour and the camera shutter speed is 1/200 second, then:

$$I = \frac{250 \times 10^6}{3,600 \times 1,500 \times 200} = 0.23 \text{ millimeters}$$

which is equivalent to 0.34 meters (13.4 inches) on the ground. Obviously, the shutter speed must be substantially increased in order to produce sharp images. Certain 70 mm cameras have been developed for low flying aircraft. These cameras have faster shutter speeds than regular survey cameras designed for 23 × 23 centimeter (9 × 9 inch) photographs, and will cycle faster to ensure stereoscopic coverage as well. An example of 70 mm photography is shown in figure 14.2.

Errors in the determination of flying height will cause corresponding errors in the scale of the photographs. For example, if the flight height is 400 meters (1,312 feet) above mean ground level, the scale on a hill only 20 meters (66 feet) higher will be 5 percent greater. In order to determine an accurate flight height above ground, a radar altimeter must be used which is designed to measure and record on each photograph the height above ground at the photograph center, unaffected by the forest canopy.

An alternative solution for precise scale determination is the use of two 70 mm Hasselblad cameras with synchronized shutters mounted 4.57 meters (15 feet) apart on a boom slung beneath a helicopter. The preferred operational flying height for this equipment is 100 to 200 meters above ground. An added advantage for stereographic pairs taken simultaneously, as with this system, is that object motion, such as wind-sway of trees, does not affect the parallax measurements used to determine heights of objects from stereoscopic pairs of aerial photographs.

Improved measurement and computing methods for use with both the above systems include measurement by photogrammetric plotter, such as the Wild A-40, equipped to provide output in the form of punched computer cards. In this system, the x, y, and z coordinates for

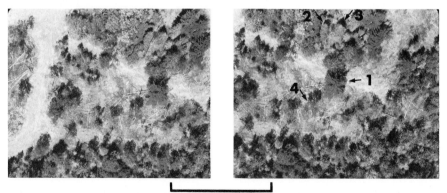

50 meters

Figure 14.2. Large-scale 70 mm stereo pair taken June 6 over central New Brunswick, altitude 275 meters (902 feet); lens 152.4 mm, panchromatic film. Species indicated are (1) white pine, (2) spruce, (3) balsam fir, and (4) white birch. (Canada Centre for Remote Sensing)

the top, base, and crown width of a tree are measured in the plotter and fed to a computer programmed to calculate tree height and volume for all trees identified within a sample plot of specified size. A number of such plots may be located in a cover type to determine the composition by species, height class, and volume of the type.

Although this method provides more information than the stand volume table method, many detailed surveys require information which can only be determined on the ground so that the aerial tree volume table method must be limited to less detailed surveys.

Advanced Imagery for Forestry

All the techniques discussed so far were developed primarily for black and white photography. Until the latter part of the 1950s, experiments with aerial color film were largely unsuccessful. Even for many later developments, especially measurement procedures, black and white films have been preferred because of availability, cost, and film characteristics. For species identification and other qualitative interpretation, however, color film has been found to be superior.

Today, the usefulness of color films in forestry is generally accepted. The cost of color film, processing, and prints is considerably more than for black and white film, but the overall cost in forest inventory compares favorably. Flying costs are about the same, but interpretation costs may be greatly reduced. Interpreters often find they can work faster and with more confidence by employing color photography in their investigations. Also, more information may be available, especially in species identification where color is a major characteristic.

Color film is so commonly used by amateur photographers that it may not be regarded as ''advanced'' imagery at all. Color infrared film is less commonplace, and its use in forestry involves a much greater technical understanding. Multispectral photography, although not based on especially advanced photographic techniques, leads to highly sophisticated, automated interpretation. Multispectral photography bridges the gap to nonphotographic remote sensing techniques which may or may not be designed to provide photographlike imagery for a particular application. Radar, thermal infrared sensors, and optical-mechanical sensors fall into this category. A variety of such sensors have been tried for forestry purposes, and some are being used on an operational basis, while others are still experimental.

Color Film Techniques

As was mentioned, natural color and color infrared films are especially useful for aerial photograph interpretation. Examples of each are shown in plate 17. Both use a color negative system so that multiple copies in color or black and white can be made. With full color reproduction, the human eye can obviously differentiate a much greater number of shades than with the same scene compressed into gray tones. Color film is composed of three layers sensitive to overlapping bands of the visible spectrum in the blue, green, and red zones respectively; that is, it is sensitive to the whole visible spectrum and, if correctly exposed and processed, reproduces natural colors faithfully.

Color infrared film at first glance, with its completely false colors, appears confusing, but for interpretation of vegetation, it has some very useful characteristics. This film also has three light-sensitive layers, sensitive to the green, red, and near infrared bands respectively. Blue light must be subtracted from incident light by a yellow colored filter such as the Kodak Wratten No. 12.

If the film is developed as a positive, the green-sensitive layer develops to yellow, the red-sensitive layer to magenta (a mixture of blue and red), and the infrared-sensitive layer to cyan (a mixture of blue and green). Because the film is what is known as a color reversal film, exposure and processing result in the greatest color dye formation in the least exposed layers. Thus, if the exposing light is mostly infrared, most of the dye will be formed in the green- and red-sensitive layers, with little being formed in the infrared-sensitive layer. The dyes are color-subtractive in viewing; yellow subtracts blue light, cyan subtracts red, and magenta subtracts green, with the result that when a film is exposed mostly to infrared, for example, the dyes formed (yellow and magenta) subtract blue and green from the white viewing light, and red is transmitted to the viewer. If this film is developed as a negative, dyes of the complementary colors to those mentioned above are formed. When a positive print is made from the negative, the colors are reversed to those described.

In nature, healthy broadleaf trees, because of active chlorophyll development, reflect infrared light strongly and appear red or magenta color on infrared film. Healthy needleleaf trees appear more purplish because of the leaf shape. Dead or dying trees of either kind usually appear bright green, while red autumn leaves appear yellow and yellow leaves appear white. Objects painted green to resemble trees show as blue or purple, and water is nearly black. A distinct shade may often serve to distinguish a particular species, especially during one season of the year.

Forest Damage Surveys

Damage may result from insect attack, disease, or a number of nonbiotic factors such as flooding, fire, frost, and air pollution. It is manifested by defoliation, malformation, or foliage discoloration. Foliage discoloration may be detected most easily on color infrared and natural color photographs, while detection of defoliation and malformation is usually best detected by stereoscopic examination of large scale photos, either black and white or color.

Damage types may be classified in greater detail by consideration of such things as the degree of defoliation or malformation; the kind of discoloration, its location and extent within the tree crowns; the distribution of damaged trees; the species of trees affected and their topographic location. With such an analysis, it is possible to prepare a photo-interpretation key to the probable cause of damage.

Vegetation Surveys

Because of the useful characteristics of color and color infrared films in species identification, these films have proven especially useful in forest and vegetation surveys at all scales. Because color infrared film is always exposed with a yellow filter (Wratten 12 or similar) to eliminate blue light, and blue is the chief color component of haze, color infrared films are particularly useful for high altitude photography where natural color is often affected by haze. A wildland vegetation and terrain survey in California using high altitude CIR aerial photographs taken at a scale of 1:120,000 has been compared with a survey of the same area using black and white photographs taken at 1:16,000 scale.[7] It was found that the results were comparable, but the small scale CIR was twice as efficient, that is, the work was completed in half the time. This was partly because 3 photographs covered the same area as 78 at the larger scale, resulting in much less handling.

In a study using 70 mm photographs at 1:160,000 scale, it was found that the information provided was sufficient to revise conventional 1:15,840 forest cover type maps. In this test of color infrared, a continuous strip—from Richmond, near Ottawa, Ontario, to Moosonee in James Bay—was flown and interpreted for regional forest classes. It was suggested that scales as small as 1:250,000 would provide useful information. At that scale, one 23 × 23 centimeter (9 × 9 inch) aerial photograph would cover 3,370 square kilometer (1,300 square miles)![8]

Forest Site Surveys

Commonly, forest site classification is based on a biophysical classification involving the recognition of vegetation associations, landforms, drainage, and other characteristics. Mapping such a classification involves detection and delineation of the boundaries of the classes, as well as identification of the classes. While various types of aerial photographs by no means exhaust the potential or variety of remote sensors available, they provide a broad spectrum of features for identification and delineation of both the biological and physical components of the classification. Other sensors may emphasize specific characteristics to a greater degree, whereas color aerial photographs provide a combination of many factors as well as high resolution, which can be particularly useful when considering units based on diversified characteristics.

Multispectral Techniques

Multispectral techniques may be photographic or nonphotographic. The essential feature of the technique is that two or more cameras or sensors are used simultaneously to record reflected or emitted energy from a scene in two or more bands of the electromagnetic spectrum. The basis of this approach is that more information is available from two or more different spectral views of the same scene than from a single view. For example, photographic materials are sensitive in a band of the electromagnetic spectrum including wavelengths from about 0.4 to 1.0 micrometers (the portion from 0.4 to 0.7 micrometers comprises the visible spectrum, the remainder is the near or photographic infrared band). With appropriate films and filters loaded in a multilens camera or in a group of cameras mounted together, it is possible to take several simultaneous photographs, each in a separate band of the photographic spectrum.

One arrangement might be to take four photographs in the blue, green, red, and infrared bands respectively. A given object would usually reflect with different brightness in each band, producing a different tone on each photograph. If the brightness is measured by measuring the tonal density (darkness) of the images on the different photos on a scale of ten, for example, the result might be a "tonal signature" or "spectral signature" of 2, 4, 2, 5 for the object. Other objects might reflect with the same brightness in three of the bands, but if one band is different, the objects can be differentiated by their tonal signature.

Unfortunately, in nature, a host of variations, such as light intensity, direction and spectral composition of illumination, atmospheric conditions, and film processing, affect the tone of images, not to mention the natural color variations in individual objects of the same kind or class. The result is that spectral signatures overlap and become nonspecific to such a degree that probability statistics must become a tool of the interpreter.

Multispectral Sensor Systems

Multiple camera mounts or multiple lens cameras with as many as nine lenses have been used for multispectral photography. The lens axes are parallel and the shutters synchronized so that identical and simultaneous coverage is obtained in the chosen spectral bands by the use of appropriate film-filter combinations.

Nonphotographic, multispectral systems have been developed in the form of optical-mechanical scanners. In such scanners, a rotating mirror directs energy from the terrain to detectors which translate it into electrical impulses. As the sensor platform (aircraft or satellite) moves forward, a series of contiguous scan lines records a strip of terrain. The impulses can be recorded on film in the form of low-resolution photographlike images of the terrain or on magnetic tapes for computer processing. They may be transmitted to a ground receiver for recording and processing. Detectors can be designed for various bands of the spectrum, both photographic and nonphotographic. Such systems offer great flexibility for multispectral sensing and are suitable for unmanned satellites. Nonorthogonal imaging (because of the scanning system) and lower resolution may be drawbacks in certain applications.

Interpretation Systems

With multispectral energy data recorded on photographs, either from cameras or optical-mechanical sensors, an unaided interpreter would be forced to locate, measure, and compare the density of identical images on several photographs. Clearly this would be a slow and inefficient way to interpret the imagery. Some degree of machine interpretation is required. Such a system has been developed in which black and white photograph transparencies are projected in common register through colored filters onto a rear projection screen. By varying the filter color combinations, a variety of false color images can be created at will by the interpreter to enhance the image contrast of a particular class of objects (pine trees, for example) and to optimize image search and recognition. The technique is well suited to the location of isolated specimens of a particular species of tree or to early detection of insect damaged trees,[9] and illustrates a combined man-machine approach.

A more sophisticated and automated interpretation system involves the use of electronic density scanners to digitize photographic image patterns for subsequent data analysis by analog computers. Optical-mechanical scanner output in the form of compatible tapes, instead of photographic imagery, provide a more direct link from sensor to computer. This is the case with space-borne (satellite) sensors which must transmit sensor output to ground receiving stations.

The digital output from sensors on the Earth Resources Technology Satellite (Landsat) was recorded on computer compatible tapes, and both colored and alphabetically coded computer maps were prepared. Landsat multispectral scanners record data in four spectral bands—green, red, and two infrared bands—which provided enough digitized spectral data to differentiate five cover classes—agricultural land, coniferous forest, deciduous forest, bare soil, and water. Information from three dates was used to prepare three single-date maps and a multidate combination of the three. Accuracy of the single-date maps compared with ground truth ranged from 67 percent to 81 percent. The multidate map was 83 percent accurate, only slightly better than the best single-date map.[10]

Thermal Infrared Imagery

Thermal infrared imagery is a good example of scanner imagery from a nonphotographic band of the electromagnetic spectrum. Heat energy, mostly from the sun, is stored in the crust of the earth and continuously emitted at varying rates which are determined by the intrinsic emissivity of an object and its temperature. (Emissivity refers to the radiant efficiency of a target as compared to a perfect radiator, such as a black body.) The spectral band usually classified as infrared extends from wavelengths of about 0.7 micrometers to wavelengths of 1,000 micrometers (700 to 1,000,000 nanometers). The atmosphere blocks much of this band, but certain wavelengths are passed through atmospheric "windows" and can be detected by airborne and space-borne sensors.

In thermal sensors, heat radiation from the ground is reflected by a rotating mirror to a heat-sensitive detector which produces small electrical impulses in proportion to the amount of radiation received. These impulses are amplified and used to modulate a light source to produce a photographlike image which superficially looks something like a black and white strip photograph. It suffers, however, from low resolution and the same kind of scanning distortion as the optical-mechanical scanners.

Although the imagery resembles a photograph of reflected light energy, it is an image record of emitted heat energy and, as such, can be recorded in darkness or through smoke and dust. Relatively warm objects are recorded in light tones while cooler ones appear dark. Night and day comparisons of the same terrain are very revealing. Bare areas, especially rock outcrops or paved roads, heat up rapidly under the rays of the sun and appear lighter than the surrounding area in daytime imagery, but at night they cool rapidly and appear darker. Water and moist areas maintain a fairly uniform temperature, and their relative tones are reversed—dark by day and light at night.

Other types of thermal detectors can record absolute temperatures along a narrow strip of terrain in the form of a trace plotted on a strip chart. Simultaneous thermal imagery is usually taken to help determine the exact location of the temperature strip, and aerial photography may also be helpful in delineating the thermal imagery.

Thermal infrared imagery has proved useful in forestry because of the moisture and local temperature relations revealed. The following two applications illustrate its use.

In forest fire suppression, small smoldering fires from lightning strikes or abandoned campfires can be located before they spread by thermal infrared detectors carried on regular aircraft patrols. Once a major fire is underway, the whole fire area is usually shrouded in smoke, preventing assessment of the direction of spread, location of burned out areas, and new outbreaks or "hot spots."

Infrared imagery taken from helicopters or light aircraft, can be delivered quickly to ground crews to help in their efficient deployment or to guide them away from dangerous locations where they might be overtaken by the advancing fire.

A less dramatic application involves studies in connection with the protracted battle against spruce budworm, the most destructive insect in the fir-spruce forests of eastern Canada.[11] Build-ups of heavy insect infestation are known to be associated with periods of relatively clear, dry summer climate for several successive years. It was assumed that during these periods, certain localities or epicenters provide optimum climatic conditions for budworm development because of aspect, slope, and other factors. Mapping climatic variations by conventional methods in the upper canopy of the forest, where the budworm feeds, would involve erecting a large number of instrument towers throughout a variety of topographic areas.

As an alternative, remote sensing with infrared sensors offered a technique by which the upper canopy temperature regime could be extensively sampled. This was carried out with an airborne infrared sensor (infrared thermometer) which produced a strip chart showing a trace corresponding to temperature variations along a 1 meter transect. Thermal imagery, to determine the transect location, was simultaneously recorded. Temperature variations were successfully mapped by this technique as a first step in elaborating the epicenter concept.

Radar Imagery

Radar was originally developed to detect the approach of aircraft beyond the range of sight and sound. It has been refined to the point where SLAR (side-looking airborne radar) units are able to produce imagery of a wide swath of terrain on both sides of the flight path. Radar is unlike the other sensors discussed in that it produces its own energy radiation. For this reason, it is said to be an "active" sensor system, as opposed to a "passive" system like photography which images naturally produced energy.

The portion of the electromagnetic spectrum in which radar operates includes wavelengths from just under 1 centimeter to about 1 meter. Because these wavelengths have great ability to penetrate cloud, fog, haze, and dust in the atmosphere, and because the system produces its own energy, radar is an all-weather, day or night sensor. This feature and the wide swath coverage of SLAR are economically attractive for reconnaissance surveys of large, remote areas.

Drawbacks to the SLAR system include image distortion and low resolution. Distortion is complex, but is greatest vertically beneath the aircraft. Conversely, the outer limits of the image swath are least distorted.

Interpretation of radar imagery requires knowledge of the microwave reflection characteristics of the scene to be interpreted. Microwave energy, that is, energy of the wavelengths previously mentioned, is transmitted at right angles to the aircraft track, in the case of SLAR equipment, and is reflected from ground objects according to their microwave reflectivity and surface angles relative to the direction of the incident beam. Various textures and tones will result, depending upon the nature of surface cover—grassland, forest, bare land, and so on. Size, shape, and arrangement of objects will also help in their interpretation. At low depression angles with the earth's surface, radar may not be able to "see" behind hills or into valleys; radar "shadows" will result.

A reconnaissance forestry and soils mapping project of 210,000 square kilometers (81,081 square miles) in Venezuela for development and planning illustrates the use of radar imagery in forestry.[12] It was done in connection with a development planning project. Radar imagery was taken in a westerly direction along north-south lines spaced at 18 kilometer (11.2 mile) intervals to give a 60 percent overlap. Navigational control of the aircraft was provided by the radar altimeter, autopilot, and inertial guidance platform. The summary report included twenty-one radar mosaics with overlays depicting hydrology, geology, forestry, soils, and radar profiles.

Because of the 60 percent overlap, it was possible to compile the imagery interpretation by stereo viewing of radar mosaics prepared at a scale of 1:200,000. Ground observations along some major navigable waterways, some aerial photography, and observations from other expeditions provided basic data and ground truth for interpretation.

The interpretation was designed to provide information on productive and nonproductive agriculture and forestry areas. Table 14.2 shows the forest and vegetation classes used.

TABLE 14.2
Classification of Vegetation Units

1		Lowland forest
	1a	Periodically flooded or riparian forest
	1b	Permanently inundated or wet forest
2		Upland forest
	2a	Highland riparian forest complex
	2b	Highland montane forest complex
3		Summit and Tepui vegetation
	3a	Montane savannah
4		Savannahs
	4a	Dry phase savannah
	4b	Wet phase savannah
5		Marsh and aquatic vegetation

Conclusion

Because forestry deals with a large, widespread resource, often difficult to survey on the ground, various types of remote sensing methods have been tested over the years. Aerial photography, because of its versatility, high resolution and relatively straightforward interpretation, is in general use. Other remote sensing techniques have been tested as they became available, and some have been used operationally as experience with their use was gained. Inventory, at all levels of detail, is the most important application of remote sensing in forestry, but other applications, including cut layout, road location, forest fire control, damage assessment, recreation planning, site classification, management, and planning, have all been developed. Ground survey is often a necessary accompaniment to remote sensing in forestry because of the level of detail required, but considerable research to lower costs by reducing ground work is carried out by government agencies, universities, and private industry.

NOTES

1. R. Thelen, "Aerial Photography and National Forest Mapping," *Journal of Forestry* 17 (1919):155-22. For further information on the history of aerial photography and mapping, the following are suggested: D. Landen, "History of Photogrammetry in the United States," *Photogrammetric Engineering* 18 (1952):854-98; S. H. Spurr, "History of Forest Photogrammetry and Aerial Mapping," *Photogrammetric Engineering* 20 (1954):551-60;

D. W. Thomson, *Skyview Canada, A Story of Aerial Photography in Canada* (Ottawa: Dept. of Energy, Mines and Resources, 1975).

2. American Society of Photogrammetry, *Manual of Photographic Interpretation* (Washington, D.C.: American Society of Photogrammetry, 1960), pp. 463-66.

3. See also: T. E. Avery, *Interpretation of Aerial Photographs,* 2d ed. (Minneapolis: Burgess Publishing Co., 1968), chap. 11; S. H. Spurr, *Photogrammetry and Photo-Interpretation,* 2d ed. (New York: Ronald Press, 1960), chap. 22.

4. H. C. Ryker, "Aerial Photography Method of Determining Timber Species," *Timberman* 34 (March 1933):11-17.

5. V. G. Zsilinszky, *Photographic Interpretation of Tree Species in Ontario,* 2d ed. (Toronto: Ontario Dept. of Lands and Forests, 1966). Summaries of some of the other material available are given by Spurr, *Photogrammetry and Photo-Interpretation.*

6. Examples of several types of stand volume tables are given by T. E. Avery, *Forester's Guide to Aerial Photo Interpretation,* Agriculture Handbook 308 (Washington, D.C.: U.S. Dept. of Agriculture, Forest Service, 1966), pp. 33-37; Avery, *Interpretation of Aerial Photographs,* pp. 201-02; and G. M. Bonner, *Provisional Aerial Stand Volume Tables for Selected Forest Types in Canada*, Publication 1175 (Ottawa: Canada Dept. of Forestry, Forest Research Branch, 1966).

7. D. T. Lauer and A. S. Benson, "Classification of Forest Lands with Ultrahigh Altitude, Small Scale, False Color Infrared Photography," In *Proceedings Symposium IUFRO S6.05,* International Union of Forest Research Organizations, Freiburg, Germany, pp. 143-62.

8. U. Nielsen and J. M. Wightman, *A New Approach to the Description of the Forest Regions of Canada Using 1:160,000 Color Infrared Aerial Photography,* Information Report FMR-X-35 (Ottawa: Dept. of Environment, Canadian Forest Service, 1971).

9. D. T. Lauer, "Multispectral Sensing of Forest Vegetation," *Photogrammetric Engineering* 35 (1969):346-54.

10. Z. Kalensky, "ERTS Thematic Map from Multidate Digital Images," in *Proceedings, Symposium on Remote Sensing and Photo Interpretation,* International Society of Photogrammetry, Commission VII, Banff, Alberta, 1974, pp. 767-85.

11. R. B. B. Dickison and D. O. Greenbank, "The Spruce Budworm Problem and Topographic Variation of Forest Canopy Radiation Temperatures," in *Proceedings of the First Canadian Symposium on Remote Sensing,* Ottawa, 1972, pp. 111-19.

12. P. M. Addison, "SLAR Imagery Interpretation Techniques and Procedures for Reconnaissance Forest and Soils Mapping," in *Proceedings of the First Canadian Symposium on Remote Sensing,* Ottawa, 1972, pp. 483-92.

15

Application of Remote Sensing in Urban Analysis

Gary K. Higgs
Mississippi State University

Introduction

Urban Characteristics and Problems

Urban areas, the scenes of the most intensive and complex human activities, exist in a wide variety of sizes, shapes, and localities. Great masses of people occupy space and interact within urban areas. The intensity of the activities, buildings, land use, and movement, along with the degree of concentration of people and the other physical and natural characteristics of cities, has given rise to many serious physical and social problems. Remote sensing is one source of information about urban areas, and it is becoming an important tool for understanding and solving many of the problems of cities and their suburbs.

Many of the modern-day problems of urban areas may, in part, be traced to: (1) the reasons for the formation of cities and the course of their evolution, (2) the competitive and conflicting nature of the arrangement and rearrangement of land and functions within cities, and (3) the size, complexity, and concentration of urban activities and areas.

Urban problems which develop from the causes of city formation and evolution generally stem from the more prominent reasons for the foundation of urban areas, such as protection, provision of services, and improvement of quality of life. Many city dwellers may feel that the level of protection, services, and general quality of life which should have been attained has not been reached. These inadequacies are viewed as serious issues. Complicating these problems or inadequacies have been those related to the intricate and conflicting nature of land use and arrangement of cities. Conflicts and complications have often resulted in wasteful misuse of urban land, or destructive, conflicting, and detrimental use, such as the placement of noncompatible land uses in close proximity. For example, problems result from the construction of factories, bars, and gasoline stations in quiet residential areas. Frequently, such conflicting land use results in disadvantages to everyone. The factory, bar, or service station, therefore, may not have adequate parking space, utilities, or services, or delivery access may be hampered. At the same time, the residents in these areas must tolerate smoke, noise, heavy traffic, or other problems associated with industrial or commercial activities.

Often a third group of problems complicates solutions related to city formation and land use competition. Those problems are related to the intensity and concentration of urban land use, such as crowding, congestion, and shortage of space. These manifest themselves in the form of interference in activities, delays in travel of people and the transporting of products, and higher costs of living.

Understanding the circumstances of urban areas and becoming familiar with the elements which make up such areas are necessary steps that will lead to solutions to many urban problems. Remote sensing provides a unique perspective and broad spectral sensitivity which is a vital way of gathering certain types of information about urban areas.

Urban Information System

Generally, information in the form of raw data concerning urban phenomena—that is, about land use, buildings, roads, and other urban characteristics—is not particularly useful in problem solving or even in asking the right questions. Data must be organized in some meaningful systematic manner in order for planners, decision makers, citizens, and public and private officials to obtain a picture of urban areas and conditions. For this reason, the vast amount of information about urban areas, whether derived from remote sensing or other sources, is often organized in the form of an urban information system (UIS). The information which makes up an urban information system may be viewed and studied separately or in combination with other data to provide understanding of and possible answers to the problems of urban areas. An urban information system usually contains a wide variety of information, and remote sensing can provide a large portion of the data.

Remote sensing data about urban areas which may be used in a UIS falls into two general classes:

1. Data on static phenomena
2. Data about dynamic phenomena

The *dynamic phenomena* about which data may be obtained by remote sensing techniques consist of things that often cannot be directly observed, either because they change from time to time, or because they are simply not physically visible, such as population statistics, traffic flow data, and socioeconomic conditions. By contrast, *static phenomena* data include such things as: size of city; number, configuration, and capacity of streets; size and type of buildings; and types of neighborhoods—industrial, residential, and commercial.

Information about dynamic phenomena are available because of their visibility in the landscape. For example, population density of a city is a statistic which tells much about the character of an urban area. Although population density is not visible, it can be calculated from features which are visible. The procedure for estimating population begins by ground truthing the number of single family dwellings and the number of units in multifamily structures. The total can be multiplied by the average family size to yield an approximate value for total population in an area. The area of the neighborhood can be measured, and then divided by the total population to give a value for population density per unit area, such as people per square mile or kilometer, per block, or per acre, depending on the mapping unit used to measure the neighborhood. The formula for determining the population density of an area is therefore:

number of single family dwellings + number of residential units in multifamily dwellings = total number of dwelling units (assumed total number of families) × average family size = population of area ÷ total area of neighborhood in square miles or kilometers, acres, or blocks = population density in square miles or kilometers, acres, or blocks

There are several sources of error inherent in this method of population estimation and density count when applied to remote sensing imagery, and to use this or any procedure properly, the investigator must be aware of the sources of error which exist. Among the more critical potential errors in this estimation procedure are:

1. Miscounting of number of structures, dwellings, or residential units in multifamily dwellings
2. Possible vacant units counted as occupied
3. Deviation of average family size in the study area from the United States average

Despite these potential errors, it has been found that this procedure for estimating population and population density by use of remote sensing imagery is fairly reliable.

The application of this remote sensing analysis can be illustrated in a residential area. Assume that a residential area imaged on an aerial photograph contains 80 multiresidence structures and 57 single family dwellings. The multiresidence units are complicated by being of different sizes; that is, they contain different numbers of dwelling units. The number of individual units in these multifamily structures may be estimated in several ways. For example, 2.5 parking spaces usually are allocated for each residence unit; therefore, a count of parking spaces divided by 2.5 yields one estimate of the number of units in the complex.

Another method for obtaining the number of units in an area is based on the assumption that in townhouse structures, each unit usually has its own entrance. The number of entrances or walkways should equal the number of units. A third, and sometimes less reliable, method of estimation is to count roof and structure divisions because frequently individual dwelling unit designs are reflected in roof layout. This is not possible if no unit division appears on the roof.

Assume that, using the walkway/entrance method for an estimate, the count indicates that within the multiunit area there are 43 structures with 7 walkways each, 35 with 5 walkways, and 2 with 2 walkways. This amounts to 480 walkways and suggests that there are 480 dwelling units. A count of parking spaces indicates that there are 1,250 parking spaces. At the rate of 2.5 spaces per unit, this would suggest 500 dwelling units. The number of residence units in this example is between 480 and 500; however, there usually are a few extra parking spaces in multifamily complexes, so 480 is more likely to be the correct number of units.

The above example appears to include 480 dwelling units in the multifamily complex plus the 57 single family dwellings, or 537 units. This value multiplied by the average family size of 2.4 indicates a population of 1,289 people. Assuming the area of this study is approximately one-third square mile, the population density can be calculated by multiplying the population by 3 to give an estimated density of 3,867 people per square mile.

The population density figure is purely a statistical value and is not physically visible, but it is an intangible characteristic of the neighborhood. Obtaining this estimate on the ground in door-to-door interviews might take days and yield answers of less reliability at a much higher cost. By use of remote sensing, it is possible to obtain a variety of nonvisible or nonphysical pieces of information about a section of a city in a rapid, relatively accurate, and easy manner.

Procedures of Analysis in Urban Areas
and the Urban Information System (UIS)

Urban areas are enormously complex when viewed from either remote sensing or ground perspectives (fig. 15.1). The result of this complexity is that the understanding and analysis of urban areas is very difficult unless an orderly procedure is followed. In attempting to study and understand urban areas, whether by remote sensing or any other method, it is generally

Figure 15.1. The complexities of the urban area can be seen in this high altitude aerial view of a part of San Francisco, taken on September 11, 1964. (U.S. Geological Survey)

necessary to decide: (1) what an urban area is, (2) where it begins and ends, and (3) of what it consists.

An urban area is land covered with urban activities. It is possible from the remote sensing perspective to draw a line around the boundary of urban areas by simply looking at an image of the city and deciding where continuous urban development stops and rural activities begin. The process of urban sprawl, however, has made it increasingly difficult to specify precisely where a city ends and rural areas begin. In other parts of a city there is a gradual decrease in the intensity of urban activities—density of houses, commercial structures, and roads—and a compensating gradual increase in open areas from the city's center to outlying areas. Thus there is no clearly discernable line which marks the end of a city.

Because many cities have similar patterns, the concept of the urban fringe has been used to refer to an area between partly urban land cover and purely rural land cover. This urban fringe zone is characterized by a mixture of rural and urban land uses, and by having significant amounts of open space that appear to be transitional between rural and urban uses. In a quantitative sense there are fewer structures per mile than in the city and very few or no active farm fields.

Plate 18 illustrates interstate highways and their focal points in the south and west portions of a city, the bounding of an urban area by terrain features, and the change of building type and density associated with different distances from the center of the city.

Urban Spatial Structure and Setting

Many things can be learned about urban areas by considering their spatial structure and context; that is, how is a city laid out, how is it related to the location of surrounding communities, and which cities and urban concentrations in a region are larger. Cities, for example, tend to be arranged in a hierarchical pattern, with one large city being surrounded by several smaller cities located at some particular distances. Clusters of smaller communities form urban agglomerations. The overall urban structure and the urban region in which a city is situated can be best viewed on a small-scale, large-area image of a major city/urban region.

Plate 19 is a Landsat image of the Milwaukee-Chicago area (southeastern Wisconsin-northeastern Illinois). It illustrates the urban spatial structure of the region. This structure can be interpreted in the light of the map in figure 15.2. Lakes, rivers, and cities can be identified and the hierarchy of urban places discerned. Milwaukee is the second largest city, followed by an urban complex made up of a string of cities and towns situated in the Rock River Valley just south of Lake Koshkonong. This system of cities includes Janesville and Beloit, Wisconsin, and Rockford, Illinois, along with numerous smaller communities—Afton and Shopiere. North of Lake Koshkonong on the Rock River, Fort Atkinson, Jefferson, and several smaller communities can be seen. Other systems of urban places in the hierarchy include the Waukesha area group of cities and towns, the North Lake Shore system of cities, the Rockford group, and the Fox River system of cities. The map in figure 15.2 is based on the observed spatial relationships of plate 19.

Further insight into the overall urban structure of a major city and the general structure of the region can be gained through the use of the image in figure 15.3, a thermal scan image showing the area from Racine, Wisconsin north along the Milwaukee shore. A number of urban areas which are not visible on the Landsat image, plate 19, can be identified on figure 15.3. These zones stand out on the thermal image because of the pronounced temperature contrast

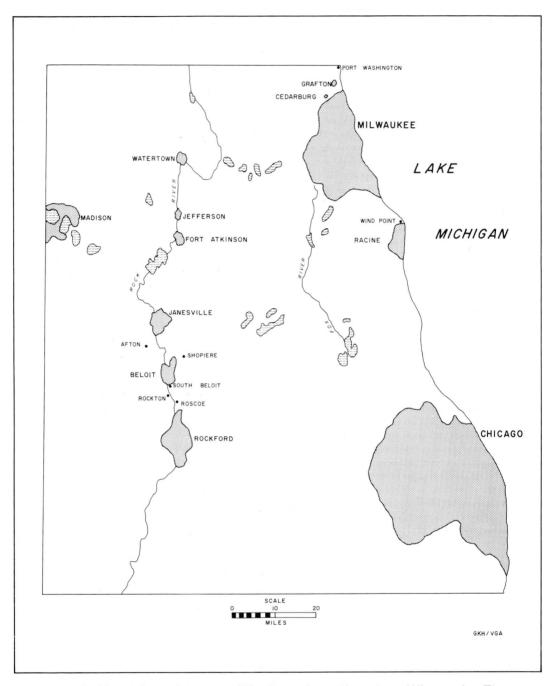

Figure 15.2. Map of northeastern Illinois and southeastern Wisconsin. The area mapped corresponds to the area imaged on plate 19.

Figure 15.3. Thermal scan of southeastern Wisconsin, north and west of Racine, made in September, 1972. Lake Michigan is to the right in the image. (NASA, Johnson Space Center)

between urban land use and rural, open, and naturally vegetated areas. Among the more obvious thermal contrasts are several small cities south and west of Milwaukee. Some of the transportation network characteristics, waterways, and neighborhood differences within the city of Milwaukee emerge on the thermal scan image. On a national basis, the overall urban pattern is illustrated by a classic nighttime image of conterminous United States and adjacent areas of Canada and Mexico (fig. 15.4). This image also serves to illustrate the patterns and sheer magnitude of urban energy consumption.

Urban Subregions and the Internal Structure of Cities

Once the urban region has been defined and delineated and its spatial structure and setting considered, remote sensing application at the next level of urban analysis is possible—that of identifying, classifying, and inventorying features and regions within the city. The objectives of urban subregion and city structure studies are to provide information on the characteristics of different parts of cities, and to understand the roles, functions, and activities associated with such areas.

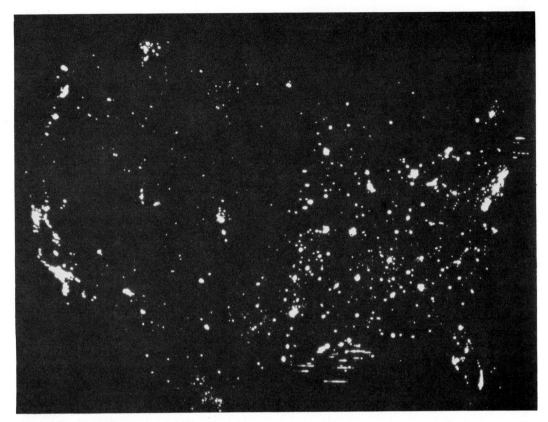

Figure 15.4. Nighttime image of conterminous United States and adjacent areas. Urban areas are located by their lights. (U.S. Geological Survey, produced by Defense Meteorological Satellite, and used by courtesy of Space Science and Engineering Center, University of Wisconsin-Madison)

Urban Subarea Analysis

The steps in analyzing subareas of urban areas and identifying their features are:

1. Locate major features, structures, and areas with basically distinct patterns, textures, and appearances. These areas are primarily:

   a. *Central business district (CBD).* This feature is present in almost all major cities. The CBD is usually the most easily seen and frequently the best place to begin an urban analysis because of its unique appearance and normally central location.

   b. *Interior residential community.* The residential district is the older residential area of the city. It is generally located adjacent to the CBD and is usually characterized by small lots and small houses, often with either very little vegetation or large, older trees.

   c. *Suburban residential communities* (plate 18). The suburban residential communities are the newer areas of the city and are generally located on the outer rim of the cities adjacent to rural lands. Frequently these residential areas are in the urban fringe and are characterized by relatively larger residences located on larger lots. Often there is substantial open space in and around these developments, and new vegetation and closely cropped lawns are mixed with some older established vegetation.

d. *Interior subcommercial areas.* These are the smaller neighborhood centers (usually older areas), strip commercial areas, shopping centers, and isolated individual establishments. These areas (with the exception of the isolated individual establishments) are distinguished from surrounding areas and buildings by being larger and more clustered, by having associated traffic patterns and congested parking areas, and by being located either centrally within a city or at focal points of the urban transportation system. Plate 20, for instance, shows the focus of major transportation networks within the city of Houston, Texas. Strip commercial areas are located along many of these routes, and neighborhood centers are located at intersections of such major routes.

e. *Suburban commercial complexes.* These are usually located in peripheral areas of cities, are surrounded by large parking lots containing hundreds of parking spaces, and large, usually single-story buildings centered in the parking areas.

f. *Industrial complexes.* Their location depends on the type and size of operation. Generally heavy industry, which primarily involves processing raw materials, is located in peripheral areas of cities, while light industry and related activities are usually located in more central areas. By contrast, final products processing, warehousing, and related finishing activities (fig. 15.5) tend to be located either near

Figure 15.5. Portion of Omaha, Nebraska. Final products processing establishments, warehouses, and related finishing activities. (Remote Sensing Applications Laboratory, Department of Geography-Geology, University of Nebraska-Omaha)

business districts to be close to customers, near large airports, or on main truck and railroad transportation routes as in plate 19.

2. Classify features and areas on the basis of character or pattern so as to group similar features and areas into subregions. Plate 20 portrays a grouping of similar areas in the city of Houston.

3. Count and/or measure the static and physically visible characteristics and features of urban subareas, such as size, total area, height of buildings, number of structures, and roads.

4. Gather and generate data concerning dynamic and other urban area phenomena that, while not always present or visible, leave evidence of their natural character, shape, location, volume, and arrangement in visible deductible substitute features, such as facilities, structure, and land uses that are relatively permanent. These nonphysical, nonvisible, or nonpermanent urban phenomena, activities, or characteristics are frequently among the most important parts of urban areas and life because they are directly related to or reflect human activities. For this reason, understanding them is necessary for a more complete awareness of urban areas.

The steps in this surrogate process of understanding the dynamic elements of an urban area are:

a. Selection of a surrogate or representative feature which is consistently associated with the temporal, nonphysical, or nonvisible urban characteristics or phenomena that are under study. The number of lanes on a highway, for example, tend to indicate the volume of traffic the highway is designed to carry, that is, its maximum capacity.

b. Study of the urban area on imagery to identify and count the surrogates, that is, the number of lanes of traffic on a section of highway.

c. Establishment of the potential relationship between surrogates and the features they represent, that is, the number of cars a highway can safely contain with proper spacing at certain speeds—the maximum traffic load. These values are given in table 15.1.

d. Calculation and measurement of surrogates, that is, size, capacity, and height. In the case of figure 15.6 the main street(1) stretches 840 feet. At a speed of 30 miles per hour, the street can accomodate 18 vehicles but has only 12. It could accomodate 6 more.

TABLE 15.1
Potential Road Capacity

Assume: Length of road 840 feet
        Average length of vehicle 15 feet

Speed	Space between Vehicle	Space per Vehicle	Number of Vehicles per Lane—Theoretical
50	50	65	13
40	40	55	15
30	30	45	18
20	20	35	23
10	10	25	33

Figure 15.6. The street (1) in this illustration stretches 840 feet between the intersections A and B. Count the 12 vehicles traveling on the street. (Remote Sensing Applications Laboratory, Department of Geography-Geology, University of Nebraska-Omaha)

    e.   Checking of relationship with ground truth.

    f.   Formulation of inferences and conclusions about the phenomena under study. Presently the street (1) in figure 15.6 is under capacity and can accomodate up to 15 more vehicles at a presumed speed limit of 10 miles per hour (6 more at a speed limit of 30 mph).

5. Calculate representative indices and derive information about the dynamic or nonphysical visible characteristics of urban subareas, such as number of structures per acre; amount of land occupied by transportation-related uses—roads, parking lots, public transportation facilities, and rights of way; and amount of land available for development or suitable for open space.

6. Compile and organize the information on static variables and phenomena that were counted and dynamic variables and phenomena that were derived as surrogates in a systematic manner in the form of urban information systems.

Interpretation and Use of Urban Subareas

    The information about presence or absence, size, density, capacity, data, land use/land cover type—however acquired—becomes an important part of the UIS and can function as a data base for decisions. If a street is frequently used to capacity, future street improvement programs possibly should involve widening or the use of alternate routes. This decision naturally depends on the time the data are acquired. For instance, if figure 15.6 were taken at 5 A.M. Sunday, the conclusions would be quite different than if the time were a weekday evening rush hour.

Most commonly the UIS and its associated data bases are portrayed on maps, but increasingly computers are replacing or supplementing maps as sources of stored urban information. Whether on the map or in the computer, the information on urban areas consists primarily of facts indicating the presence or absence of "phenomena," size, location, capacity, statistical characteristics (such as density, concentration, percentages), types of land use/land cover, and any other pertinent data.

From these basic components of the UIS, a number of very important relationships and facts may be derived. These provide insight not only into the nature and character of the urban settings and human activities but also into the proper and reasonable course for future development to solve urban problems. Such elusive characteristics and facts are: estimates of energy use, estimates of population density, rates of change of population density with distance between central city and suburban areas, and estimates of the predominant functions of different areas in cities (residential, commercial, and industrial). These estimates actually may be more than simple extensions of knowledge by surrogates because they deal with complex relationships and characteristics.

Estimates of rate of change of population density with distance from the central part of a city to a suburban area (population lapse rate) require selection and calculation of population per unit area. Assume an aerial photograph with a scale of 1:12,000 (1 centimeter = 120 meters or 1 inch = 1,000 feet). Imagine that the image is divided into eight squares, each of which is one inch on a side. Assume also that beginning with the first square beyond the fiducial mark, the number of residences or parts thereof in each square inch is as follows:

First square inch—13   Fifth square inch—15
Second square inch—33   Sixth square inch—11
Third square inch—33   Seventh square inch—32
Fourth square inch—24   Eighth square inch—1

The diagram in figure 15.7 shows the population lapse rate, or change, along a transect for the area. Such information as population lapse rate is valuable in planning the availability of social and other public services, including public transportation routes, new streets, street improvement, mail deliveries, trash collection routes, and police patrol plans. Such service policy decisions and the types of derived and directly observable information on which they are based are among the most useful for learning about urban areas. For that reason such complex relations as density of structures and people per given area are frequently used in the detailed analysis of the sections of urban areas, that is, residential, commercial, and industrial sections.

Residential Area Analysis

One of the basic functions of urban areas is the housing of people. In most contemporary urban areas the residential function is both the most intensive and the most expensive land use and occupies the greater portion of the city area. Residential areas are easily distinguished on a

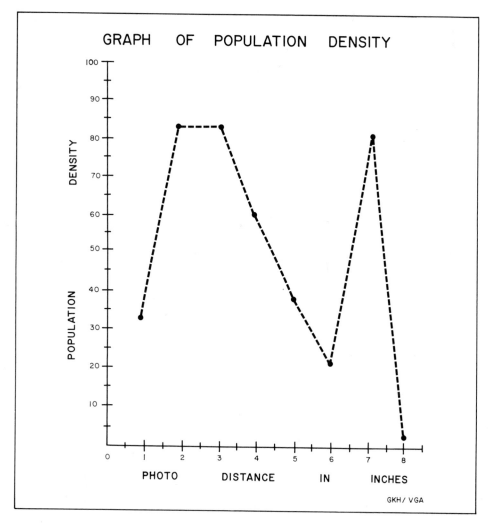

Figure 15.7.

variety of types and scales of imagery, and their analysis begins, as does most other area analysis, by a simple definition, identification, and delineation of the areas involved. Certain "keys" (typical features and characteristics) may be used for this at virtually any scale or on any image type (table 15.2).

Many of the items in the key are directly observable, while others require some calculation, manipulation, or inferences before they can be used to evaluate residential areas. Regardless of whether they are easily seen or require some special adjustment, the items are not only important indicators of residential areas in themselves, but also of the different types of dwelling zones; single or multiple family, or low, medium, or high income. With these keys and the classification of residential type, it is also possible to gain some ideas about the quality and nature of life, type of problems, and possible solutions in specific areas of a city. Generally, the

TABLE 15.2
Keys for Identifying Residential Areas

	Single Family Residence	**Multifamily Residence**
Large Scale Imagery Only		
Directly observable	Doghouses Alleys Fences Narrow streets Trees overhanging streets Garages, automobiles Possibly small commercial structures Sidewalks	Size and shape of structure Height of structure Balconies Fire escape Incinerators Skylights Spacing of entryways Roof divisions Chimneys Number of sidewalks Porches Arrangement of windows Number of parking spaces Street parking

Calculable	Percent of land covered by
	    structure
	Percent of area covered by
	    streets
	Average frontage
	Percent of land covered by parks,
	    playgrounds, pools
	Utilities
	Architectural type
	Age of buildings
		Percent of land covered by
		    structure
		Percent of area covered by
		    streets
		Average frontage
		Percent of land covered by parks,
		    playgrounds, pools
		Utilities
		Architectural type
		Age of buildings

Suitable for Both Large and Small Scale Imagery		

Directly observable	Relative position in city
	Green belt
	Urban fringe
	Absence of industry, railroads,
	    commercial structures
	Sidewalks
	Street pattern
	Separated from surrounding
	    areas by differences in tone
	    and texture
	Moderate amounts of
	    vegetation
		Relative position in city
		Absence of industry, railroads,
		    commercial structures
		Sidewalks
		Parking lot
		Larger open green areas
		In dense cities: no green space
		Relatively large structure
		Mixture of land uses

TABLE 15.2 (Continued)

	Single Family Residence	Multifamily Residence
Suitable for Both Large and Small Scale Imagery		
Directly observable	Trees usually separated Shrubs Yards Open areas between buildings Relatively small structures Moderate density Absence of large structures Absence of parking facilities	

most valuable indicator of urban problems is residential quality of life. The desirability of an area as a place to live, therefore, can be categorized into five groups: house characteristics, yard characteristics, transportation factors, locational characteristics, and neighborhood characteristics (table 15.3).

The process of evaluating residential area quality of life usually consists of six steps: (1) outlining, (2) observing, (3) counting features, (4) measuring areas, (5) calculating some indices and relationships, and (6) forming conclusions about the nature of problems and their solutions and the quality of life and conditions of the neighborhood. The steps for evaluating residential quality and the appropriate indicators are summarized in table 15.3.

The process of evaluating the quality of life may be illustrated by analyzing and comparing several images of residential neighborhoods which have different characteristics. Figure 15.8 shows closeup views of five types of residential areas: (A) is a mobile home park; (B) is a low income public housing complex; (C) has mixed residential—apartment buildings and single-family, middle-income homes which are closely and regularly spaced on small lots; (D) shows an upper middle class residential area of single-family, large-lot suburban dwellings on rolling, well landscaped, and tastefully configured streets; and (E) shows blighted housing.

Qualitatively, there are substantial differences in apparent quality of environments and living conditions in the various types of residential neighborhoods illustrated in figure 15.8. In a qualitative sense one of the easiest and most obvious evidences of upkeep and general neighborhood atmosphere is to observe residential roofs. On figure 15.8D the appearance of each roof is generally clean, indicating similar roof conditions throughout, as well as fairly new roof materials. In figure 15.8C, however, differences in tone and appearance among house roofs are apparent, indicating porches or new additions to a structure. The most striking contrast in roofing materials in this image appears in two of the multifamily structures where definite, straight lines are found between roof materials of different tones and colors. It appears that certain sections of these units have been reroofed.

By contrast, roof materials in figure 15.8B indicate a general overall mottled appearance with no sharp boundaries typical of new material. Frequently such tones reflect uneven surface conditions and moisture content variation which can indicate areas of poor repair or leaks. In a

TABLE 15.3

Indicators of Residential Quality and Neighborhood Decline

	Observe	Count	Measure	Calculate	Conclude
House Characteristics (Usually large scale imagery)	Age of structure Soundness or quality of construction Style of architecture Structure of roof Condition of fences Condition of accessory buildings Condition of drives and walks	Structure in poor repair Roof in poor repair Number of drives and walks		Building size, shape, and height	Living area Overall quality and condition of structures
Yard Characteristics (Large to medium scale imagery)	Presence of play space Presence of trash and litter Presence or absence of trees			Size and shape of yards	Lot area Play space Upkeep
Transportation Factors (Usually medium to small scale imagery)	Presence of or proximity to highways Traffic volume and bottlenecks Parking on streets Parking lot usage Maintenance condition of streets and sidewalks	Vehicle and pedestrian count	Traffic speed	Number of cars vs. road capacity	Accessibility Congestion Adequacy of parking

	Observe	Count	Measure	Calculate	Conclude
Locational Characteristics (Small scale imagery)	Relative location to other residential areas		Distance to: CBD industrial district commercial area rural area railroad highways Institutional land use		Relative location within cities
Neighborhood Characteristics (Small scale imagery)	Spatial structures Street pattern: radial parallel dendritic Land uses Presence of unattractive land uses New construction Prevalence of single family homes Green space Commercial area	Mixture of land use Junk yards, heavy industries Commercial vs. residential Number of single family homes vs. area of residential neighborhood		Population density Housing density Percent of land crowding Land use budget	Shape, size, concentration Traffic flow potential Homogeneity of neighborhood; differences from surrounding areas Future trend of neighborhood Crowding

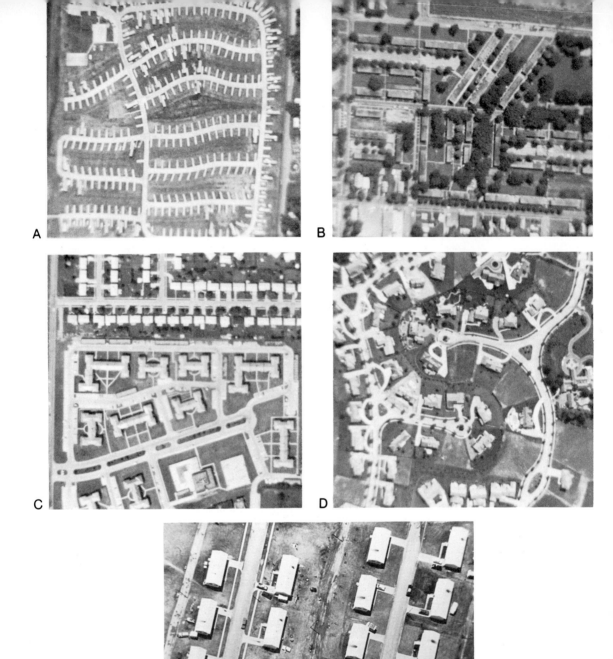

Figure 15.8. Residential types used to illustrate quality of life in cities. A, mobile home park; B, low-income public housing complex; C, mixed residential—apartment buildings and single-family, middle-income homes; D, upper middle-class, single family houses; and E, blighted housing. (Remote Sensing Applications Laboratory, Department of Geology, University of Nebraska-Omaha)

qualitative sense there is a substantial difference in apparent quality of environment and living conditions in the various types of residential neighborhoods of figure 15.8.

In a quantitative sense certain objective dimensions of the quality of life in the different types of neighborhoods can be assessed by calculating the land use (or land cover) budget. The land use budget of an area is basically a calculation of the amount of land area which is used for, or covered by, specific types of functions, such as residence structures, automobile space, vegetation, agriculture, water, commercial activities, industry, open land suitable for development, and open land unsuitable for any other use. One of the principal values of the land use budget is that it can be used with qualitative evaluation of a neighborhood to indicate similarities and differences among neighborhoods. Sometimes the land use budget provides a measure of the quality of life (and types of problems and potential solutions) as indicated by neighborhood characteristics.

In developing a land use budget for comparison of residential areas and understanding the quality of life, one of the main factors to be considered is dwelling unit differences. Furthermore, one of the most basic quantitative house measurements is size of structure. In addition to being of value in identifying neighborhood differences, size of living units is important in quality of life estimates because it often reveals much about potential living space per unit and cost of structure, which in turn indicates something about community economic health.

Additional land use/land cover budgets can be made for other elements—such as open grass, automobile space (roads, drives, parking areas), and tree cover—in order to provide insights into such important but vague neighborhood characteristics as the availability and size of green space and play areas, degree of crowding, and general community atmosphere. Among these elements, transportation constitutes a very large percentage of the land cover, including such specific categories as the length and width of roads, driveways, and parking areas, as well as the total area devoted to transportation.

Evaluation can be made of yard characteristics also, as evidenced in figure 15.8E, by counts of litter and junk cars, the adequacy and upkeep of transportation facilities, and the accessibility of various portions of the neighborhood to major transportation routes. Following the outline of table 15.3, therefore, it is possible to evaluate a residential area or neighborhood, such as those in figure 15.8, and to formulate conclusions relating to the character and livability of an area, including its types of problems and their potential solutions.

Commercial Areas

Commercial areas may be subjected to land use and land budget analysis in much the same manner as residential areas. The analysis of commercial areas not only gives information about their characteristics and role within a city, but also about the city itself. Commercial functions vary greatly within and among cities in both amount of area covered, type of activity involved, and general location. Commercial activities range in size from very large shopping centers containing several hundred individual stores to single retail stores scattered among residential or industrial land use districts.

The amount of land occupied by various commercial activities is frequently a roughly representative indication of their worth in terms of (1) dollar production for tax purposes for local government, (2) retail sales, (3) provision of goods and services, and (4) employment of people. Size of physical area, therefore, is important for understanding the role and function of a commercial area. The classification most frequently employed for commercial establishments and complexes is based on the variety and volume of goods sold, which is normally closely related to physical size—that is, the bigger the store or group of stores, the more goods are stocked and sold. Large centers with high volume and a wide variety of items available are termed first order centers, while progressively smaller and less well stocked shopping areas are termed second, third, and fourth order centers.

Types and Locations of Commercial Activities

In urban areas the oldest and largest commercial complexes (frequently over 100 stores) are located in the central portion of cities. Usually the CBD is a first order commercial complex. Smaller neighborhood centers, consisting of about 12 to 20 establishments, normally are located in older residential areas surrounding the central portion of a city. Depending on their sizes, these neighborhood centers may be second, third, or fourth order complexes. Individual shops and strip commercial complexes are often scattered in such areas. In the newer suburbs, the very large and expensive suburban shopping centers are first or second order complexes and are generally located at the intersections of major transportation routes. The principal keys used to identify and delineate the CBD and other commercial areas are listed in table 15.4.

As a first step, the analysis of commercial activities using remote sensing requires the identification and spatial definition of the areas. This is normally easy because the commercial structure of a city stands out very brightly on aerial photographs and space imagery. Plate 20 illustrates the appearance of the commercial complexes of the city of Houston.

Following the CBD in order of importance in size and sales within a given city are the major second order suburban shopping centers. These centers can be identified on plate 20. They are distinguished by their size and situation (on major transportation routes), and by the bright spectral response indicating broad and nearly equal reflection. Further detail on the characteristics of these major complexes, of small commercial sites (small neighborhood centers), and of individual establishments is generally not visible on satellite multispectral imagery. Additional information, however, can be obtained by small-scale, high-altitude imagery, such as plate 20.

Estimating Commercial Activity

A variety of analyses of commercial activities can be performed on remote sensor imagery to provide information in support of planning, policy making, and problem solving, as well as to support the evaluation of existing circumstances. For example, on plate 20 the areas of white color are primarily industrial, commercial, and transportation zones. By calculating the land budget and the percentage of land cover in white it is possible to determine what percentage of Houston is involved in these activities.

Finer measurements concerned only with commercial land use can be made on small areas with a high degree of accuracy, as in the case of the commercial/office facilities shown in a portion of the downtown of a large city (fig. 15.9). The typical setting and context of the CBD is obvious in the image, with major high rise buildings located in the upper right surrounded by parking lots, low buildings of commercial character, and some industrial activity in the upper left and right center. The volume of CBD activity is obviously quite high as evidenced by the

TABLE 15.4
Keys for Identification and Classification of Commercial Areas

Central Business District
      Large number of stores
      Large multistory buildings
      Parking lots or parking structure
      Central location within city
      Absence of general vegetation
      Presence of traffic-restricted open mall
      Selected landscaping
      Absence of residential areas
      Absence of industrial areas
      Presence of mass-transportation facilities or vehicles
      Little bare land/large proportion of land covered
      Possible evidence of renewal or conversion of land use
      Focal point of transportation network
      No or very little on-street parking

Older (Inner) Neighborhood Centers
      Dozens of trash containers
      Corner newsstands
      Street intersection traffic lights
      Street traffic lane markings
      Medium-sized buildings, 1 to 3 stores
      Small parking lots associated with individual structures
      On-street parking
      Located in areas adjacent to central part of city
      Located in neighborhoods of residential land use (usually older homes—single, multiple, or mixed dwellings)
      Centered at focus of transportation network or intersections of relatively major roads
      Absence of general vegetation; large percent of area covered with man-made materials
      Little landscaping
      Small shops and stores
      Parking meters and trash collection
      Signs over street and sidewalk
      Billboards
      Little evidence of land use renewal and reuse
      Some vacant land scattered in patches
      Public transportation

TABLE 15.4 (Continued)

Small isolated stores or centers

Estimated 1 to 6 stores

Small structures or converted residences

Usually surrounded by residential land use

Frequently indistinguishable from residences except for signs, parking facilities, modification

On-street parking

Presence of general vegetation

Not located at transportation network focal point

No trash containers

Possibly no public transportation

Suburban Shopping Centers

Usually large open areas with large central structures containing major retail establishment and many smaller shops and stores

Extensive parking areas marked off in traffic lanes and car divisions

Located in settled suburban area with large lot residence subdivision surrounding it

At major focal point of transportation network or on major highway

Absence of natural vegetation

Landscaped and planned grounds

Public transportation facilities

No on-street parking

---

traffic, congestion, and parking lot usage. By delineating the CBD, it is possible to calculate the number of square feet of land in the CBD. Also, it is possible to calculate the number of square feet in parking lots, streets, or covered by buildings.

Probably the most common and pressing problems of neighborhood commercial areas are the serious threats to their existence posed by the competition with large suburban shopping centers. Studies of older areas have indicated that the most significant factors troubling them are: (1) traffic, congestion, and parking, and (2) poor selection of goods, price, and low number of establishments. Obviously these neighborhood centers cannot match either the CBD or the very large suburban complexes in terms of alternatives.

Remote sensing cannot provide much insight into the issues of prices and selection of goods in such areas, but it can provide perspective and possible solutions to problems of store numbers and size, congestion, and related issues by comparison with other centers. The contrast between shopping centers is clearly illustrated by comparing figures 15.6 with figure 15.10, both in Omaha. It is possible to compare these areas on the basis of parking, size, and, therefore, presumed selection and volume of goods.

Small individual establishments scattered in residential or other areas, whether they are drug stores, beauty shops, gas stations, or other small retail outlets appear distinctive, and at large scales enable very detailed analysis not only of square footage, but also of specific function and degree of usages—traffic, sales, volume, and specific types of activities that occur with specific areas as evidenced by patterns of stains, spills, and wear marks on the different areas.

Figure 15.9. Low altitude vertical aerial photograph of a part of the downtown of a large city. Detailed measurements of commercial land use can be made at this scale. (NASA, Johnson Space Center)

Figure 15.10. West Roads Shopping Mall, Omaha, Nebraska. (Remote Sensing Applications Laboratory, Department of Geography-Geology, University of Nebraska-Omaha)

In the very large suburban shopping centers (figs. 15.10 or 15.11A, B, C), not only can accurate estimates of space and retail volume be obtained but also estimates of such dynamic phenomena and patterns as to which types of establishments or establishment arrangements attract the greatest number of customers, as well as when most customers enter. For instance in figure 15.10, parking lot usage is obviously much below capacity. It is significant to note, however, that it is much higher in certain portions of the lot than in others. This pattern raises questions, such as where is the best store location in the building—the one with the greatest customer accessibility? What types of stores or establishments should be there? If the mall were to be enlarged, would new land have to be purchased for parking? Answers to these questions can be obtained in part from figure 15.10.

Industrial Areas

Industrial areas occupy an important and somewhat large amount of the total area of most cities (fig. 15.12). Table 15.5 summarizes some of the basic characteristics of different industries in terms of those features which can be observed on remote sensor images. For a detailed discussion on applications of remote sensing in industrial analysis refer to Chapter 17.

Locational factors and the spatial distribution of industries adjacent to an urban area can be seen in figure 15.12. The typical setting of a major heavy industrial concentration can be seen in the area between the residential area and the river. The industrial area is characterized by large buildings, large outside storage yards, roads, railroads, water transportation facilities, and small areas devoted to parking.

Fabricating and assembly industries usually are located adjacent to the central business districts or between the CBD and inner residential areas. Finished products and warehouse ac-

Figure 15.11. Features of suburban shopping centers. (Remote Sensing Laboratory, Mississippi State University, NASA grant NGL-25-001-054, Applications of Remote Sensing to State and Regional Problems)

A   B   C

Figure 15.12. Heavy industrial complex. Observe the large buildings, storage yards, and associated transportation linkages. New residential areas have developed near the industrial area. (NASA, Johnson Space Center)

tivities tend to be located in still more peripheral urban sites or among other (often older residential) land uses (plate 18).

Transportation Systems and Traffic Flows in Cities

Urban areas are probably the most changed part of the cultural environment, and the transportation system of cities is one of the most changed elements of the urban scene. Technology in transportation has had a significant influence on two components of the patterns and systems of transportation of urban areas: (1) the static component—such as transportation characteristics, including the location, arrangement, and connection of roads and (2) the dynamic characteristics, including flows and obstructions of traffic, goods, and people.

TABLE 15.5
Factors for Distinguishing General Classes of Industry

Industry	Parking Lots	Materials Storage	Transportation	Building Size, Equipment	Site Size
Primarily processing of products	Generally small lots and few employees	Large open area	Main highways and railroad lines	Large buildings Large outdoor equipment	Large site
Fabricating and assembling of products	Medium size	Little or no open storage Storage in medium-sized buildings	Primarily loading docks and trucks Occasional railroad for large products	Medium to large buildings No equipment outside	Medium lots Often grouped with similar industries
Finishing of products	Small lots	No external storage	Usually loading docks and trucks	Medium buildings—often multistory No equipment outside	Medium to small lots Grouped with similar industries

The Nature of the Urban Transportation System

The characteristics of the transportation network of a city reveal much about an urban area. The network can be analyzed to produce information about the urban structure and the nature of its problems and their possible solutions. For instance, in both figure 15.1 and plate 20, the transportation structure of a city stands out. Plate 20 shows a large city with an extremely complex transportation network which has no clear focus except for the intersections of primary access routes and main streets. The outstanding characteristics of this network are seen in the contrast of configuration in the orientation and density of the transportation subsystems of different neighborhoods. In plate 20 the focus and character of the transportation network is indicated by the converging routes and the circumferential and crosstown throughways.

Static Elements of Transportation Systems

At a much larger scale, a variety of high detail analyses of the static elements of a transportation system become possible. One of the most important of these analyses for understanding cities is the determination of the land use budget. Within the context of the static elements of transportation and traffic flows of cities, the principal component of the land use budget, as has been stated earlier, is automobile space, that is, the amount of land used by and allocated to automobiles. Another component is the influence on the road network of environmental factors, such as hills, valleys, rivers, and swamps.

The static analysis of automobile space is obtained from the total area of streets and parking lots in a city or neighborhood. These data are obtained by measuring the length and width of each street, drive, or lot, and then calculating the area of each. The results can be totaled and the percentage of automobile space in each image, or part of an image, calculated. On this basis, the approximate amount of automobile space for the commercial area of figure 15.10 is 59 percent, and for the industrial area of figure 15.12, about 8 percent. Thus, in terms of the static elements of their transportation and traffic flow systems, the two images are quite different.

Dynamic Elements of Transportation Systems

The dynamic aspects of the accessibility within a city can be approached on both a general level and a specific level. Assuming that delays and obstructions in travel occur mainly at intersections, it is possible to count intersections along the shortest route between sets of points and arrive at a number which reflects the estimate of accessibility between these two points. By counting intersections between certain points in a city, it is possible to find the center of accessibility for a city.

Parking, Traffic Movement, Congestion, and Obstructions

The dynamics of parking and traffic flows are among the most significant problems facing major cities today, and the availability and capacity of parking areas are clearly visible on remote sensing images, such as figure 15.9. Often this information can be used for planning and decision making regarding changes in urban land use which are reflected in the land use budget. In figure 15.10, there appears to be excess parking capacity, and planning new developments to utilize certain portions of the parking lot might be reasonable. Caution, however, must be exercised in making such decisions using remote sensing because of the time at which the data may have been gathered. For example, if the image in figure 15.10 were taken at 6 A.M. on a Sunday morning, the interpretation of the situation would result in a much different conclusion than if the images were taken at 2 P.M. on a Tuesday, 2 P.M. on a Saturday, or on December 24.

Measures of the static and dynamic characteristics of a transportation network when combined with certain other dynamic characteristics of transportation and highway systems within urban areas—such as an assessment of which areas are traffic generators at which time of day, where traffic bottlenecks are and their duration, and what the speed and density of traffic flows are or should be—provide important pieces of information for solving some of the most irritating urban problems. The information may be used not only to understand urban areas and their dynamics but also as a basis for policy decisions on such detailed issues as transportation route expansion and relocation, speed limits, and road maintenance levels.

Information on which areas are traffic generators at different times of the day are easily available from images such as figures 15.6 and 15.9. If these had been imaged at the close of a business day, thousands of vehicles would be seen leaving the central business district on the main highway (2) in figure 15.6. Their destinations would be the residential areas of the city, and the bridges, intersections, and feed-ons would show major bottlenecks in the flow of traffic to the residential areas. The absence of traffic and obstructions in figure 15.9 reflects the primary difficulty in transportation systems—the fact that during much of the day roads and other automobile spaces are vacant or largely underused, while during very short periods of time they are subjected to heavy use.

These conditions complicate the policies, investments, and decisions involved in road planning and design, and are part of the reason that information from all sources, including remote sensing, must play an important role in transportation system design. For example, in figure 15.9, a midday photograph, it is obvious that parking facilities are near or at capacity, while roads show a high degree of under utilization. Similarly, figure 15.13, which shows a parking lot at capacity and nearly totally filled on-street parking, reflects transportation problems during a busy period when traffic on crowded and congested streets flows into parking areas.

The essence of congestion is revealed in the bottleneck of figure 15.14 where a recent automobile accident has closed a roadway and completely blocked traffic flow. Figure 15.15 showing part of San Francisco and the approaches to Golden Gate Bridge, illustrates the potential for traffic congestion and bottlenecks. During rush hours, traffic on several multilane (5 lane) roads bunches up and creates congestion where roads converge and diverge.

Figure 15.13. Parking lot almost filled to capacity. (Remote Sensing Applications Laboratory, Department of Geography-Geology, University of Nebraska-Omaha)

Figure 15.14. An accident blocks traffic flow. (Remote Sensing Applications Laboratory, Department of Geography-Geology, University of Nebraska-Omaha)

Figure 15.15. Bottlenecks are created in traffic flow by the way roads are laid out, as when several thoroughfares converge at the Golden Gate Bridge in San Francisco. (U.S. Geological Survey)

In using these types of images to solve traffic flow and congestion problems, several questions must be raised: (1) Where is the congestion?, (2) Why is the congestion there?, (3) Where is the traffic going?, and (4) How can the flow be improved, bottlenecks avoided, and congestion dispersed?

A fundamental consideration in traffic planning is traffic speeds, and these can be estimated closely. For example, suppose a van is observed travelling on the image of a highway. Also suppose the oblique distance or effective flying height of the aircraft from the highway is approximately 2,000 feet and that the camera focal length is 3 inches. The scale of the image at the highway, therefore, is 1:8,000 or 1 inch = 667 feet. If the images were taken 15 seconds apart and the image of the van moved approximately 1 3/8 inches on the photographs, or 915 feet on the ground, then the speed of the van was about 41.5 miles per hour. If this speed represents the approximate rate of traffic flow along this route, the capacity of this section of road is one car every 55 feet (its own length of 15 feet + 10 feet for every 10 mph of speed, or 40 feet). By determining the length of the road and knowing the need for 55 feet per car, it is possible to calculate the number of cars the stretch of road can accommodate. The speed of vehicles is only one of the important factors influencing traffic and transportation characteristics in cities. Considered in light of some other characteristics, such as congestion, parking, and neighborhood traffic characteristics (source or destination), speed often stands out in summarized flow pattern studies, and, as a result, it has received a great deal of attention.

Natural Disasters

Natural disasters, such as storms, earthquakes, floods, and other violent happenings, have occurred throughout time and in many places. They generally attract the greatest interest when they affect or modify human behavior. These events are almost always dynamic in character. Sometimes they last only a few seconds; however, evidences and results of their existence are static and can be used to determine what happened and what can be done to relieve suffering in the future.

Cities occasionally become the scene of these violent natural happenings. Figures 15.16A and 15.16B show damage from a 1975 tornado in part of Omaha. In such a disaster, remote sensing becomes a valuable tool for damage assessment, insurance loss evaluation and reimbursement, planning for short- and long-term emergency disaster aid, and general disaster planning. Aid in the form of shelter, food, and medical care is the first concern, as well as security to prevent looting of the damaged and abandoned areas. Images such as these can provide estimates of the number of people and amount of area requiring aid. For example, among the 14 residences in figure 15.16A, 6 units appear to be more or less destroyed; thus, shelter and other aid must be provided at least six families, or approximately 15 persons. In addition, 5 residences appear to be partly damaged and only 3 appear to be relatively undamaged. Of the 44 units in figure 15.16B, 15 appear to be totally destroyed and only 3 show no damage. For the neighborhood in figure 15.16B, 38 persons, and possibly other individuals, will need food, clothing, shelter, and other forms of aid.

An additional long range need in such natural disaster areas is general planning. This is often quite important since such natural disasters frequently result in long-run improvement by

A   B

Figure 15.16. Tornado damage in Omaha, Nebraska, 1975. (Remote Sensing Applications Laboratory, Department of Geography-Geology, University of Nebraska-Omaha)

causing the removal of old or undesirable facilities and their replacement with newer, better planned, and more suitable activities and structures.

Conclusion

The urban scene and setting, composed as it is of dynamic and static elements and evidences of human occupance, is an extremely complex situation. The forces, pressures, and general intensity of activities—building, movement, land use and resources, and competition for space—generate stresses and difficult situations which further complicate the problems of life in cities. Despite the intensity of forces pushing cities to disorder and decline, the remote sensing perspective indicates clear evidence not only of order but of a successfully functioning system. To a very large extent the present condition of cities and urban areas is one of natural evolution.

Seldom has there been effective shaping of urban activities, processes, and land uses through planning, and even less often has shaping and decision making occurred under conditions of relatively complete information.

The perspective afforded by satellite and airborne remote sensing imagery has provided a great deal of information not only on the static elements of urban areas, but also on the dynamic activities. The value of this information lies not only in its volume, the great detail of facts, and its objective and repetitive nature, but also in its availability, low cost, and ease of use. For these reasons such imagery is increasingly being employed in urban studies as part of the input into urban information systems. By itself, remote sensing can provide only part of the information necessary to understand urban areas and solve urban problems, but remote sensing data, because it provides the basic structure of the city, is probably the most important part of any set of information about a city.

16

Application of Remote Sensing in Regional Planning

Lee Guernsey and Paul W. Mausel
Indiana State University

Introduction

Remote sensing for regional planners will benefit planning operations in many ways. This chapter includes information about the progress which has been made by using Landsat data for regional land use/land cover studies, a state land use inventory, surface hydrology studies, coastal zone studies, and vegetation studies and forest inventories, along with an analysis of surface mining reclamation activities.

Studies conducted by the writers support the unmatched capabilities of remote sensing with respect to recording large area land use/land cover trends and the flexibility of various classifications. This chapter emphasizes refinements of the remote sensing techniques toward a more reliable data base for geography and planning activities in improved resolution in remote sensing imagery.

The chapter also attempts to answer questions about relative costs of using Landsat satellite data in comparison with other methods of obtaining land use/land cover data. The cost information included should assist geographers and planners in deciding upon the feasibility of using satellite multispectral data for their studies. First, however, to enable the reader to appreciate fully the importance of remote sensing to regional planning, the chapter provides information on the role of governments and planning at all levels—federal, state, and local.

Background

Before the middle of the nineteenth century, American land developments were simply the products of subjugation. The overall development was haphazard and random until, in 1862, the Homestead Act set the stage for a more systematic stewardship of our land. The national concern for a more organized growth which followed developed against a backdrop of the emerging importance of cities and urban regions as the open areas steadily receded. A few scattered ordinances were passed that forbade "nuisance" land uses from developing except in certain districts designated for such uses.

Selected ordinances regulating building height and land use appeared in some United States cities by the turn of the twentieth century. During the next decade many cities passed local ordinances dividing real estate into districts restricting some land uses. This system provided planners with a process containing a wide range of options with which to achieve a consensus of land uses. In 1926, the Supreme Court gave its blessing to these zoning practices.

Land use regulations were generally considered to be mostly urban problems. A Standard Zoning Enabling Act of 1922 delegated the responsibility for zoning to city governments. Variations of the law were adopted during the 1940s and 1950s. These variations resulted in an increased number and kinds of zones. Later, greater flexibility was introduced through open space ratios, floor plan ratios, and performance standards.

While the Standard Zoning Enabling Act was a state enabling act, it was only an enabling act and was directed at delegating land use control to the local level. It has become increasingly apparent that the local zoning ordinance has become a primary means of land use control in the United States during the last half century.

The nature of the innovations in land use regulations varies from region to region, and from one city to another within a region. Some of the devices employed are old ones, others are new in concept and design. The commonality, however, is a land resource orientation which attempts to preserve and protect the land for the use of the region as a whole.

An increasing dissatisfaction with existing land use patterns is manifesting itself in a trend toward providing a more viable interest for local and regional governments in land use management. The trend is similar to the wave of ecological concern that has aroused recent national attention. The idea that people do not own the land in the same sense that they own structures or commodities which they build or create may appear self-evident to geographers and planners. In the context of land use traditions, however, it is a changing concept. The existing land use regulations were created by investors in real estate interested in maximizing the value of land as a commodity. Subdivision regulations have encouraged uniform lots to be divided into tradeable units. Urban and regional land use regulations generally made little attempt to conserve land for particular purposes or to direct it into a specific use, but only sought to prevent land from being used in a manner that would depreciate the value of neighboring land. To achieve this purpose they sought to restrict those land uses which adversely affected the price of neighboring land by concentrating them in specific areas.

Land use regulations have traditionally been restricted to urban areas. Generally, there were few regulations of land outside urban areas where reductions in value were not likely to occur. Recently, there is an increasing recognition that the purpose of land regulations should go beyond protecting commodity values of land. A realization is growing that important social and environmental goals require additional specific regulations and land uses.

This recognition is seen by the changing role of regions in land use regulations and also in changing actions of urban governments. Modern zoning ordinances typically rely less on prestated regulations. They require developers to work more closely with city administrative officials in designing developments that conform more closely with surrounding neighborhoods. Similarly, recent actions tend to encourage large scale developments in which various land uses are arranged and designed according to comprehensive plans rather than lot-by-lot developments.

As planning perspectives have evolved during the past century, so have the techniques for gathering the resource base information which is prerequisite to intelligent land use regulation and planning. Remote sensing techniques have a long history as valuable data collection tools, from mountain-top surveys of the landscape to today's cameras and more sophisticated sensing devices, such as airplanes, satellites, boats, balloons, and cars. The unique perspective and scale which these techniques provide make them both essential and invaluable to planning at all governmental levels.

The Role of the Federal Government

The federal government has affected land use planning by various federal grants and aids that support cities and regions. Many federally aided public facilities must conform to local and regional land use plans. In states where there are significant federal land holdings, the federal government also has had a direct influence on land management.

Federal Programs

The federal government has had a significant influence because of a host of water control projects, federal installations, and the vast array of federal aid to city and regional governments and private individuals. Federal programs support highway construction, airports, urban renewal, sewer and water facilities, and open space acquisition, as well as the housing mortgage insurance and subsidy programs designed to aid specific families. For projects assisted by these programs, remote sensing methods can provide useful information pertaining to site selection, projections of project impact on the environment, and monitoring of impact and land use modifications during and after project completion.

Environmental Protection Agency and Council on Environmental Quality

With a growing concern for the environment, the role of the federal government in influencing land use has come under increasing scrutiny. The establishment of the Environmental Protection Agency and the Council on Environmental Quality has provided more concrete assurances that federal programs will consider regional impacts on land use. The federal water quality control "208" studies are a case in point. Some of these studies are making use of remote sensing capabilities for detecting area and point sources of water pollution, both natural (e.g., sedimentation) and man-made (e.g., thermal). Another example is the study published by the Public Land Law Review Commission under the title *One Third of the Nation's Land,* completed in June 1970 at a cost of more than $7 million, where 137 separate legislative recommendations were made. Its preparation required an evaluation of nearly two centuries of legislation relating to public lands.

Several recommendations are included in the report. For example, the Commission recommends that public lands should be sold when the greatest public benefit would result from such action and that "management of public lands should recognize the highest and best use of particular areas of land as dominant over other authorized uses." There should be an "effective" role for regional and local planning in federal land use planning, and the Commission recommends that no public land use be permitted which conflicts with local zoning. It also urges that environmental quality be recognized by law, as an important objective of public land use management.

The National Environmental Policy Act of 1969 requires that approval and design of federal or federally aided projects include an impact study on the environment. Land use consequences are an important criterion in this evaluation. In its first annual report to the President, the Council on Environmental Quality identified the regulation of land use as an important element within broader environmental concerns.

Considerable federal legislation has recently been passed that would attempt to enhance or maintain the natural environment while fostering controlled development which follows proper ecological principles. Essentially, the basic premise behind the major land use legislation is that a greater degree of coordination of policies and programs would solve many of our current economic, social, and environmental problems and reduce existing waste of natural resources.

The Role of the State

States have become concerned with land use. Their recent concerns have been directed to land use problems. States have directed their attention toward decisions involving important state interests in areas of critical concern, while retaining local control over the great majority of matters that are primarily local.

Areas Important to States

Defining all areas having important state interests is difficult; however, two key criteria are appropriate for defining the areas. One is to designate sensitive areas that have unique environmental values, delicate ecological balances, or unique resource characteristics. The other criterion consists of land uses that have a widespread impact, not because of their uniqueness or sensitivity, but rather because of their magnitude.

The first category includes swamps and marshes, lake shore beaches, wilderness areas, and historical and archaeological sites. The second category includes mined lands, floodplains, open spaces, highway interchanges, airports, parks, and recreational areas. Most of the land features and/or uses in these two categories can be readily delineated by remote sensing techniques. Furthermore, these techniques can be applied to monitor fluctuations in the areal extent as well as in the vegetative, hydrologic, and geologic characteristics of both the critical areas and their neighboring lands. The geographical distribution is often the critical factor in designating areas of state critical concern. For example, developments such as parks or airports will have an impact far beyond the area where the facility is located. The withdrawal of large acreages from private uses is also of state concern.

Critical Areas

To identify the areas of critical state concern, specific criteria are needed. The criteria should be as explicit and as quantifiable as possible to avoid arbitrary decision-making and fluctuating standards. Such criteria might include a specified minimum size, unique or rare phenomena, flood-prone areas, unique topography and geological formations. A certain number of scenic areas where urbanization is expected to alter future natural conditions and areas to be acquired for flood-control reservoirs may also be included.

After criteria are established, a compilation of critical areas should be identified on a use basis in as small an areal unit as possible and with as much separation as remote sensing techniques permit. The approach to using remote sensor imagery is especially appropriate for use in the development of a data bank of state information about these areas. They should be examined to assure that only the clearly critical ones are included with cognizance that the state's programs should single out the most acute areas and not include all environmentally valuable areas.

Once the initial inventory has been remotely classified to meet the purposes of the program, specific delineations of the land use zones should ensue. These should be minimum and maximum zone sizes; a minimum size to embrace the relevant environmental features and maintain state significance; and a maximum size to keep areas at a manageable level as well as to take into account boundaries of political jurisdictions. The delineation of zones should include the area immediately adjacent to critical areas where developments could significantly affect the environmental quality of critical areas but where lesser control over developments may be appropriate.

Once critical environmental zones have been designated, development standards and criteria for each zone should be established. As much as possible, the standards should be stated in performance terms to permit greater flexibility. By using remote sensing technology, states may acquire the data needed to protect and preserve areas that are of critical concern. By also providing a framework for coordinating public improvements, states can further insure protection and preservation of critical areas.

Inventory and Preservation of Open Space

Another role of the state in regulating land use is to identify and preserve an appropriate amount of open space. Of the rationale, those relating to conservation, recreation, and shaping future urban growth are most commonly accepted. Open space can be used by states to preserve historical, archaeological, and geological sites; to conserve wildlife habitats, water supplies, valuable forests, prairies, and agricultural land; and to minimize water runoff, soil erosion, and flood damage.

Of the many uses of open space, recreation has received the most emphasis. Most states have a recreational situation which is characterized by too little land, overcrowding, and lack of adequate programs to acquire open space for recreation.

There are also approaches toward influencing future urban growth by open space. One approach is to preserve open space so that developments are shaped around it. Another approach is to preserve open space temporarily until it is considered ready for a planned unit development. The open land may be purchased by the state and temporarily used for recreation or it may be preserved. Urban growth may be regulated on a large scale by having greenbelts surround new towns, or it may be on a small scale by maintaining a green strip between a highway and a community.

A major barrier to developing a positive strategy for keeping land open is a general lack of information about where the best undeveloped acres are located. Multispectral imagery and aerial photo interpretation have proved to be useful for developing open land resource inventories. The inventories can also identify areas for agriculture, forest, and recreational open use. These inventories can provide states with data for developing appropriate policies to assure wide open space management.

The open space inventory would enable the states to begin setting priorities on open land. Obviously not all land presently used for agriculture, forest, or open space could stay that way permanently. Economics of food production, for example, will continue to concentrate farming on fewer acres, with or without state intervention.

Several states have passed acts that provide for the creation of conservation districts which may acquire, preserve, and maintain open space. The principles of preserving scenic and aesthetic assets and protecting natural resources are specifically supported by county governments that may levy countywide taxes to support the program. The significance of these new conservation districts is not so much in the granting of new authority as it is in providing a new emphasis and impetus to a coordinated program of preserving open space.

Vulnerable Environments

The conversion of vulnerable environments to intensive land uses and the degradation of environmental quality are creating a steady decline in available public opportunities to use and enjoy areas that are essential to man's well being. These include areas that have value not only in the esthetic sense, but also can contribute to a better understanding of ecology and man's relationship to his environment. Consequently, most states have such objectives as (1) to protect the vulnerable environments by restricting developments, (2) to enhance the educational and scientific value of sites that are preserved, (3) to strengthen an appreciation of natural history, and (4) to foster a greater concern in the conservation of the state's natural heritage. Remote sensing techniques can assist in the identification and delineation of these environments, an obvious prerequisite to satisfying the aforementioned objectives.

Vulnerable environments may be classified into two categories. One category includes those areas that have a significant cultural, scenic, and ecological interest or possess high-risk development potential. A second area of vulnerable environments includes those land uses necessary to health, such as landfills and wastewater spray irrigation lands. Remote sensing technology can help identify optional sites. In short, a program of site predetermination and reservation is needed at state levels to assure the availability of these lands to meet our projected needs.

The Role of Regions

Regional Planning

The concept of regional planning is a relatively new approach in coordinating planning activities during recent years. Planning regions are normally composed of several counties that may be linked by common problems, resources, and opportunities. Economic and social interdependence bind the people of a region together. The regional planning approach provides the means for tackling mutual problems and taking advantage of joint opportunities. It allows for an economy of scale by pooling regional resources. Though the definition of regional planning in the United States is changing, the fact remains that regionalization in land use planning is on the upswing in the United States. Regional planning has both the weaknesses and strengths of confederation. Regional planning is normally advisory and is subject to termination, succession, and unevenness. Its principal tools of implementation are education, influence, and leadership.

Support for Regional Planning

In 1966, the federal government called for increased cooperation by establishing planning regions. Certain advantages and benefits seemed inevitable. Regional planning would provide the framework for state-wide developmental decisions. Implementation of regional planning programs meant a better coordination of federal, state, and local programs.

In 1969, the Bureau of the Budget issued Circular No. A-95 which further promoted regional planning by intergovernmental cooperation. The regulations promulgated under Bureau of the Budget Circular No. A-95 were aimed at promoting more effective coordination of planning and development activities that were assisted by funds from the federal government. The major device of the regulation was encouragement of systematic communications among federal, state, and local governments in carrying out planning activities. The regulation also required that applicants for federal assistance for regional planning identify related regional planning activities and demonstrate how they will coordinate their own activities with them. In order to comply with this regulation, regional planning agencies will require a comprehensive regional resource inventory as well as an up-to-date record of land use. Remote sensing methods not only can acquire a large portion of this information, but machine processing for analyzing remotely sensed data can then inexpensively and quickly map this information at different scales to fit the user's needs.

Circular A-95 promulgates regulations which provide, in part, for:

1. The establishment of a project notification and review system to facilitate coordinated development planning on an intergovernmental basis for certain federal assistance
2. Notification, upon request, of governors and state legislatures of grants-in-aid made under federal programs in each state
3. Coordination of federal development programs and projects with state, regional, and local development planning

Circular A-95 is an expansion of an earlier notification and review system created by Section 204 of the Demonstration Cities and Metropolitan Development Act of 1966, which created areawide clearinghouses for data banks.

Regional planning has developed at all levels of government. Federal legislation stipulates regional planning as a prerequisite for loans and grants for public works projects. State governments are finding regions to be an effective spatial unit for planning various programs. Local governments realize that their problems do not end at their jurisdictional boundaries and that cooperation is a necessary step toward the solution of regional problems.

Council of Governments

Another growing area of regional planning is the Council of Governments. The Council of Governments deals with regional problems by coordinating plans and policies, but it is a voluntary organization composed of elected officials. By bringing together elected local officials into a regional discussion of common problems, a better understanding and awareness of each other's difficulties will develop and regional goals will be established with priorities set for the optimum allocation of land uses.

Councils of government lack the authority to implement regional policies and decisions since the organization is voluntary. However, the Council is normally composed of influential elected officials who are working for the common good of the region. This involvement and trust helps promote and implement land use plans.

The Role of Local Planning

Land Use Guidelines Formulated by Local Government

An effective land use management program must involve all four levels of government: federal, state, regional, and local. In the urge to act innovatively states must not lose sight of the fact that the majority of land use management tools still remain in the hands of local government. New ways, however, should be discovered by which these techniques and tools can be improved.

Local regulations of land use have been in existence for many years in urbanized areas. These local regulations have traditionally regulated small-scale developments in urban areas, but, at a time of increasing demands for space and land, the value of traditional controls is being reexamined.

Most land use guidelines are actually formulated and enforced by local ordinances. Most states hand local governments a skeleton outline of land use, and local governments fill in the body of information about that land use.

Most local governments, however, have only that power lawfully vested in them by the state. As the state moves toward a system that is not weighted heavily towards development, it is increasingly necessary to merge both state and local regulations into a single system with specific roles for both state and local governments defined more clearly.

Zoning

Zoning is the primary tool that is used by local governments to regulate land use. With almost no exceptions, zoning is a legal device enacted by local governments to exert control over an area's land use. It is a regulation that is the responsibility and prerogative of local governments; however, authority to adopt the controls is delegated to local governments by state enabling acts.

A distinguishing characteristic of zoning is the division of jurisdictional areas into zoning districts with uniform regulations throughout each district, but with differing regulations for different zones. Zoning ordinances can be based in part on soils studies, clear delineation of flood plains; and accurate measurement and reasonable forecasts of the major categories of land uses that can be classified and mapped by remote sensing techniques.

The traditional role of zoning is that of dividing areas into districts most suitable for residential, commercial, and industrial activity and restricting all future development to an appropriate district. This restricted concept has been expanded in recent years to become a device for excluding nuisances, for arranging land uses that are not nuisances, and for establishing height, lot size, floor space, and bulk standards. It also has been applied to protect floodplains and steep slopes and to preserve historical sites, marshes, and wetlands. Zoning can be used to preserve prime agricultural lands, greenbelts, and scenic open spaces from being absorbed by sprawling urbanization. Many of the zoning measurements can be made remotely.

Subdivision Regulations

Subdivision control ordinances are another tool that can be used by local governments to regulate land use. The rationale for subdivision regulations is simply that the subdivided and developed land has a vital and lasting effect upon the entire community. The private developer seeks the benefit of a systematic recording of his lots for sale. Land subdivision regulations arose from the need of local governments to assure rational patterns of streets as new areas were developed. Controls normally are applied through specification or approval of such criteria as reasonable regulations regarding size, scale, and other plat details; orderly development of the area; coordination of streets; adequate provisions for drainage and flood control, light, and air; storm and sanitary sewers, as well as the other utilities provided.

The most beneficial placing of subdivisions in an urbanizing area is obtained by powers of zoning and subdivision control in conjunction. The placing of structures within the subdivision is affected by the community's desires for a specific street layout; lot size; dedication of widening strips along existing boundary streets.

The placing of subdivisions may be more stringently regulated if unique factors exist. For example, a subdivision located on especially steep or rocky slopes or on low-lying ground may have unique sets of requirements or design standards which can be applied to it. A subdivision that will create parking, traffic, or transportation problems should similarly be subject to conditions which will ameliorate these problems. Therefore, if unique situations demand a certain flexibility in subdivision control ordinances, the local governments need a certain ability to deal with the community's best land use interests.

Protective Covenants

One of the oldest local land use regulations is the deed restriction. Local governments enforcing zoning and subdivision regulations may require developers to file covenants designed to ascertain the existing land use of a given area. They generally go beyond the requirements of health, safety, and welfare of zoning and subdivision codes. In addition, they provide insurance against possible adverse effects on nearby surroundings caused by new developments. Title companies, banks, and other institutions may insist upon the protection provided by well-thought out deed restrictions.

Other Regulatory Tools

There are other land use regulatory tools in addition to zoning laws, subdivision controls, and protective covenants. They also have an effect upon local land use patterns. Official mapping is aimed at preventing developments in the beds of streets. As applied to local highway programs, this tool normally enables the purchase of rights-of-way at prices more nearly approximating raw land values. Setback ordinances are designed to prevent construction on that portion of a tract of land abutting existing streets and local roads for safety and to enable the more reasonable acquisition of these lands when the widening of streets becomes necessary. In addition, ordinances forbidding construction of homes on land not served by public streets and ordinances specifically forbidding building development where the terrain is too rocky for sewer and water installation, too low to be healthful, too steep, or too prone to flooding to be safe are all possible tools to aid in the accomplishment of local land use plans.

Local governments may affect land use by adopting building construction codes that restrict the height of buildings and materials used, provide inspection, and condemn buildings deemed unsafe for fire hazards. Cities also may build and regulate all functions connected with water transportation, storage, commercial activity, and waste disposal within the local area.

Land needed by cities for alleys, streets, sidewalks, parks, and public building sites may be obtained by eminent domain proceedings. Local governments may build and manage public parks and levees, obtain public utility easements, and even mow the grass and send the owner the bill.

Remote sensing techniques provide vast and ever increasing amounts of information concerning atmospheric conditions and earth surface and sub-surface features. These techniques can be, and to a limited extent have been, applied at all governmental planning levels to assist in policy formulation and problem solving on a variety of scales, from aerial photo interpretation for large scales to machine processing of multispectral data for smaller scales.

To date, this information has been used in many ways, including resource base inventories, project site selections, impact studies, and land use regulation and planning programs. Perhaps the greatest collective advantage of remote sensing techniques, however, lies in their capacity for monitoring (1) land use modifications by nature and man, both individually and in concert, and (2) the effectiveness over a period of time of planning policies and projects through repeated analysis of project areas. At a time when intelligent resource management and effective land use planning are more crucial than ever before, it is of paramount importance that this watch dog capability of remote sensing be utilized to its fullest.

Data for Regional Studies

The planning process involves selecting data to be analyzed; gathering data on physical, economic, and human resources; determining the demand for resources; and a plan design. An overview of the kind of data useful to conduct urban regional studies is fundamental to planning at all levels of government. In 1968, Congress added to the federal highway legislation a clause requiring that plans for a federally aided highway project submitted by a state highway department must certify that it has considered the economic and social effects of such a location, its impact on the environment, and its consistency with the goals and objectives of such urban planning as has been promulgated by the community. The National Environmental Policy Act of 1969 goes even further by requiring every federal agency to assess in detail the environmental effects of proposed legislation and activities. These reflect the growing concern that our environment be conserved and the use of our natural resources planned.

Much of the analysis necessary prior to developments can be avoided if a detailed inventory and analysis of environmental resources are included within initial planning. The main purpose of such an analysis would be to determine the capacity for development and, at the same time, to determine which areas should remain in open space. Thus, for greatest benefit, reliable information about the environmental resource base should be collected on an areawide basis. Obtaining and maintaining an information base is fundamental to urban regional planning studies.

Fifty years ago, the planning process, because of its dependence on time-consuming field studies and laborious field mapping for the necessary resource base information, was severely limited in its scope and effectiveness. With the advent of aerial photograph interpretation and its associated mapping capabilities, the entire planning process of data retrieval, analysis, program design, and implementation was not only greatly accelerated but was also made more accurate and effective for larger areas.

Today's satellite technology is having a similar effect, helping expand planning effectiveness to a regional basis as resource base inventories become more readily compilable and comprehensive. With the continued refinement and combined application of field and remote sensing techniques, the planning community can now develop, utilize, and maintain inventories necessary for intelligent and effective programs with various purposes and at various scales.

Soil

An enormous amount of useful data may be obtained from soil surveys, soil association maps and soil information and interpretation studies. The results of soil surveys are a series of maps that depict data about soil type, thickness, slopes, drainage, elevations, and vegetation cover. The Soil Conservation Service was mandated by the United States Congress in 1966 to make their surveys "in connection with community planning and resource development" in addition to agricultural interpretation.

Modern soil surveys provide the following types of data: (1) soil capabilities for common cultivated crops, crop yield estimates, suitability groups, and crop adaptation; (2) capability of the different kinds of soil to sustain natural cover for birds and animals; (3) soil suitability for lawns, golf courses, playgrounds, parks, and open space reservations; (4) identification of areas subject to flooding, stream overflow, ponding, seasonally high water table, and concentrated runoff; and (5) soil properties influencing engineering uses.

Planning interpretations about soil suitability for various land uses can then be derived for the following:

1. Suitability ratings for potential residential, commercial, industrial, transportational, natural and developed recreational and agricultural land uses
2. Suitability ratings for use as a source material for road base, backfill, sand or gravel, topsoil, and water reservoir embankments and linings
3. Ratings with respect to flooding potential, watershed characteristics, susceptibility to erosion, and susceptibility to frost action
4. Suitability for wildlife habitat and habitat improvement, lawns, golf courses, playgrounds, and parks and related open areas requiring maintenance of vegetation

These data may be combined on a series of interpretive maps showing suitability for such land uses as: (1) agriculture, (2) industry, (3) transportation facilities, (4) housing with sewerage, (5) large-lot housing without public sewerage, (6) small-lot housing without sewerage, (7) recreation, and (8) wildlife preserves.

Topography

Most topographic maps are developed through aerial photograph interpretation, itself a remote sensing technique. Because of scale and time limitations, however, these maps do not separate, among other things, forest and vegetation types, agricultural land uses, urban building types or functions, residential district types, nor changes over time. These differences and more can be readily identified through machine processing of remotely sensed data, for areas much larger than those covered by aerial photographs or topographic maps, and in much less interpretation time. Consequently, the best utilization of these data sources is through a combination of their respective strengths. Remote sensor data analysis can definitely supply vast amounts of information quickly and cheaply, but such analysis in turn requires "ground truth" information which can be supplied by aerial photographs, field inspections, and topographic maps.

Topographic data may be obtained from the U.S. Geological Survey topographic quadrangle maps at scales of 1:24,000. Topographic information is fundamental to urban and regional planning because hills and valleys, ridges, height and degree of slope all affect the intrinsic suitability of the land. In addition, the large-scale topographic map reveals information about transportation networks and facilities, vegetation, drainage, and settlements.

Remote sensing techniques can be applied at various times during the data collection process to verify and/or augment the information which is being supplied by some other method, such as the delineation of flood prone areas as identified in a soil survey. Perhaps more useful to the planning process, however, is the capability of remote sensing techniques to provide constantly up-to-date land use maps. Such maps can be used in conjunction with the aforementioned suitability maps to identify planning priorities: (1) areas where the current use is so alien to the land's suitabilities that immediate planning attention is required to correct mismanagement and avoid environmental damage; (2) areas which currently exhibit a marginally suitable use pattern that should either be modified or at least carefully monitored; and (3) areas where current use patterns are sufficiently in accordance with suitabilities that they constitute areas of intelligent conservation.

Through proper interpretations of the USGS topographic quadrangles, data can be gathered about the amount and direction of slope, bedrock and surface features, waterways, erosion patterns, landslides, slumps, and other related physical and cultural features. Since these all affect the type and cost of existing and potential land uses, the information is advantageous to everyone and every agency involved in urban and regional planning.

The pattern of historical settlement can be correlated with topographic limitations since early settlements occurred on level or rolling land that was not excessively wet. Buildings are more easily and more cheaply built on flat lands than on steep lands. Low, poorly drained areas result in unstable foundations and wet basements; and early settlements in floodplain areas have occasionally been swept downstream relinquishing their site to floodwaters.

With some dramatic exceptions, large scale developments have not occurred on steep lands. In more recent years, however, some housing has been built on hillsides to take advantage of better views. Builders of residential structures in some areas have utilized steep sites, but the historical preference for level lands for development resulting from the relative ease of construction is a factor that remains unchanged today.

As a rule, steep slopes, those of 10 to 15 percent or more, are more easily eroded than level lands; the extent of erosion during construction and prior to soil stabilization is substantially increased on steep slopes. Septic tanks installed on steep slopes are more subject to failure than similar installations in more level landscape. Where provision is to be made for public water and sewage collection systems, the difficulties and costs are significantly greater on steep slopes. In addition, the acreage requirements for roads and structures increase with increases in slope. Efficiency is related to cost, and some costs of developing steep lands have to be borne by the public, especially when local units of government must maintain roadways or other utilities or when erosion and resulting stream sedimentation occur.

Rock Structure

The USGS Geological Portfolios are excellent sources of data about rock structures. The USGS has excellent maps that include interpretations for urban and regional planning. The rocks are sources of raw materials. They also serve as reservoirs for ground water supplies and as waste disposal sites. In addition, they are important in supporting building foundations and contain deposits of important mineral resources.

Basic data included in the Geological Portfolios are the distribution, thickness, orientation, structural detail, composition and geologic history of the rocks. They also include interpretation of basic geologic data for engineering purposes. This interpretive data include such engineering properties as; (1) foundation-bearing quality and supportive strength; (2) tunneling and boring capabilities affecting transportation and utilities; (3) potential construction hazards; (4) physical limitations and costs; (5) depth to bedrock; (6) fault movement potential; (7) workability; (8) ground water supplies; and (9) suitability for septic systems and refuse disposal. Further interpretation can yield information about the capacity of the subsurface to absorb waste and on the interrelationships between climate, topography, vegetation, soils, and water supply.

Remote sensing techniques are best applied in research covering large areas, sacrificing fine detail in exchange for a regional perspective. The types of geologic information available through the application of such techniques include (1) indications of fault lines, folds, and plate tectonic activity, and (2) indications of rock type, structure, surface form, and particle size. To collect such information, use may be made of a variety of remote sensing systems, such as, color infrared photography, ultraviolet sensors, radar which can image through clouds, and thermal scanners which image through some vegetation. The full utility of this information is only realized when it is combined and analyzed with data supplied by other sources.

Among physical requirements for developments are the designation of subsurface conditions that are adequate to support foundations of buildings. Structural limitations are a factor to be used in determining the intrinsic suitability of areas for development. If inferior areas are developed, careful design and construction are required. Although severe rock limitations from the standpoint of supporting foundations may, in some instances, be overcome through design, there exist areas where the probability of structural failure due to unstable conditions is sufficiently high that development should not be recommended.

Minerals

The United States Bureau of Mines distributes maps, tables, and reports on mineral resources. A most complicated aspect of a plan is the problem of assuring adequate supplies of minerals for the future and assuring their availability in areas where they are most needed. The mining methods used have often degraded water supplies, contaminated the land, caused subsidence of land surfaces, and created ugly landscapes. There are many conflicts between mining and other land uses because the mining process requires a technology that is inconsistent with maintaining a high quality environment demanded by other land use developments.

Urban and regional plans often neglect to identify the nature and location of mineral deposits. The result is that recoverability of low-value but essential minerals, such as the limestone, sandstone, clay, sand, and gravel used heavily in construction of buildings and highways, is often endangered by encroaching urban developments. Long-range planning of mineral resource extraction can avoid both of these problems. Recognizing the necessity for maintaining the recoverability of these resources, some urban and regional planning agencies, as well as many of the producers, have attempted to minimize the detrimental effects of mineral extraction and to make provision for site reuse after extractive operations are concluded. Since sand, gravel, and other mineral resources are extremely valuable, proper regulations should be enacted to assure that these minerals can be extracted without significant negative effects. When high-value minerals, such as coal, oil, and gas are found in metropolitan areas, they need to be identified so their sites can be reserved for existing or future extraction.

Remote sensing techniques can not only aid in site identification, but more importantly can also be used to monitor mining operations once extraction begins, to insure compliance with regulations, and to better understand the effects on and the evolution of the landscape from pre-extraction to final reclamation. In areas already mined and abandoned, with varying degrees of reclamation taking place, remote sensing techniques can quickly delineate the extent of mining effects and the nature of the reclamation so that agencies responsible can plan for continued rehabilitation and subsequent use of the land.

The geologic surveys include a survey of the location, distribution, amount, and composition of the various minerals present. They include resource geologic maps and supporting text and tables. Interpretive maps showing mineral quality and quantity, depth and type of overlying soil and rock, and mine refuse areas can be produced from this material. An interpretation of such data to determine the cost of reclaiming strip mines, worked-out sand and gravel deposits, and other mineral extraction sites is important to urban and regional plans.

Water

Water resources are located either as surface waters in lakes, rivers, ponds, and streams or as ground water. A knowledge about how to maintain the quality and quantity of our water resources is fundamental to urban and regional planning. Sources of data about a state's water resources include State Departments of Natural Resources, Geological Surveys, Water Resources and Stream Pollution Control Reports, Reports from the Public Health Service and the United States Army Corps of Engineers.

In the areas of water supply and water supply forecasting, remote sensing methods can provide tremendous amounts of information. Some of the aspects of water supply which can be measured and kept up to date by repeated monitoring with a variety of remote sensing instruments and techniques are: (1) snow conditions, including water content, depth, and areal extent; (2) ice cover and break-up on lakes; (3) water level fluctuations in lakes, rivers, and reservoirs, and associated fluctuations in ground water levels; (4) the effects of weather conditions before, during, and after their occurrence; (5) the loss of surface waters through seepage lines (detectable by vegetation and temperature); and (6) the effects of irrigation, both on the source of the irrigated water and on the area irrigated.

In studies of water quality, remote sensing techniques have been applied and have provided information on: (1) point and area sources of pollution; (2) water temperature variations and the delineation of their effects; (3) salinity levels and delineation of zones of mixing; (4) sediment transportation and deposition rates; and (5) coastal zone management of estuarine environments.

The use of water can be increased through proper planning. Variable amounts of water occur because of rainfall fluctuations. In addition, such factors as time of occurrence, condition of the soil, slope of the land, ground cover, and intensity of rainfall affect the amount of water from a particular rain which will soak into the ground or run off into nearby streams. A detailed analysis of runoff data may also be used to evaluate the supply of water that could be used and also to indicate the probability of flood occurrences.

Water supply and water quality are in a state of constant change, even from hour to hour. They change as runoff reaches a lake after a storm, as a riverside industry discharges its wastes, or as the tide rises and falls. Thus frequent reevaluation of the state of the resource would be very valuable. Remote sensing techniques can provide such information to an extent never before available, enabling man to respond more intelligently and efficiently to subtle changes in conditions as well as to emergencies.

Areas for Preservation

When a planning policy is being adopted, particular land uses should be singled out for immediate special concern and protection. The areas should be excluded from development except in cases of overriding public interests. These critical areas include selected swamps and marshes, lake shore beaches, unspoiled wilderness areas, historical and archaeological sites, and other unique or fragile areas. In the preservation of these and adjoining areas, mere observance of existing air, water quality, and other controls is not sufficient guarantee of retaining the character of these areas of state-wide significance. Thus, a method for identifying areas of critical environmental concern and a better system for preserving such critical areas and restriction development in and around them should be established. Such a system of regulation could be integrated with the usual zoning concepts and procedures already in practice throughout most areas.

Swamps and marshes represent a type of habitat that, once destroyed, cannot be replaced. Their significance derives from the fact that they satisfy the habitat need of the most diverse and numerous wildlife aggregations within a region. With proper management, some swamps and marshes can enhance the quality of life both near and far from developed areas. Many species in wetlands have specialized feeding or breeding requirements satisfied only within narrowly defined habitats. Whether any one wetland satisfies the requirements of a particular species of bird or mammal, or whether it plays an important role in the fisheries of a stream or lake depends upon many factors. These include the chemical composition of the water, the vegetative cover of the wetland area, the presence or absence of flowing or open water, and other interrelated variables.

The areas that have supported plant or animal populations characteristic of presettlement times or that provide habitat for rare species take on more value as their number and size diminish. Once wilderness with pockets of settlement, most areas have felt the impact of people at one time or another until now only occasional remnants remain of the wilderness ecology. These natural areas are of value to the scientific community, for they provide living evidence of past ecological conditions.

An attempt to identify natural areas and important plant habitats can be aided by monitoring significant albedo effects in individual Landsat band densities and ratios. Extracting information on such basic features as sinkholes, springs, bluffs, and rivers can also be done by sensor systems adjusted for best interpretation. These and other natural features have significance to the scientific community and are utilized for educational programs. Since these natural areas represent evidence of past geologic events and processes, they become a part of our National heritage and are worthy of preservation.

Much biophysical environment and land use data can be obtained by remote sensing techniques. The various laboratories for remote sensing can be of immense help in obtaining these data. The ability in short time spans to identify and record various surface characteristics at different times of year and over large areas fits into the needs of urban and regional planners. The importance of Landsat to a data system can become formidable. The Landsat program can provide: (1) regional-urban photomaps, (2) land use maps updated for short time periods, (3) data and maps about the distribution of the earth's physical features, and (4) data and maps about many of man's activities. The development of studies about soils, topography, rock structures, ore deposits, water surfaces, vegetation, and features that could be included in a program of area preservation can be enchanced by Landsat data.

Landsat Data

A large share of data used by regional planners is oriented toward the identification of basic spatial patterns of earth surface features. Very specific information concerning features such as individual dwellings, population distribution, commercial-industrial characteristics, and transportation patterns are particularly important for geographers and planners analyzing urbanized areas. Generally, this type of detailed information must be acquired from on-site enumeration or intensive interpretation of large scale aerial photographs. The data needed for interpretation by regional planners, even in areas with large amounts of urbanization, are most frequently concerned with levels of generalization much less specific than those used by urban planners. A regional planner requires frequent and large amounts of information which provides insight into the distributional characteristics of earth surface features. Among these are classes and subclasses of water, agriculture, natural vegetation, soil, disturbed lands, wetlands, and selected cultural features. Often this information is valuable when developed at map scales as small as 1:250,000; the need to acquire information for an individual dwelling, drainage ditch, or a single industrial building is absent for most analysis associated with regional planning.

The most common source of regional land use or land cover data has been through interpretation of medium or small scale (1/20,000-1/100,000) aerial photographs. There are, however, numerous problems associated with acquisition of regional land use and land cover information from photography. For example, recently a six county Regional Council of Governments in Indiana-Kentucky required a contemporary basic land use/land cover map of its 5,957 square kilometer (2,300 square mile) area. A study was made of existing aerial photographs that could be interpreted to develop the land use/land cover map required for the region. It was found that every county had useable photo coverage, but for virtually every county the photographs had been taken in different years. The photographs of some counties were nine years old while in other counties they were only three years old. This lack of temporal uniformity created concern among the regional planners with respect to developing a truly accurate and contemporary land cover map of the region.

Another problem, as is the case in many agencies responsible for regional planning, was the lack of professional expertise in aerial photograph interpretation within the Regional Council of Governments. Consequently, an analysis of aerial photographs would have to be accomplished by available personnel, or new professional interpreters would have to be hired. Few small and medium sized regional planning agencies have budgets which are sufficiently flexible to hire additional high cost photo interpreters, consequently a majority of photo interpretation conducted by such agencies is done by the existing planning staff. As in the case of the

Regional Council of Governments, the reliance on individuals skilled in planning but not trained to develop accurate and consistent land use/land cover maps from aerial photographs is often necessary but is less than ideal. The estimated cost of acquiring new aerial photographs and interpreting them by the regular regional planning staff in the six county area totaled more than $1.40 per square kilometer ($3.50 per square mile). New aerial photography would provide recent data acquired at the same time; however, the quality of interpretation of that data and the high cost of analysis would remain problems.

Since 1966, computer technology has been applied to analyzing digitized spectral data. Computer analysis of digitized photographs and digitized spectral responses recorded from earth surface features by aircraft optical mechanical scanners has proved useful for many specialized research uses, such as corn blight detection and detailed thermal mapping of land and water. Computer assisted processing of photographs or digitized multispectral data obtained from aircraft scanners can provide regional land use/land cover data but the cost of analysis generally is prohibitive. Only with the advent of the first Earth Resources Technology Satellite has it been both technologically and economically feasible to machine process multispectral data for regional land use/land cover information of extensive areas.

The two Landsat satellites, orbiting more than 900 kilometers (360 nautical miles) above the earth's surface, obtain spectral information in four bands that range from .5 to 1.1 micrometers. Each satellite obtains spectral information from the earth in 14 separate swaths of about 185 kilometers wide. A given area is scanned every 18 days; however, acquisition of useful surface information is dependent on cloud free conditions. About 60 meters by 80 meters is the minimum area from which spectral information is obtained.

The 80 meter resolution of Landsat 1 and 2 images may appear to be disadvantageous for acquiring earth surface feature information; however, this resolution is satisfactory to use for developing a large portion of the land use/land cover information required by regional planners. The assets of the Landsat system for acquiring regional land use/land cover data far outweigh any limitations associated with resolution or other characteristics of the sensors. For the first time regional planners and other scientists concerned with the more macro characteristics of the surface of the earth have access to massive amounts of contemporary information which can be acquired cheaply, and theoretically at a nine day interval from a specified area.

Information developed from machine assisted processing of Landsat multispectral data has the advantage of being generated in a similar manner. It may be processed by a single analyst for thousands, or tens of thousands, of square kilometers thereby reducing subjectivity and inconsistencies which are often induced by manual analysis by several interpreters. Land use and land cover classification data developed from machine processed Landsat multispectral data are collected in a computer compatible format which can be combined with information acquired, and stored in computer compatible format with other sources and matched in a versatile geographic information system. The cost of developing regional land use and land cover inventory information displayed at a scale of 1:24,000 from computer analysis of Landsat multispectral data is less than that of aerial photograph interpretation for large areas. Generally, machine processed Landsat multispectral analyses are less expensive than photo interpretation when 3,000 or more kilometers are included.

In the Indiana-Kentucky Regional Council of Government example, which covered 5,957 square kilometers, it was found that machine processing of Landsat multispectral data by the

Indiana State University Remote Sensing Laboratory (ISURSL) cost $0.54 per square kilometer ($1.50 per square mile) to develop a ten category basic land cover inventory at a scale of 1:24,000, while the estimated cost of obtaining new aerial photographs and interpreting them by the regular staff was approximately $1.40 per square kilometer, (about $3.50 per square mile). Recently, the ISURSL has developed computer programs which permit development of a ten category land cover inventory at a scale of 1:24,000 for less than $.36 per square kilometer ($1.00 per square mile) when thousands of square kilometers are to be analyzed.

There are limitations in the current capabilities of Landsat for land use/land cover data acquisition. The information provided from machine processing of Landsat multispectral data often is insufficient for use needs that require very specific data because of the 80 meter resolution characteristics of the system. In regional planning there are a few types of data which may be difficult to obtain through Landsat analysis, although overall a majority of regional land cover information can be developed. The most limiting factor, however, is the fact that currently there still are relatively few places where multispectral data from optical mechanical scanners can be analyzed effectively by computers which have sophisticated programs specifically designed to convert digitized spectral information into meaningful earth surface feature data. There are also few analysts who are capable of conducting basic and applied research using machine processing techniques to generate data needed by regional planners.

Regional planners use information obtained from a variety of sources that include on-site acquisition, published materials, aerial photo interpretation, and analyzed Landsat data. The traditional sources of information for regional planners should be retained but the proportion of usable data these sources generate can decrease in proportion to the amount of Landsat data introduced. Because of the cost advantage, the vast amount of data continuously available, and other advantages discussed previously, it is virtually certain that eventually a large share of basic regional land use information will be acquired through machine processing of multispectral data obtained from earth resources satellites. The future is bright for application of satellite multispectral data to regional planning and other resource management issues. Since the knowledge of machine processing of Landsat data, or other multispectral data, applied to regional planning and associated environmental resources problems is somewhat limited in the user community, an emphasis on this form of remote sensing is a major focus in this chapter. On the other hand, other types of remote sensing and data acquisition are important and valid methods of obtaining regional land use and land cover information.

Basic Principles of Machine Processing of Multispectral Data

Various approaches to machine processing of multispectral data are possible. Often the availability of computer equipment and computer programs determine the specific method of analysis. Some analysis systems stress single band density slicing (adjustment of spectral responses to a condition in which desired earth surface features are highlighted) and color enhancement. Other systems rely on more complex multiband density slicing which may also use color enhancement techniques. Even more sophisticated and complex is the use of computer programs designed to statistically analyze the spectral response characteristics of designated earth features in n dimensions (n = the number of separate bands of data used).

The more sophisticated types of analysis use computer programs that develop spectral information analogous to a multiband spectral signature. The numerical characters of classification programs are designed to use these spectral data for identifying earth surface features. There are other approaches to classification which are related to those described in this chapter; however, basically all machine processing of multispectral data is concerned with identification of earth surface features by analyzing several bands of spectral information with the aid of computer instrumentation. The less complex, but generally also somewhat less versatile, approaches to this type of analysis use a mini-computer system. The cost of a mini-computer system is at a level which many larger planning agencies and smaller universities can afford, thus it is likely that the largest number of future multispectral analyses will use a mini-computer-based system rather than large general purpose computers.

Machine processing of multispectral data was pioneered primarily by two very large and well-equipped remote sensing laboratories which had access to large digital computers and computer packages designed to process multispectral data. These two laboratories are the Laboratory for Applications for Remote Sensing (LARS)/Purdue University and the Environmental Resource Institute of Michigan (ERIM)/Ann Arbor. The set of computer programs designed to analyze multispectral data at LARS has been designated LARSYS. LARSYS and programs similar to LARSYS are the most used computer packages in the world for analyzing data.

It is unlikely that many large facilities such as those established at LARS or ERIM will be developed in the near future; however, remote terminals with access to their technology are beginning to be developed. For example, the Indiana State University Remote Sensing Laboratory (ISURSL) has access to most LARS facilities at a remote terminal in Terre Haute, Indiana. In addition, Indiana State University has additional computer programs designed to process multispectral data which work within the remote terminal LARSYS environment. Through mini-computer systems and remote terminal access to the large remote sensing laboratories, regional planners and other regional scientists have the opportunity to acquire much of the multispectral data they require for their regional information system.

Computers with access to programs of varying degrees of sophistication for analyzing Landsat multispectral data are supplying solutions to regional land use planning problems. The Indiana State University Remote Sensing Laboratory is a facility which specializes in the machine processing of Landsat multispectral data specifically for developing regional land use and land cover information for planning and environmental agencies. The ISURSL, through its remote terminal of LARS and unique independently developed approaches to multispectral analysis, is a facility which has access to current state-of-the-art remote sensing technology. Numerous examples of Landsat data which were analyzed by the ISURSL for regional planners or environmental agencies in response to a need for general earth surface feature information will be presented later in this chapter. These examples will provide insight into many of the current uses to which machine processed Landsat multispectral data are being applied in the fields of regional land use and regional environmental analyses.

One approach, briefly outlined herein, was used to develop the earth surface feature information for applied research projects. Other approaches to LARSYS analysis of multispectral data also have been used successfully by the ISURSL. Figure 16.1 outlines the sequence of selected procedures used for a majority of the Landsat analysis examples used in this chapter. The example, which is simplified for presentation, describes a basic low-cost approach used in machine processing of multispectral data.

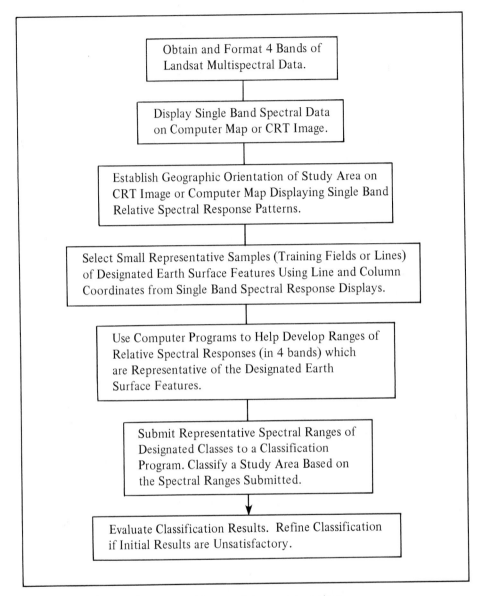

Figure 16.1. Procedures used in machine processing.

Step 1. Multispectral Landsat data tapes are purchased from EROS Data Center. Each data tape contains 4 bands of digitized spectral response information for more than 7,000,000 80 meter by 60 meter units (ground resolution unit). Each ground resolution unit (GRU) is identified by a specific line and column coordinate which is retained throughout the classification procedures. The multispectral data tapes are put into a format to be computer compatible with LARSYS.

Step 2. Relative spectral response values (in digital format) of a study area or subarea are displayed a band at a time for one or more bands on computer maps or on a cathod ray tube (fig. 16.2). This display shows patterns of spectral reflectance/emittance which can be

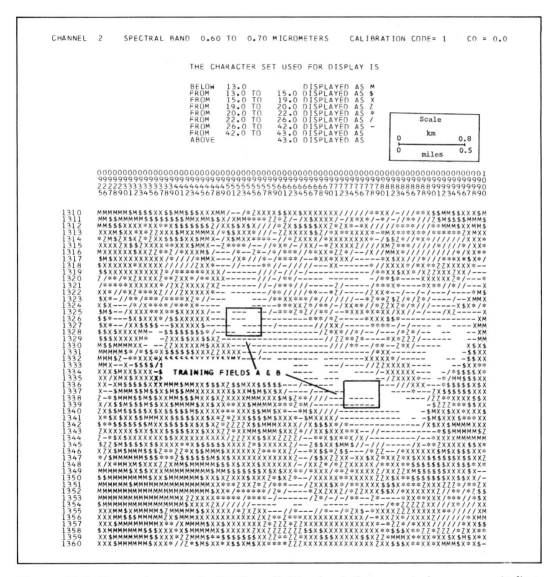

Figure 16.2. Computer map indicating digitized relative spectral response (reflectance) values in the .6-.7 micrometer band of a rural western Indiana area. Samples of bare strip mine spoil are outlined. Relative spectral response values in four Landsat bands which are representative of these strip mine spoil samples are provided in table 16.1. Multispectral data used in analysis of this area were obtained from the June 9, 1973 Landsat pass. (Source: ISURSL)

associated with selected earth surface features to the extent that geographical orientation by line and column coordinates can be achieved.

Step 3. The displays of single band relative spectral responses are examined. The resulting patterns on the displays are compared and associated with patterns of known earth surface features to establish geographical orientation between spectral data and selected features on the surface of the earth (figs. 16.2 and 16.3).

Step 4. Small representative samples of earth surface features (training fields or lines) designated for classification are selected by analyzing ground information (i.e., aerial photography, recent topographic map) from a small subarea of the total study area. These representative samples are identified by line and column coordinates which are indicated on the computer compatible data tapes and all displays of spectral information which use the digitized data from those tapes.

Step 5. A variety of computer programs are implemented to develop information concerning the spectral characteristics of the representative training fields. The dominant ranges of digitized relative spectral responses of each feature designated for classification are determined through this analysis. An example of spectral ranges in all four bands for the candidate training

Figure 16.3. A 1974 photograph of the study area displayed in figures 16.2 and 16.4. (Agricultural Stabilization and Conservation Service)

samples indicated in figure 16.2 are summarized on table 16.1. The first three bands of Landsat data have 128 possible digitized relative spectral response values while 64 values are possible for the last band.

TABLE 16.1
Relative Spectral Response Ranges for Two Samples of Mine Spoil

	Band 4 (.5-.6 micrometers)	Band 5 (.6-.7 micrometers)	Band 6 (.7-.8 micrometers)	Band 7 (.8-1.1 micrometers)
Sample A	38-46	40-47	42-53	32-39
Sample B	40-49	41-46	44-54	33-41

Step 6. Spectral ranges characteristics of each feature designated for classification are submitted to a computer program designed to seek out and identify unclassified ground resolution units. Ground resolution units which have the spectral range characteristics of a representative training sample will be identified as the feature whose spectral characteristics were used for training. An example of the classification of mine spoil in the area indicated in figure 16.2 which is based on the spectral information from samples A and B is provided in figure 16.4. Theoretically it is possible to use spectral statistics from samples A and B (and preferably from a few other samples) to identify mine spoil over an entire Landsat frame (more than 30,000 square kilometers).

Step 7. Classification results from step 6 are compared with known ground information. Refinements or even reclassification of a study area may be made if this evaluation reveals that major problems in the classification results exist.

Land Use and Land Cover Classification Systems

The types of earth surface feature information which urban regional planners and macro environmental analysts need is highly variable. A majority of these data are concerned with the more macro aspects of large regions. Very specific earth surface feature information is also required by regional planners; however, analysis of large quantities of basic land cover data over multicounty areas is the focus of many planning or environmental projects.

No single land cover and land use classification scheme is fully suited for use by regional planners. Each region has its own set of land use and land cover features which to a greater or lesser degree it is possible to analyze through remote sensing techniques. The variety of earth surface information which regional planners and environmentalists request from remote sensing specialists is very large. Regional problems vary from area to area and the type of information requested from remote sensing analysts is rarely identical. Even though no two regional analyses require exactly the same types of data, it is nevertheless true that there are many categories of land use and land cover information which are commonly requested.

Many classification schemes designed to subdivide major land use/land cover classes into a logical framework are available to urban regional planners. The land use/land cover classification system which is being considered most widely by regional analysts is the one developed by

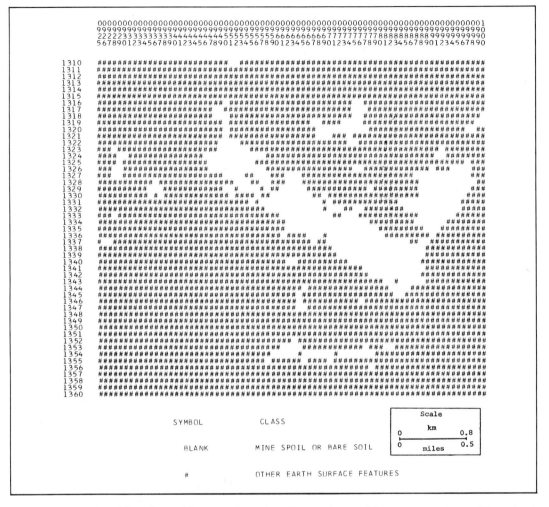

Figure 16.4. Identification of bare strip mine through machine processing of Landsat multispectral data. Samples of bare strip mine spoil similar to those delineated in figure 16.2 were analyzed and used in classification of this feature. (Source: ISURSL)

Anderson, Hardy, Roach, and Witmer entitled "A Land Use and Land Cover Classification System for Use with Remotely Sensed Data" (U.S. Geological Survey Professional Paper 964). The abstract from this Professional Paper provides insight into its characteristics and objectives. The abstract states:

The framework of a national land use and land cover classification system is presented for use with remote sensor data. The classification system has been developed to meet the needs of Federal and State agencies for an up-to-date overview of land use and land cover throughout the country on a basis that is uniform in categorization at the more generalized first and second levels and that will be receptive to data from satellite and aircraft remote sensors. The proposed system uses the features of existing widely used classification systems that are amenable to data derived from remote sensing sources. It is intentionally left open-ended so that Federal, regional, State, and local agencies can have flexibility in developing more detailed land use classifications at the third and fourth levels in order to meet their particular needs and at the same time remain compatible with each other and the national system. Revision of the land use classification system as presented in U.S. Geological Survey Circular 671 was undertaken in order to incorporate the results of extensive testing and review of the categorization and definitions.

It is evident from the abstract that this classification system was designed to meet a specific classification need. The approach is designed to use remotely sensed data. This classification system may prove to be less than adequate as remote sensing technology and approaches to regional land use analysis change. As with all classifications the inclusion of some categories may be questionable, while the exclusion of other categories may be a center of objection. Considering that the classification, however, was designed to provide an overview of land use and land cover throughout the entire country it seems obvious that this system provides a very suitable framework which is applicable throughout the nation. The consideration and integration of remotely sensed data makes this classification particularly suitable for use by regional planners and environmental scientists who are relying on sensors aboard Landsat, high altitude aircraft, and other platforms for acquiring land use and land cover data. Table 16.2 indicates the Level I and Level II land use and land cover categories of the USGS classification system. Details concerning the classification are not presented here, but it is strongly suggested that all persons interested in regional land use/land cover classification read this important paper.

The examples of applied research presented in this chapter emphasize developments of regional land use/land cover information from machine processing of Landsat multispectral data that generally use the USGS classification as a framework. The classes of land use/land cover features that were ultimately developed rarely coincided exactly with the USGS system because the data requirements of the regional research required specific types of remotely sensed land use and land cover information which were not specifically identified in Levels I and II.

Landsat Analysis Applied to Regional Land Use and Land Cover Issues

The discussion in this section focuses on specific examples of applied research in regional land use/land cover analysis conducted by the ISURSL and Department of Geography and Geology (ISU) in response to user needs for earth surface feature information from Landsat data. In all examples presented, the earth surface feature information was developed from analysis of Landsat multispectral data using machine assisted processing techniques within a LARSYS computer environment. Land use/land cover information has been acquired by the ISURSL for (1) The State Planning Services Agency, State of Indiana; (2) U.S. Forest Service; (3) Environmental Protection Agency; (4) Southwestern Indiana and Kentucky Regional Council of Governments; and (5) Oklahoma Water Resources Institute.

Development of Basic Land Cover and Land Use Inventories for Regional Planning

Frequently the distribution of the most basic land use/land cover features of an area is among the information most desired by urban regional planners. Accurate basic land cover information acquired for the same date over large areas is virtually impossible to obtain except through analysis of Landsat multispectral data. This basic land cover data acquired from Land-

TABLE 16.2
Land Use and Land Cover Classification System for Use with Remote Sensor Data

Level I	Level II
1 Urban or Built-up Land	11 Residential
	12 Commercial and services
	13 Industrial
	14 Transportation, communications, and utilities
	15 Industrial and commercial complexes
	16 Mixed urban and built-up land
	17 Other urban or built-up land
2 Agricultural Land	21 Cropland and pasture
	22 Orchards, groves, vineyards, nurseries, and ornamental horticultural areas
	23 Confined feeding operations
	24 Other agricultural land
3 Rangeland	31 Herbaceous rangeland
	32 Shrub and brush rangeland
	33 Mixed rangeland
4 Forest Land	41 Deciduous forest land
	42 Evergreen forest land
	43 Mixed forest land
5 Water	51 Streams and canals
	52 Lakes
	53 Reservoirs
	54 Bays and estuaries
6 Wetland	61 Forested wetland
	62 Nonforested wetland
7 Barren Land	71 Dry salt flats
	72 Beaches
	73 Sandy areas other than beaches
	74 Bare exposed rock
	75 Strip mines, quarries, and gravel pits
	76 Transitional areas
	77 Mixed barren land
8 Tundra	81 Shrub and brush tundra
	82 Herbaceous tundra
	83 Bare ground tundra
	84 Wet tundra
	85 Mixed tundra
9 Perennial Snow and Ice	91 Perennial snowfields
	92 Glaciers

sat analysis are generally intended to provide a base for many initial regional land use planning and environmental decisions. Generally the Level I categories of the USGS classification system (table 16.2) are developed to provide the most basic regional land cover data bases.

Several Level I regional land cover inventories have been conducted by the ISURSL for regional, state, and federal agencies. More than 100,000 square kilometers (38,610 square miles) of basic land cover inventories have been developed by ISURSL for various regional planning agencies utilizing machine processing of Landsat multispectral data. These classifications ranged from as few as 6 categories of land cover (forest, water, agriculture, soil, wetlands, and built-up) to as many categories as 10 (deciduous forest, mixed forest, row crop, small grain/pasture, commercial/industrial, older residential, newer residential, soil, silty water, nonsilty water). The number and types of categories selected were largely determined by the individual agency contracting for the land cover information. A larger number of other classes, not requested by these agencies, also could be developed.

Each land use inventory in this basic type of analysis was developed to meet the needs of a specific planning region at a low cost. In all cases, the contracting agency required contemporary and similar quality land cover information ranging from thousands to tens of thousands of square kilometers. Additionally, the data were to be developed within a three to five month period depending on the specific analysis. The planning or environmental issues facing the contracting agencies were of a nature that distributional information concerning a few basic categories of land cover was sufficient to help provide solutions to land related problems at a regional scale of analysis.

The specific uses made of the basic land use and land cover developed for state and regional planning agencies by the ISURSL are enumerated. However this list is not intended to indicate all possible applications of these data. The uses made of the ISURSL developed basic regional land use data are proliferating, but among the current applications are:

1. Assistance in development of flood hazard boundary maps
2. Surveying extent of strip mined lands
3. Evaluation of land use characteristics on strip mine lands
4. Providing basic land inventory data to up-date Conservation Needs Inventory information
5. Providing basic land inventory data for use by tax assessors
6. Providing data to help develop a regional open space plan
7. Providing basic earth surface feature information which is incorporated into existing data for use in a variety of regional planning projects
8. Identification of basic land use changes (from rural features to urban features) in metropolitan fringe areas

More detailed land use/land cover inventories have been developed by the ISURSL. These inventories contain more than 10 land use/land cover classes. Usually subclasses of water and more detailed differentiation of forests and separation of agriculture into row crop and small grain/pasture (or small grain and pasture) are developed. These more detailed land use/land cover inventories can be used for applications previously cited. In addition, they are currently being applied to evaluate pollution and erosion characteristics of watersheds. An evaluation of the surface water quality associated with the erosion and surface runoff in watersheds is also being applied by using the more detailed classification of land use and land cover features via analysis of Landsat data.

The most frequently requested land use/land cover data are alphanumeric computer maps that are geometrically corrected at a 1:24,000 scale. Smaller scale computer maps at a scale of 1:250,000 have also been commonly requested by users but the maximum amount of detail is obtained at a 1:24,000 scale which accurately overlays 1:24,000 USGS topographic maps. Classification displays at this scale also serve as bases upon which additional earth surface feature data are plotted.

Statistical summaries of land use/land cover categories within a township, county, watershed, or other geographical/political area can be determined by a computer during the development of a classification. Virtually all regional planners request these statistical summaries of classification results by individual categories of land use and land cover.

Black and white or color coded digital display (CRT) images of classification can be generated from the machine processing of multispectral data (plate 21). These photographiclike displays are very effective in providing classification results both at a small scale and in a form which is easily interpreted. The CRT displays of single band spectral responses or classification results are visually pleasing and for some purposes are the most effective way to present information. The large to medium scale computer maps, however, still serve the regional planner best for a detailed land use analysis. The display of computer maps showing a single class at a time (see examples in the following surface hydrology section) improves the ease of interpretation but increases the number of maps which must be analyzed.

Classification results from LARSYS analysis are obtained less frequently in the form of contour maps and electrostatic printer maps (figs. 16.5 and 16.6). When these two types of classification displays become more readily available, they will undoubtedly increase in popularity because of their ease of interpretation and flexibility of scale.

Examples of results from land use and land cover classifications developed for state and regional planners are presented (figs. 16.7 through 16.10). Each example shows results from a small subarea which contains a few of the more general land use/land cover categories used in classification. Photographs of the areas used as classification examples are provided in order to give insight into the quality of basic land use and land cover information which machine processing of multispectral data typically provide.

A determination of classification accuracy of land use/land cover features developed in the various analyses conducted for regional planners and regional environmentalists has been made. Table 16.3 lists the land use/land cover classes which have been developed from Landsat data by the ISURSL for regional planners or regional environmental analysts. The list would be greatly expanded if analysis by all Landsat analysts throughout the country were included. The accuracy of identification of the features indicated on table 16.3 generally ranges from approximately 80 percent to 98 percent correct.

Water and many of its subclasses always are correctly classified from 95 to 98 percent of the time. Classification accuracy of forests generally has been near 90 percent. A majority of the other features identified on table 16.3 typically are correctly classified with an accuracy of 85 percent to 90 percent. Particularly difficult classification problems lower classification accuracy of selected features indicated on table 16.3 to approximately 80 percent in the worst instances. Overall the classification accuracy of a typical basic land use/land cover inventory

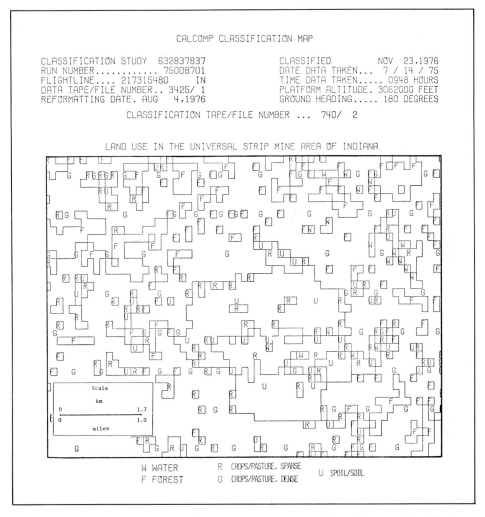

CALCOMP CLASSIFICATION MAP

CLASSIFICATION STUDY 632837837   CLASSIFIED   NOV 23,1976
RUN NUMBER........... 75008701   DATE DATA TAKEN... 7 / 14 / 75
FLIGHTLINE.... 217315480   IN   TIME DATA TAKEN..... 0948 HOURS
DATA TAPE/FILE NUMBER.. 3425/ 1   PLATFORM ALTITUDE. 3062000 FEET
REFORMATTING DATE. AUG 4,1976   GROUND HEADING..... 180 DEGREES

CLASSIFICATION TAPE/FILE NUMBER ... 740/ 2

LAND USE IN THE UNIVERSAL STRIP MINE AREA OF INDIANA

W WATER   R CROPS/PASTURE, SPARSE   U SPOIL/SOIL
F FOREST   G CROPS/PASTURE, DENSE

Figure 16.5. Display of earth surface feature classification results (Universal strip mine, western Indiana) in contour map format. The classification displayed by contour map was developed at the Laboratory for Applications of Remote Sensing at Purdue University and ISURSL.

developed from Landsat data averaged approximately 90 percent. Other information used in conjunction with Landsat improves the quality of information which can be used for planning purposes. The urban-regional planner or environmental analyst should not attempt to utilize Landsat data solely for the development of basic land use/land cover inventory base data. Information obtained from analysis of Landsat data provides a good general data framework which can be applied directly to a planning issue or can be combined with other available data for application to more complex planning issues.

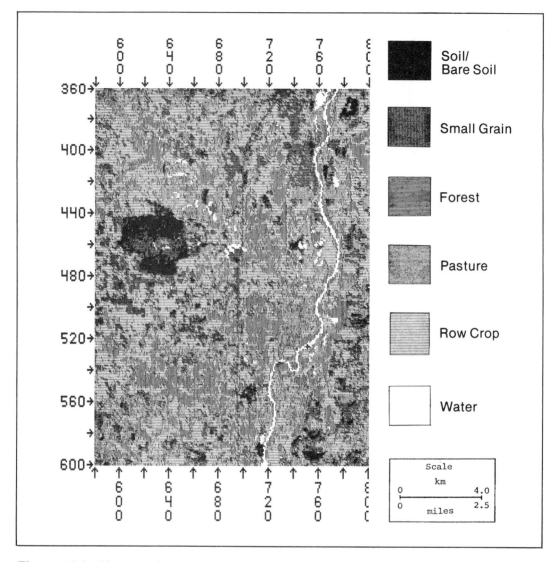

Figure 16.6. Display of earth surface feature classification results in electrostatic printer format. The classification displayed is of land cover of rural Vigo and Vermillion counties, Indiana, (Source: ISURSL and LARS/Purdue)

Development of Surface Hydrology Data for Use in Regional and Environmental Planning

Water is one of the easiest features to identify accurately through machine processing of Landsat multispectral data. More than 98 percent of all water in most areas can be located through Landsat analysis. Only very small water bodies of less than three or four acres or very

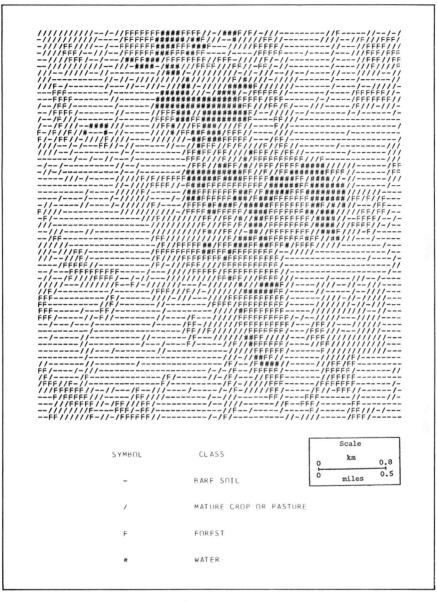

Figure 16.7. Classification of a rural western Indiana area through machine processing of June 9, 1973 Landsat multispectral data. The area displayed is from a 13-category land use/land cover classification of the entire state of Indiana. The northern part of the area displayed was mined for coal approximately 25 years ago and has since reverted to forest. The classification results have been simplified for reproduction. (Source: ISURSL)

Figure 16.8. Photograph of the area classified through machine processing of Landsat multispectral data displayed in figure 16.7. (Agricultural Stabilization and Conservation Service)

narrow water bodies of less than 80 meters wide (263 feet) are poorly identified. It is also possible to accurately identify spectral subclasses of water that are associated with physical, chemical, and biological substances dissolved or contained in suspension.

Water containing different amounts of suspended sediments have distinctly different spectral characteristics. A determination of the suspended sediment content of water can be applied as input into erosion models. Suspended sediment can be used as one measure of water quality, and often can be used as a surrogate to estimate water depth. Spectral variations in water identified through Landsat also can be used to help identify chemical differences in water. Evaluations of water characteristics associated with surface coal mining in Indiana and Illinois have

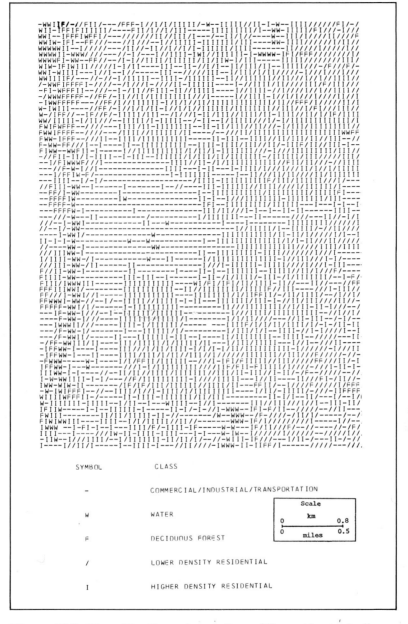

Figure 16.9. Classification of a portion of Terre Haute, Indiana, through machine processing of June 9, 1973 Landsat multi-spectral data. The area displayed is from a 13-category land use/land cover classification of the entire state of Indiana. (Source: ISURSL)

Figure 16.10. Photograph of Terre Haute, Indiana, area classified through machine processing of Landsat multispectral data displayed in figure 16.9. (Agricultural Stabilization and Conservation Service)

TABLE 16.3

Land Use and Land Cover Features Commonly Developed by Machine
Processing of Landsat Multispectral Data in the Midwest

Water	Pasture
Water, high in suspended sediments	Bare soil
Water, medium amount of suspended sediments	Mineral soil, high organic content
Water, low in suspended sediments	Mineral soil, low organic content
Forest	Muck soil
Deciduous forest	Strip mine spoil
Mixed forest	Bare rock
Coniferous forest	Sand
Forest, high density	Commercial/industrial
Forest, low density	Inner city
Forest and pasture mixture	Older residential
Agriculture	Newer residential/suburban
Row crop	Forested wetlands
Small grain	Nonforested wetlands

used Landsat data. Variations in water pH and chemical elements in solution detected from one strip mine lake to another have been identified using Landsat. ISURSL water research has indicated that Landsat analysis of water might provide an insight into the degree of eutrophication of lakes and ponds. It has been found that lakes and ponds adjacent to many suburban developments in metropolitan Indianapolis, Indiana, are spectrally distinct from seemingly similar water bodies nearby which have not experienced residential development. Preliminary analysis suggests that these spectral differences are partially related to indirect human input of materials (i.e., fertilizer, waste from septic systems) which have strongly altered the normal biotic characteristics of the water. These biotic differences cause spectral changes to occur in the water in one or more of the four Landsat bands.

Data about the distribution of water without considering spectral subclasses also are vital for many urban-regional planning and environmental planning programs. The basic distribution of water is important to regional planners as one of the major categories of land use/land cover classifications. There are many practical applications of basic water distribution information in addition to conventional uses. For example, the frequency of Landsat data is ideal for monitoring change. Major changes in land use/land cover can be identified several times a year if required. The dynamics of land use/land cover change as urbanization encroaches on rural lands or as changes associated with surface mining modify rural landscapes. These are among the uses to which the monitoring capabilities of Landsat can be applied. The monitoring of surface hydrology conditions through analysis of Landsat has great potential value for regional and environmental planners. Variations in spatial patterns of water are often much more frequent than other changes in land use/land cover features; consequently, the frequency of monitoring for many surface hydrology problems is greater than that for other uses.

An example in which frequent Landsat monitoring of surface water provides valuable information for regional land use planners and environmental analysts is illustrated by recent ISURSL research on selected areas of the Wabash River floodplain. A several date Landsat analysis of the Wabash floodplain was begun in an attempt to accurately identify spatial patterns of wetness. This research required that six or more multispectral analyses of the Wabash floodplain be completed each year for a minimum period of five years. These analyses are conducted at various times of the year in order that most characteristic surface water and land use conditions on the floodplain can be observed. In this long term study, it has become evident that frequent long term monitoring of surface hydrology and land use conditions in floodplain areas has the potential to provide information on which decisions can be made about the optimal human use of floodplains.

Frequent monitoring of floodplains provides insight into dynamic land-water relationships of flood-prone areas. Lands which are flooded first are easily identified through this analysis, but equally important is the ability to identify the spatial patterns of flooded land following floods of different intensities. Also, it is possible to identify the recession of water from the land following a flood. If enough dates of observation are made, it is possible to develop zones or areas of wetness on floodplains. For example, parts of the Wabash study area appear to be virtually permanently free from surface water with the exception of a very short time period associated with extraordinary high flood stages. Other areas were free from surface water more than 10 months each year but appear to have standing water for a short period every year. This type of land can be considered an intermittent wetland which is very similar to the best drained and least flood-prone portions of the floodplain but has some standing water every year. Other areas tend to be covered with water more than 8 months of the year; however, these areas were dry for short periods. This type of wetness pattern is characteristic of intermittent wetlands which are nearly permanent water features. Several classes of intermittent wetlands between these two extremes can be discriminated through machine processing of Landsat multispectral data. The extreme of surface wetness is represented by permanent water features. These features, as indicated previously, are among the most accurately identified through Landsat analysis. The water associated with the intermittent wetlands and permanent water features can be further analyzed to provide indications of siltiness, eutrophication, and other foreign substance influences in the water. Identification of water saturated land which technically is not covered with water can be made through Landsat analysis.

The surface hydrology information and associated land use/land cover information derived from a long-term multidate monitoring of a floodplain has numerous applications to regional planners, hydrologists, and others concerned with environmental issues. Among these applications are:

1. Delineation of floodplains.
2. Providing information of floodprone lands for use as input into Federal Flood Insurance programs.
3. Developing a map indicating regions of wetness which include (a) permanently dry; (b) several classes of intermittent wetlands; and (c) permanently wet.
4. Comparison of existing patterns of land use as determined from Landsat analysis and other sources with patterns of wetness. Through analysis of this information it is possible to evaluate the degree to which land is being used effectively (environmental and economic aspects). For example, does underuse of land occur in intermittent wetlands because land owners overestimate the delitereous affects of seasonal flooding and ponding? Is it possible that land use activities such as row crop agriculture are occurring on intermittent wetlands which flood and pond too frequently to be economically feasible in a modern agricultural economy?
5. Developing models of suggested good land uses on floodplains that are based on wetness characteristics which maximize economic value within a framework of proper environmental safety.
6. Analyzing water quality patterns on floodplains through time.

Two examples of research using machine processing of Landsat multispectral data which are totally or partially focused on analysis of surface water are presented. Figures 16.11 and 16.12 are two examples extracted from the Wabash Floodplain monitoring research previously discussed. These two examples indicate the distribution of wetlands and various classes of water on two of the many dates used for analysis. The two examples selected are representative of floodplain conditions where extensive flooding of lowlands has occurred (May 1973) and floodplain conditions in which surface water is confined to the river channel and permanent water bodies (September 1973). These two examples illustrate the wide range in surface hydrology conditions from one time period to another. An examination of the floodplain several times throughout the year helps provide the surface hydrology and land use information required to determine wise uses of the various types of intermittent wetlands which are very widespread on many floodplains.

Figure 16.11. Distribution of water and wetland on the Wabash River floodplain in Vigo County, Indiana, on May 4, 1973. The surface hydrology information was obtained from machine processing of Landsat multispectral data. (Source: ISURSL)

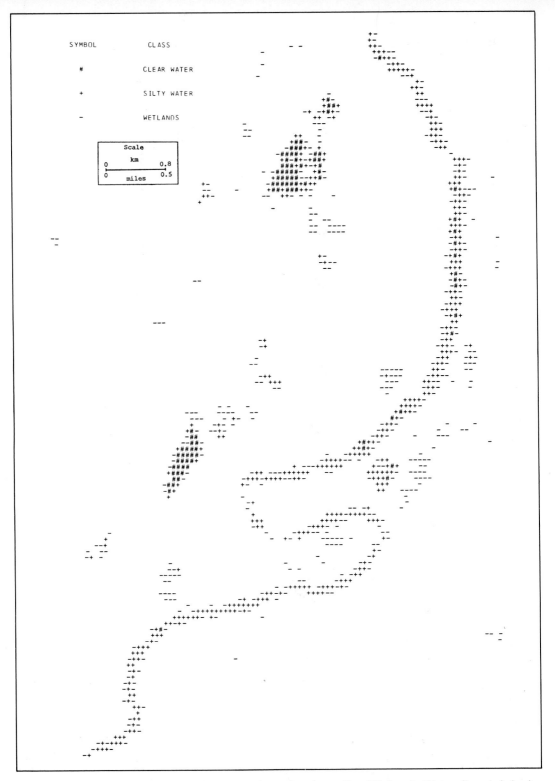

Figure 16.12. Distribution of water and wetland on the Wabash River floodplain in Vigo County, Indiana, on September 7, 1973. The surface hydrology information was obtained from machine processing of Landsat multispectral data. (Source: ISURSL)

Figure 16.13 is derived from a study of water quality in large surface water bodies of the Indianapolis metropolitan area. These water bodies were used to help supply the area with water, recreation, and a means of waste disposal. There is considerable spectral variation in the water because of variations in suspended sediments, biological life, and chemicals in solution. The example used for illustration is Eagle Creek Reservoir northwest of Indianapolis (fig.

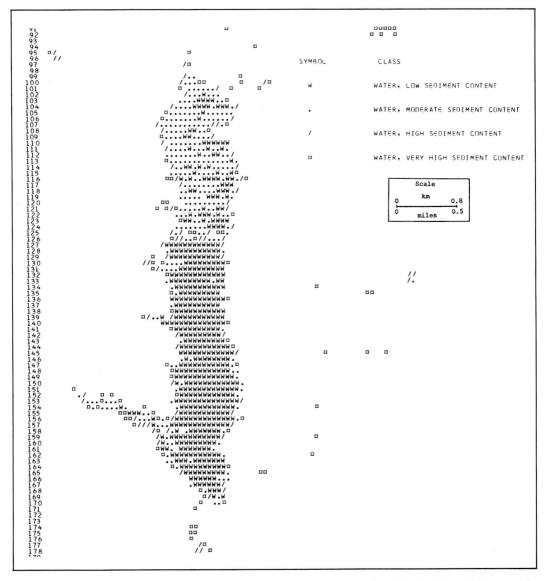

SYMBOL	CLASS
W	WATER, LOW SEDIMENT CONTENT
.	WATER, MODERATE SEDIMENT CONTENT
/	WATER, HIGH SEDIMENT CONTENT
□	WATER, VERY HIGH SEDIMENT CONTENT

Scale

km　0.8

miles　0.5

Figure 16.13. Identification of water characteristics in the Eagle Creek Reservoir (metropolitan Indianapolis) area through machine processing of June 10, 1973, Landsat multispectral data. (Source: ISURSL)

16.13). As many as five spectral classes of water have been identified in Eagle Creek Reservoir. Field research indicated that spectral variations were primarily a function of mineral particle matter in suspension. The shallower portions of the reservoir often have very silty water in contrast to the deeper and physically less disturbed portions of the reservoir where silty and clayey materials were more likely to have settled to the bottom. Figure 16.13 depicts the reservoir at a time when four distinct classes of water associated features were identified. Deep water containing small amounts of suspended sediments were found in the southern two-thirds of the reservoir north of the dam. In contrast, the northern one-third of the reservoir was shallow and it generally contains water with large amounts of suspended sediments. Very shallow water at or near the shoreline often are identified as a third and fourth class. Through this type of analysis it is possible to select water quality characteristics which play a significant role in regional planning and environmental evaluation.

Machine Processing of Landsat Multispectral Data
for Use in Coastal Zone Management Programs

Coastal zone areas currently are receiving a great deal of attention from scientists, legislators, and planners. This attention has been in response to the critical and fragile environmental characteristics of these areas. The coastal zone is a major breeding area for a large variety of aquatic life which has great ecological and economic significance. Population pressure in coastal zone areas often is great in response to the development of recreational facilities and permanent homes in surroundings made desirable by the attractions inherent in a marine influenced location.

The coastal zone has proved to be very attractive for aquatic life, land animals, vegetation, and human development. The climatic, biotic, hydrologic, and other resources of these areas make them among the most productive biological zones in the world, as well as areas which are becoming increasingly desirable for a variety of human activities. The biological wealth of the area is, in large part, attributable to an environment in which both fresh and salt water interact in the presence of tidal action which moves and mixes water of differing characteristics. Vegetation, such as the *Spartina* marsh grasses, provides breeding grounds for aquatic life.

Large amounts of nutrients in the water, contrasting currents near shore, and an abundance of micro and macro biotic life which are vital in the food chain of biologically rich areas are features which make these areas major ecological regions. Their significance to people extends far beyond the physical boundaries of the coastal zone.

The varied and plentiful life in the coastal zone is very sensitive to environmental changes. Increases in water temperature, the addition of chemical pollutants, and the introduction of foreign biological wastes alter ecological relationships. A relatively minor alteration in these relationships could result in disaster to the fishing and/or recreation industries.

The value of the coastal zone for residential use associated with permanent homes, summer cottages, tourism, and recreational developments depend largely on the varied and interesting biotic life, sandy beaches, and unpolluted waters. Great population pressure is placed on coastal zone lands which are in many ways physically unstable. Disposal of human wastes in these areas is extremely difficult and expensive if conducted properly. Inexpensive and improper disposal of wastes poses a direct threat to people via disease and an equally great threat to disrupting the sensitive ecology in the area. Construction of homes and hotels on unstable eroding sand spits or offshore bars subject to hurricanes often results in great economic loss. Access to large quantities of water and raw materials invites the development of selected types of industry and power plants. These activities also must be carefully monitored in order to prevent damage to the coastal zone environment.

All levels of government have expressed concern for intelligent use and development of coastal zone environments. Legislation and funding for land use planning and environmental impact assessment in coastal zone areas has begun. The need for earth surface feature information to help implement current planning and impact assessment projects is great. Machine processing of Landsat data is one important method of obtaining the land use and land cover information which is required for optimal coastal zone management practices.

A recent Landsat analysis of a coastal zone management area in southeastern North Carolina (Cape Fear region) conducted by the ISURSL has been completed. Seventeen land and water categories of information were developed to provide land use/land cover, hydrological, and ecological information useful in coastal zone management. Table 16.4 summarizes the categories of earth surface features which were accurately identified through machine processing of Landsat data. In consultation with ecologists from North Carolina State University and the University of Georgia several potential uses of the type of data developed in coastal North Carolina have been suggested.

Among the potential uses of the information generated by Landsat are:

1. Monitoring spectral characteristics of water influenced by effluent from the new nuclear power plant near Southport, North Carolina, with changes in micro aquatic life being anticipated to cause spectral variations in water

TABLE 16.4

Earth Surface Features Identified in Coastal Regions of North Carolina
Through Machine Processing of Landsat Multispectral Data

Salt water (more than 1 meter deep)
Salt water (less than 1 meter deep)
Fresh-brackish water
Shallow fresh-brackish water, humus-mud bottom
Shallow fresh-brackish water, mud bottom
Shallow fresh-brackish water, sandy-mud bottom
Marsh grass, medium and tall *Spartina* and *Juncus*
Marsh grass, short *Spartina* and *Juncus*
Nonmarsh grasses, dense
Dense hydrophytic forest, cypress dominant
Lowland hydrophytic deciduous forest
Dense mixed forest, well drained (maritime forest)
Sparse pine forest, well drained
Sand interspersed with grasses/shrubs (medium density)
Sand interspersed with grasses/shrubs (low density)
Sand, nonvegetated
Agriculture

2. Identification of varieties of marsh grasses with differing heights which can be used as ecological indicators and keys to biological activity
3. Identification of patterns of salt water intrusion into estuaries which provide data needed for determining the areas where many types of shellfish can be planted successfully
4. Identification of spectral variations in water which can be used to trace water movements and water mixing in estuary/tidal areas that have very dynamic surface hydrology characteristics
5. Identification of land cover features which have the best drained land in the coastal zone, these areas being the most suitable to use for construction with ecological and economic safety
6. Identification of land cover features characterized by poor drainage which consequently are not suitable for intensive commercial and residential development
7. Identification of ecological zones which provide insight into floral and fauna characteristics of coastal zone subregions
8. Monitoring of macro physical changes in land which result from wind action, wave action, and human development of the coastal zone

A sample from the North Carolina coastal zone land use and land cover classification is provided in figure 16.14. Not all 17 classes are represented in this example, but selected major features of interest to planners, ecologists, and hydrologists are presented.

A central feature in figure 16.14 is the delineation of marsh grasses (both shorter and taller *Spartina* and *Juncus*) which serve as a principal breeding ground for many types of important coastal zone aquatic life. The pattern, total amount, and height of the marsh grasses are clearly discernable from intensive Landsat analysis. Monitoring of their areal pattern and characteristics makes it possible to identify expansion or contraction in the distribution of these valuable plants as well as to note selected physiological changes. This information can be used to help identify physically or culturally induced environmental changes which are affecting the role marsh grasses play in the coastal zone ecosystem.

Figure 16.14 also illustrates how Landsat information provides insight into the dynamic and often confusing surface hydrology of estuarian waters in coastal zones. Tidal movements in and out of the estuary twice a day on an irregular daily schedule make detailed analysis of salt water intrusion (into the fresher water estuaries during high tide) and fresher water movement (from rivers into estuaries during low tide) virtually impossible for large areas by on-site methods. The tide was moving ocean water up the Cape Fear River estuary at the time of the Landsat overflight on the data used for analysis. South of the subarea shown in Figure 16.14 the tidal flats and fresh estuarian water was replaced by the intruding ocean water. North of the area shown in figure 16.14 the tidal marshes and tidal flats were greatly exposed and fresh estuarian water totally occupied the estuary.

Frequent Landsat analysis of an estuary would show the complex pattern of fresh water and salt water movements throughout the estuary because the schedule of tidal movements varies every day even though the passes of the satellites over a given point in the United States are at the same time each morning. From this type of information many of the dynamic characteristics of surface hydrology in coastal zones can be better understood and thus applied to environmental and planning issues which are related to water conditions. Evaluation of

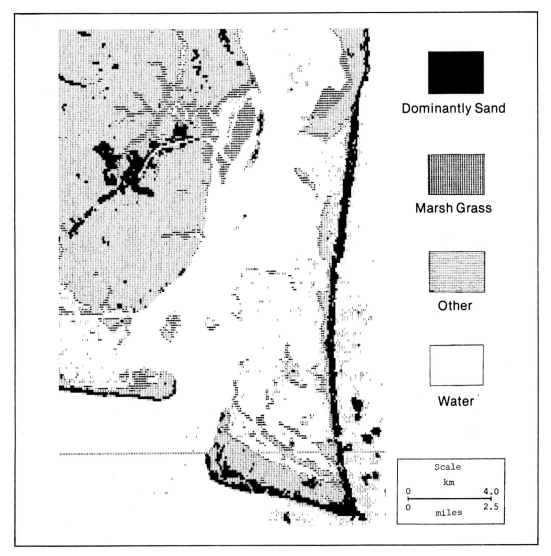

Dominantly Sand

Marsh Grass

Other

Water

Scale
km
0   4.0
0   2.5
miles

Figure 16.14. Classification of selected features through analysis of October 11, 1972 Landsat multispectral data from a coastal zone area (Cape Fear, North Carolina). A total of 17 land use/land cover classes were developed in this classification, but only 4 of these classes are displayed. (Source: ISURSL)

human waste disposal, analyzing the effect of chemical and thermal effluents on environment and establishing economically valuable organisms (*Spartina* marshes, oysters, new species of fish) in coastal zone waters are among some of the activities which need accurate surface hydrology information frequently.

Uses of Landsat Multispectral Data for Analysis of Mineral Resources

Landsat data have been used to help locate new faults or to better define existing faults and fault linaments. This information has been used to identify general areas which have high probabilities of containing valuable mineral (primarily metallic) deposits. Another, but more widespread, use of Landsat data is in the analysis of surficial mining activities, foremost of which is coal.

Surficial mining research has assumed several forms most of which are not associated with finding new mineral deposits. Research has focused on delineating areas which have been strip mined, identifying the nature of the land use which occurs on reclaimed strip mined land, identification of current land use/land cover features on land designated for future strip mining, and monitoring changes in the areal extent of strip mining.

ISURSL has been involved primarily in developing information from Landsat data which can be used to help assure wise land management practices in areas deemed to be suitable for strip mining development. Landsat data are being used to assess the current uses made of lands which have been strip mined. Frequent identification of newly developed surface mining areas are being made to help guide the course of future regional development.

Many of the types of land use/land cover information which are commonly developed for the purposes previously stated are indicated in table 16.3, but additional categories of information specifically associated with surface mining are often developed. These additional categories include specific subclasses of bare mine spoil, subclasses of partially vegetated mine spoil, and subclasses of surface mined water features.

An example of recent research conducted by the ISURSL in surface mine areas is illustrated in figure 16.15. One purpose of the Landsat analysis was to determine the extent of strip mining in a major western Indiana coal field. A second major object of the research was to identify uses made of land which has been striped mined during the past 30 years.

As indicated in Figure 16.15, it was discovered that the oldest (15 years or more) strip mine lands were densely covered with forest, while somewhat younger (8-15 years) strip lands were covered with thick grasses. More intensive analysis later provided insight into specific characteristics of the mixed and coniferous forest that occupied the older sites. In the study area, sparse to medium density grass covered the land which had been stripped within the last 3-8 years. A few recently stripped areas (2-3 years old) were discovered to have been reclaimed and planted into small grains. Land stripped within two years of the acquisition of Landsat data was dominated by mine spoils which had little if any green vegetation cover, as were other spoil areas on steep slopes. Strip mine ponds and lakes were also identified but no attempt was made to develop spectral subclasses of water in this example.

Overall, the information from this analysis provides a cross-sectional view of coal strip mine land development through time. Several similar analyses would provide land use/land cover information from which the national pattern of coal strip mine land use could be accurately ascertained. Periodic monitoring of these areas would identify new mining activity and indicate patterns of land use and land cover change on existing stripped land.

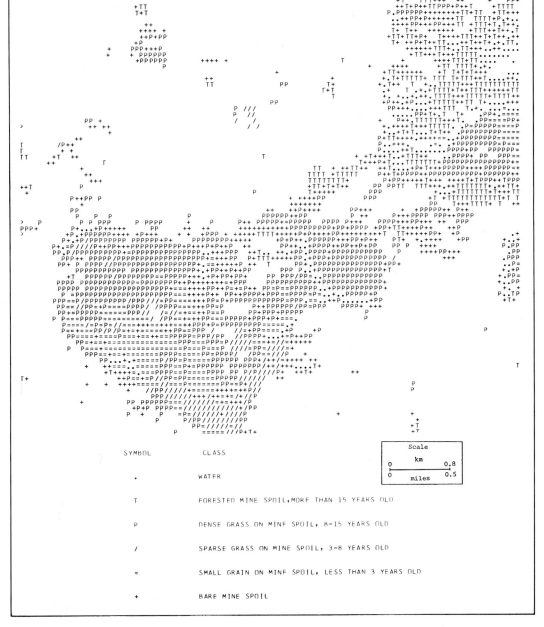

Figure 16.15. Distribution of land use/land cover features in a coal strip mine area in Clay County, western Indiana. Identification of these earth surface features was acquired through analysis of September 7, 1973 Landsat multispectral data. (Source: ISURSL)

17

Application of Remote Sensing in Industrial Analysis

James John Flannery

University of Wisconsin—Milwaukee

Introduction

Historically the analysis of industrial activity from remote sensing imagery can be traced to the period of World War II. During that time aerial photographs were used to identify tactical military targets and to select strategic military objectives for aerial attack. Commonly, the strategic military target was an important war-related industry such as an iron and steel plant or a petroleum refinery. Military aerial photograph interpreters became adept in identifying types of industries and in evaluating the effects of aerial bombing strikes.

Today, the interpretation of industrial activity from remote sensing imagery has many peaceful uses. For example, remote sensing of industry is an important source of data for people in government or industry who are engaged in such activities as planning, property evaluation, zoning regulations, inventories of raw materials stockpiles, and environmental considerations of water and air quality.

Aerial photography remains the most commonly used type of remote sensor imagery for industrial analysis. A number of factors account for this, such as the low cost of aerial photographs and the high definition of targets on them. Identification of specific types of industries, recognition of particular problems associated with industrial activity, or determination of the general class of industry requires imagery of high resolution.

Because of the need for high resolution of detail, aerial photography for purposes of industrial analysis is frequently flown at large scales, usually larger than 1:10,000, and in many cases larger than 1:5,000. Commercial remote sensing firms, in addition to using very large scale photography, also may prefer to use conventional color and infrared color photography because of the inherent definition advantages of these products over black and white imagery.

Imagery from some of the new and sophisticated sensor systems has a utility for industrial interpretation. Unfortunately, little research has been done on the application of these new systems to industrial analysis. There has been some use of near infrared and nonphotographic sensor systems in investigations of the effect of industrial activity on the quality of the environment. For example, color infrared photography has proved to be an effective tool in monitoring certain types of water pollution.[1] In addition, the feasibility of using ultraviolet, thermal infrared, and microwave sensors to detect marine oil spills has been tested.[2] It was found that all the sensors tested have considerable potential for oil spill investigations; however, radar, in particular, seems to have the best application for mapping the aerial extent of oil spills.

The application of thermography to problems associated with industrial activity has been demonstrated in a number of studies. Thermal scanned images of heat pollution in streams and the use of thermal infrared sensors to determine the precise location of subsurface fires in coal waste piles are two examples of the successful practical application of scanners to problems associated with industrial activity. Also, there has been research on the use of Landsat imagery to monitor those extractive industries, such as strip mining and forestry, which, because of their aerial extent, can be imaged by the Landsat system. Notwithstanding these developments, the aerial photograph continues to be the most valuable and commonly used remote sensing tool in industrial analysis, particularly in problems of identifying types of industries.

Industrial Interpretation from Aerial Photographs

In this chapter industry is defined to include not only manufacturing but the extractive industries and the utilities as well. Mining is the only extractive industry discussed in this chapter because forestry has already been treated in chapter 14.

The same rule applies to the interpretation of industrial activities that applies to all aspects of imagery analysis. In the case of the interpretation of industrial activity this means that the interpreter must know such things as the nature of a specific industrial process, the equipment employed, the kinds of raw materials used, the nature of the finished product, the waste products, and the types of related support or auxiliary facilities. It is important, therefore, for the interpreter to be familiar with the industry he is called upon to analyze, and to know how that industry is usually imaged on aerial photographs or satellite images.

The Identification of Types of Industries

There are many different ways to classify industries depending upon the intent or purpose of the investigation. Industries, for example, may be grouped under headings such as consumer goods industries, capital goods industries, light industries, and heavy industries. These are very broad categories, and interpreters are often called upon to be more specific than this and to classify industries on the basis of the general type of product manufactured. For example, a study may require classifications of industries as metal industries, metal fabricating industries, nonmetallic industries, chemical industries, textile industries, and food industries. The problem with these classifications is that they are not very useful in attempting to identify types of industries from remote sensor imagery.

Every industry has certain individual units or components which can be imaged on aerial photographs. Some industries, however, are difficult to identify because their functional units are not imaged in any distinctive manner. Others have unique characteristics which make their identification relatively easy. For example, all industries require some kind of structures or buildings. In some cases the industrial process may give a single character to its structures making it relatively easy to identify the industry on an aerial photograph. Also, industrial equipment may be external to the building or present in storage yards where it is easily imaged by an aerial camera. These features are found with blast furnaces in an iron and steel complex or forgings in a storage yard.

Frequently the pattern or arrangement of various units in an industrial facility will reveal the nature of the industry. The layout of the plant—that is, relative positions of structures, their sizes, interconnections among them, the location of raw material stockpiles, and finished goods storage—provide a substantial aid in the identification of the nature of a given industrial activity.

In many instances it may be quite difficult to identify a specific industry on the basis of exposed equipment, unique structures and buildings, or layout. This is frequently the case with light industries. Light industries are commonly housed in buildings which are similar in appearance. Also, nonindustrial operations, such as wholesaling and storage, may be housed in buildings similar to those of light industries.

When attempting to identify many kinds of industries, certain support components which are essential to the industrial process, and which are visible on the photograph, may provide important clues to the nature of the industrial activity carried on inside the structure. These are referred to as *corollary* components,[3] and they include such things as power stations, workshops, chimneys, ventilating equipment, pipelines, tanks, conveyor belts, overhead cranes, derricks, raw material stockpiles, finished goods storage, and loading docks. All of these are related, in fact are crucial, in one way or another to the ongoing industrial activity. The interpreter, however, should be aware that many corollary components are common to a wide range of industries; therefore, it is important to know which components are associated with which types of industrial facilities.

Specific kinds of industries, therefore, can be identified from aerial photographs because: (1) unique types of equipment, structures, or other elements are imaged and provide an indication of the type of industrial process; (2) the pattern or arrangements of the various units in the industrial complex, such as structures and locations of stockpiles, may give additional clues concerning the nature of the industrial activities; and (3) images of selected corollary or associated units, commonly in revealing combinations, also may indicate the nature of the industrial activity carried on in structures which, in themselves, are not indicative of the specific industry.

Classification of Industries for Aerial Photograph Interpreters

Chisnell and Cole have provided a classification of industrial activity which is well suited to the needs of aerial photograph interpreters, although other industrial classifications and keys for aerial photograph industrial interpretation were developed as early as World War II for military purposes (fig. 17.1). The Chisnell and Cole contribution, however, is the most comprehensive and useful for aerial photograph interpreters who deal with the identification of industries. The Chisnell and Cole system was designed to aid the relatively inexperienced interpreter—the one not necessarily highly qualified in industrial interpretation. Following their method, in many cases, the interpreter can identify a specific industry and its components from aerial photographs.

A further advantage of the Chisnell and Cole method is that in those situations where a particular industrial activity is difficult to identify, the interpreter can at least assign the in-

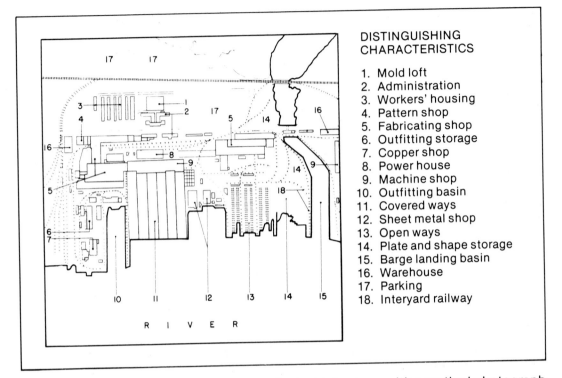

DISTINGUISHING
CHARACTERISTICS

1. Mold loft
2. Administration
3. Workers' housing
4. Pattern shop
5. Fabricating shop
6. Outfitting storage
7. Copper shop
8. Power house
9. Machine shop
10. Outfitting basin
11. Covered ways
12. Sheet metal shop
13. Open ways
14. Plate and shape storage
15. Barge landing basin
16. Warehouse
17. Parking
18. Interyard railway

Figure 17.1. A diagram of a shipbuilding plant which, along with a vertical photograph of the same facility, served as an interpretation key during World War II. A number of other industries were shown by both diagram and photograph in the *Photographic Interpretation Handbook-United States Forces,* U.S. Army Air Forces and Navy Department, 1944.

dustry to a broad category. The Chisnell and Cole classification is, first of all, based on industrial components as they are imaged on aerial photographs. By use of imaged components, all industries are classified into three major groups: *extractive, processing,* and *fabricating.* The processing industries are further subdivided into mechanical, chemical, and heat processing, and the fabricating industries into light and heavy fabrication. Once the industry has been classified, the number of possibilities needed to be considered by the interpreter are substantially reduced. On the other hand, within the groups there is still enough image variation that, hopefully, the interpreter can identify specific industries with the help of interpretation keys.

Extractive Industries

If there is one characteristic which might be considered common to mineral extractive industries, it is the presence of waste piles. Exceptions are found in those mineral industries which employ wells to extract minerals, such as the petroleum industry, salt industry, and sulfur industry where the Frasch process is used.

Another relatively consistent characteristic of the mineral industries is the presence of considerable surface disturbance. The surface disturbance may be deep, and sometimes extensive excavations can be seen, such as those associated with open pit mining, quarrying, or digging for sand and gravel. Sometimes surface disturbances may take the form of furrows of overturned earth which are common in strip mining and placer or dredging operations. The former is common in many parts of the world, including parts of the United States. If minerals are being extracted from underneath the surface of the earth, evidences of surface modification in the form of excavations or turned earth usually will not be seen. Underground mining, however, can be identified by the presence of waste piles, piles of mined materials, and a variety of corollary image units. If excavation is in process, the aerial photograph may contain images of the associated excavating equipment, such as drag lines, mechanical shovels, bulldozers, or dredges. In addition, there is the requisite transportation equipment necessary to remove mined materials from the site, such as trucks, mine cars, railway equipment, and conveyor belts.

Buildings are not necessarily of equal significance among the various minerals industries. In some mining operations the buildings might be quite large and impressive. This is the case with those minerals requiring considerable amounts of processing before they are shipped from the mine site. Copper and taconite are two ores which, because of their low grade and/or physical nature, must be processed and concentrated at or near the mine site in order to make it economically feasible to move them over long distances to smelters. Other types of mineral industries, such as sand, gravel, and quarried stone, do not necessarily require large and elaborate structures. Figures 17.2 and 17.3 are stereograms of extractive industries.

Processing Industries

Processing industries are those which modify raw materials by the employment of heat, chemical, mechanical, or kinetic energy. The most common image characteristic of the processing industries on aerial photographs is stockpiles of raw materials. Stockpiles may exist in different forms. They may be in outdoor piles, in covered buildings, in ponds or tanks, or in a variety of closed and open storage units. Raw materials must be moved from the stockpiles into processing facilities. As a result of this, there usually will be evidence on aerial photographs of transporting facilities, such as conveyor belts, mobile cranes, and railroad cars.

Frequently, the processing units themselves may be imaged on aerial photographs. Smelters, ovens, kilns, or chemical processing units, such as the cracking towers in petroleum refineries, can be identified on large-scale aerial photographs. If the processing equipment or facility is not directly visible, evidences of it may be revealed by the shape of the housing structure or by associated images of chimneys and vents. Because much of the processing equipment in mechanical industries is large, the structures which house the equipment are also large.

Processing industries employ considerable amounts of energy. The image components of these energy needs are powerhouses, piles of coal, storage tanks for oil and gas, and electrical transformer stations. Also common to most of the processing industries is the problem of massive waste disposal. There may be direct image evidence of waste materials, such as waste piles, or the presence of facilities structured to deal with the waste, including drainage channels, settling tanks, or ponds. Some finished products of processing industries are stored in the open. Open storage of finished goods appears orderly and, therefore, may cause confusion for the interpreter.

The processing industries can be subdivided into mechanical, chemical, and heat processing industries. There are instances in which two or more of these types of processing activities

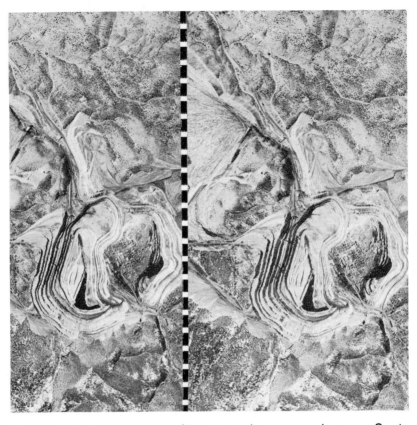

Figure 17.2. A stereogram of an open pit copper mine near Santa Rita, Grant County in southwestern New Mexico. The photograph was taken in 1937 and the mine no longer has this appearance. The terrace method is used and a large mine tailings dump is clearly imaged. Notice certain mine buildings on a high island surrounded by the pit. (Courtesy of the U.S. Department of Agriculture and University of Illinois)

are carried on in a single industrial facility. In an integrated iron and steel plant, for example, not only is heat used in the smelting process in the blast furnaces and in other aspects of the steel manufacturing operation, but there is also a chemical phase related to the byproducts coke ovens which are characteristic of modern iron and steel plants. Another example is a papermill where both mechanical and chemical processes are employed to reduce the pulpwood to the consistency where it can be manufactured into paper. Combinations of this type, however, do not interfere with the intended purpose or utility of the processing classification.

Mechanical processing industries. These include such industries as sawmilling, flour milling, meat packing, and ore concentration. All mechanical industries alter the physical characteristics of raw materials, that is, they carry on physical reduction processes. Large units of raw materials are mechanically altered into smaller forms, as in the sawmill industry, where

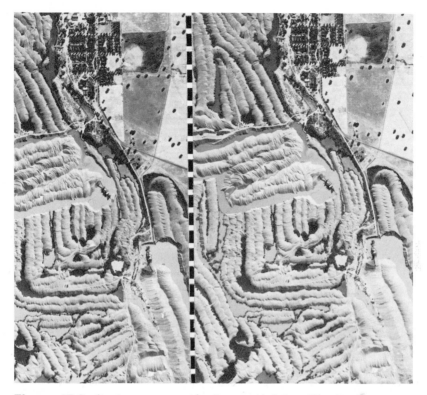

Figure 17.3. A stereogram of placer mining with dredges near Hammonton in Yuba County, California. The two dredges can be located near the freshly-turned spoil banks. They are processing both previously worked and fresh alluvial materials without evidence of bulk handling units, suggesting that a precious metal is the objective. (Courtesy of the U.S. Department of Agriculture and University of Illinois)

logs are sawed into boards. The most important identifying image characteristic of mechanical processing industries is open raw materials storage.

Along with raw materials there will be strong image evidence of equipment which handles raw materials, such as hoists, cranes, and conveyor belts. Frequently, although not always, energy units are associated with mechanical processing industries, such as in paper manufacturing where power is an important element and, therefore, a major image component. Powerhouses, along with their associated features, are the energy support components most common to the mechanical processing industries.

This category of industry differs from the other two processing industries in that tanks, pipelines, and chimneys are not numerous. Where they are found, they are usually restricted to the energy unit and are not part of the main processing unit. Many mechanical processing industries have their equipment housed in large and complex buildings. There are exceptions, however, such as in the sawmill industry where the buildings may be very small and frequently are temporary types of structures. A number of utilities such as hydroelectric power and water and sewage treatment plants image as mechanical processing facilities and are, for convenience, so classified here (figs. 17.4 and 17.5).

Chemical processing industries. These include those industries which produce such materials as sulfuric acid, ammonia, and fertilizers. One of the most common chemical industries is petroleum refining. In this group of industries raw materials are altered in terms of their chemical composition through the use of other chemicals, heat, and/or pressure.

The images which aid in the interpretation of chemical processing industries are those of units used to store either raw materials or finished products. In the chemical processing industries, storage facilities usually are in the form of tanks or other kinds of sealed or closed

Figure 17.4. The industry shown here is a paper mill. Typical of mechanical processing industries is the large area devoted to raw material storage (A); primarily stocks of pulpwood arranged in long orderly rows. Bark is stripped from logs in the woodhouse (B); they then pass through the grinding house (C) before being processed into paper in (D). (Courtesy of Consolidated Paper Corp. and Aero-Metric Engineering)

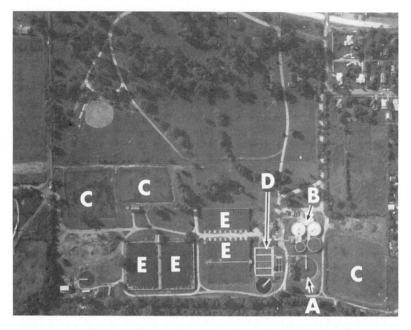

Figure 17.5. Some utilities image as mechanical processing industries. In this case the utility is a small sewage plant. Outside "storage" is a strong image component. In the sequence of the treatment process, (A) is the primary tank, (B) four anerobic digesters, (C) drying lagoons, (D) the activated sludge tanks, and (E) the filter beds. (Courtesy of University of Illinois)

units. In certain chemical industries, such as in the agricultural fertilizer industry, tanks for sulfuric acid or ammonia, or large unsealed buildings for the storage of solid raw materials and manufactured products, will be present.

Pipes and pipelines connecting tanks and processing units are another important image component of chemical industries. The processing units may be exposed units and open girder support structures for pipes. Many processing units are tall, narrow cylindrical structures, such as cracking towers in a petroleum refinery. The presence of smoke or steam plumes imaged on aerial photographs is another clue to the location of processing units. In most chemical industries the proportion of buildings to pipelines and storage tanks is not very large (fig. 17.6).

Heat processing industries. Heat processing industries are those which use heat to restructure the important constituents or to separate them from waste materials. Examples are smelting operations and the generating of electricity in thermal or steam-powered plants. Other examples of heat processing are primary finishing work, such as hot steel rolling and the making of bricks and cement. Common image characteristics associated with heat processing industries are those related to heat production and use; therefore, images of tanks, piles of coal, chimneys, and smoke or gas plumes are apparent on aerial photographs of these industries.

The heat processing industries have a significant amount of their areas devoted to the storage of raw materials, as well as areas which are devoted to the storage of waste materials

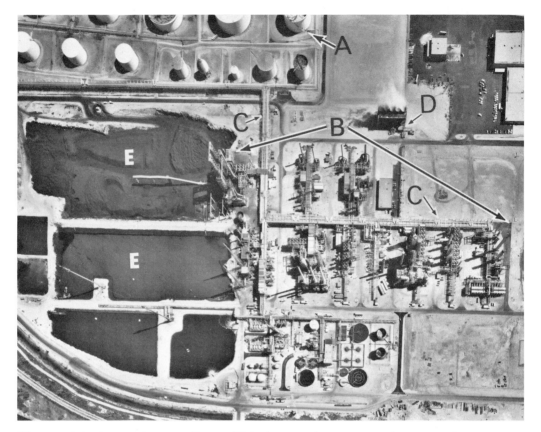

Figure 17.6. A petroleum refinery. This is a typical processing unit. The storage tanks cover a larger area than is here shown. (A) storage tanks, (B) fractioning towers and other cracking units, (C) pipelines, (D) water cooling units, and (E) residual storage. (Courtesy of Chicago Aerial Survey)

generated by the process. Blast furnaces are easily recognized because of their characteristic structure and open exposure. When processing units are housed in buildings, these are frequently designed for a specific purpose and are easy to identify. Roof design is particularly revealing. Also, buildings are designed, and sets of buildings arranged in patterns, to allow for the easy movement of processed goods through the plant, providing another clue for the interpreter (figs. 17.7, 17.8, and 17.9).

Fabricating Industries

Fabricating industries, in general, are engaged in the production or assembly of finished goods from materials usually produced by the processing industries. Individual kinds of fabricating industries are the most difficult of all to identify from aerial photographs because the buildings housing these industries do not provide many clues for identification. Also, there is a larger variety of fabricating industries than of extractive and processing industries, with most of the fabrication being carried on inside the buildings. Not only are the processing

Figure 17.7. A cement manufacturing plant typical of the heat processing industries. Raw material stockpiles of crushed limestone are located at (A). The various processing units (B), such as the crushing house, kilns, and drying house, are arranged in line for efficiency. Storage of the finished product is in elevators (C) with loading facilities; railroad at (D); and water at (E). (Courtesy of Aero-Metric Engineering)

Figure 17.8. Shown here is a thermal electric power station in Indiana on the shore of Lake Michigan. Such plants image as processing facilities. (A) is a water cooling tower, (B) the coal stockpile, (C) a conveyer to (D) the generating house, and (E) the transformer and switching yards. (U.S. Army Corps of Engineers photo)

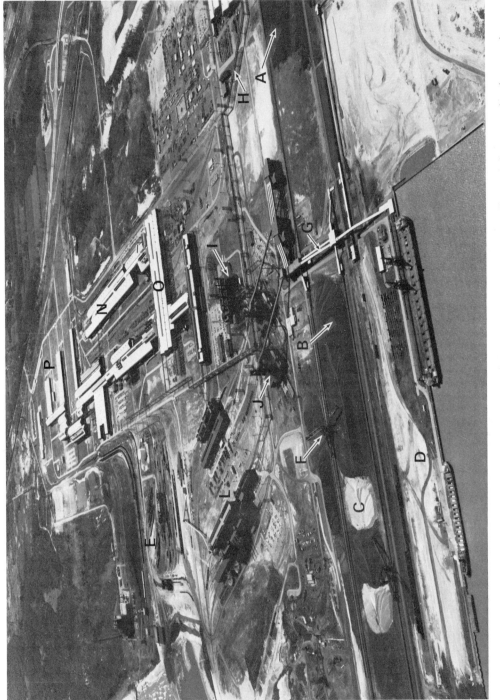

Figure 17.9. An oblique view of the Burns Harbor Bethlehem Steel plant in Indiana. It is an early view and some of the buildings now present are not shown. The integrated coke ovens are to the right of the picture near the coal piles (A). (Courtesy of Bethlehem Steel Corp.)

facilities themselves located in buildings, but as a consequence of their vulnerability to weather elements, raw materials and finished goods are often stored inside and, thus, out of view of the aerial camera. There are, therefore, not as many exposed image components to aid in the identification of fabricating industries as in the extractive and processing industries.

The two major subtypes are *light fabrication* and *heavy fabrication* industries. Of the two, heavy fabrication industries are easier to identify on aerial photographs than specific types of light fabrication.

Heavy fabrication industries. Heavy fabrication industries have more potential image components than those classified as light fabrication. Some examples of heavy fabrication include the manufacture of railway equipment, earth moving equipment, and shipbuilding. The raw materials and finished goods are large and heavy and are often stored outdoors. There is usually a greater need for energy. Overhead cranes and other heavy moving equipment are more typical of a heavy goods facility than of a light industry. The factories are usually located along railroads or bodies of water, and in many cases railroad tracks enter the buildings. The buildings themselves cover a large area and are frequently one story in height. Figure 17.10 is an example of a heavy fabrication plant, although it would be difficult from this photograph to identify precisely what is being fabricated within the buildings. Clues to what is being fabricated in this photograph are in the storage yards of raw materials and finished products.

Light fabrication industries. They include the manufacturing of textiles, garments, shoes, plastics, and light metal goods. Often the raw materials and the finished goods are stored in the buildings, perhaps in the same units where fabrication takes place. The design, size, and plans of the buildings are not distinctive and so are of little aid to the aerial photograph interpreter. Buildings may be small or large in size and single or multistoried. There is no uniformity of building structure even in the same industry, although there are exceptions (compare A and B in figure 17.11). Heavy moving equipment usually is not needed and so is not imaged on photography taken of light industries, its absence being a clue to the fact that it might be light industry.

Chisnell and Cole recommend that the interpreter proceed from the general to the specific when attempting to identify industries on aerial photographs. First, the interpreter must determine if the industry is an extractive, processing, or fabrication industry. If it is processing, the next step is to determine if it is chemical, heat, or mechanical processing. If it is a fabricating industry, is it light or heavy fabrication? By use of the Chisnell and Cole method the interpreter at least will be able to get this far in the classification process. If there is need to identify specific extractive, processing, or fabricating industries, the interpreter must use an interpretation key for that industry. If interpretation keys are not available, it is not difficult to construct one.

Structuring an Industrial Interpretation Key

Interpreters who are involved in industrial analysis in remote sensing, particularly from aerial photographs, will find the Chisnell and Cole classification of industries useful. When working with industries, however, as with any other interpretation task, it is desirable to develop more detailed identification keys. The keys are checklists of the image components of a given industry. Much of the preceeding material is a type of loosely structured or informal set of keys for certain groups of industries.[4] When developing a key for a specific industry the interpreter should study the technical literature covering the industry. In addition to a description of the industrial process, there will be photographs and diagrams of typical plants in which the functional units and support facilities will be identified. This information can be verified by field investigations after the interpreter has applied his book knowledge to the problem of identifying the industry and its important components on an aerial photograph. The key, however, is composed of *identification features*—components of the image on aerial photographs.

In addition, adjunct or supporting data should be used, some of which may not be part of the photographic image, such as geographic location. Knowing that an aerial photograph of an

Figure 17.10. A heavy fabrication plant. Products include, among others, auto and truck frames, auto control arms, and railway brake equipment. (A) is incoming steel storage, (B) and (C) are respectively truck and auto frame storage, (D) is auto frame production, as are (E) and (F). Chimneys and vents on these buildings indicate the use of heat. (G) is a paint shop. Notice the considerable railroad service facilities. (Courtesy of A. O. Smith Corp.)

Figure 17.11. Shown are two light fabricating industries. (A) is a manufacturer of light metal products for homes and hospitals. It's most distinctive feature is a heat treatment unit at (X). Other than for that, (A) could be a box factory, hosiery mill, or other light industry. (B) is a plastics products firm. Unlike most light industries it is distinctive due to the complex of roof structures, including ventilators, filters, tanks, and pipes, which are common to many plastic plants. (Courtesy of Plastics Engineering, Inc. and Aero-Metric Engineering)

open pit mine was taken in northern Minnesota makes it easier to reach the conclusion that the mine produces iron ore.

In addition to a good geographical background, historical knowledge is an aid in the interpretation of cultural features on aerial photographs. Old multistory buildings located along rivers near falls and rapids as imaged on aerial photographs of part of Massachusetts are probably old—or former—textile mills.

The essential ingredient of an aerial photograph interpretation key is the recognition of identification features. The recognition key for an iron and steel plant follows, with the underlined items being the identification features and the letters locating these features in figure 17.9—the Burns Harbor, Indiana plant of Bethlehem Steel Corporation. Because the Burns Harbor plant is relatively new, the spatial organization appears more orderly than that of older iron and steel works. The facility uses the basic oxygen process for making steel. The interpretation key for the iron and steel industry provides a good example of a key for a specific industry, and it demonstrates the importance of identification features. The major items in the key are as follows:

1. Raw material storage in *piles* consisting of *coal* (A) black in color; *iron ore* (B) dark gray on black and white photography and reddish gray on color photographs; and *limestone* (C) white or light gray. Usually on or near heavy transportation facilities such as *docks* (D) or *railroad yards. Stocks of scrap steel* (E) will usually be located elsewhere in proximity to the steel processing equipment. A large *supply of water* is also another essential raw material.

2. Transport facilities such as *overhead cranes* (F), *conveyor belts* (G), *rail lines* and a few *pipelines* for moving bulky or heavy raw materials (H).

3. *Powerhouses* (I), large buildings with many *stacks,* vents and their *smoke plumes,* as well as coal piles or other fuel sources all indicating massive uses of heat.

4. Exposed processing units such as the *blast furnaces* (J), large towerlike structures where the ore is smelted into pig iron, and the furnace support equipment including *stoves,* four cylindrical units in a row, to heat air before it is forced through the furnace, the *casting house,* at the base of the furnace, and hoist *conveyors* to load the furnace. Also in the open are the *coking ovens,* long low structures with *hoist houses* and *conveyors* to load the coal from nearby stockpiles into the ovens and *pipelines* to carry combustible waste gases from the coking ovens to heat using units.

5. Large complex buildings such as the steel making units where iron from the blast furnaces and scrap steel are heat processed into molten steel. Many modern steel plants employ the *basic oxygen furnace* (L), housed in rather tall buildings not too distant from the blast furnaces. Older plants rely on the *open hearth* furnace with a set of hearths enclosed in long narrow buildings, usually identified by a row of *evenly spaced tall chimneys,* one for each furnace, along one side of the structure. *Storage of scrap steel* (E) may be observed near these units. Other complex structures are the *rolling mills* (N), a set of long one story interconnected buildings. The steel usually enters the mill as an ingot and may be rolled either hot or cold. Other similar and related structures are *slab mills, wire mills,* and *tube mills.* A feature essential to the rolling mills is the *soaking pits.* If a milling operation requires hot steel it is kept hot in the soaking pits. *Soaking pit buildings* (O) are part of the rolling mill complex but with their long dimensions usually at *right angles* to that of the mills. They have roofs characterized by many *ventilating units.* The soaking pits are usually found at the beginning of the production line. *Storage and shipping* (P) buildings are at the end of the production line and have roof structures evidencing little need for ventilation. *Loading docks, large doorways* and *railroad tracks* entering the building are associated identification features.

6. There are waste products and processing units such as piles of *slag,* a by-product of smelting, and *water treatment* units, buildings, and settling ponds to purify the huge amounts of water used in various cooling operations.

The above identification key works best when applied to United States facilities; however, it is still applicable to aerial photographs taken of iron and steel plants in other countries. It is not necessary that keys be involved and lengthy. The length of the key above is caused by the inclusion of brief explanations of phases of the industrial process as an aid to the reader. In a real situation the interpreter would first become familiar with the process and then build a key based on those identification features which would normally image well enough on aerial photographs to become part of the key.

A large scale aerial photograph of a typical industrial facility, accompanied by notations, makes an effective "key." This is particularly true if the photograph is matched, for clarification, with a map, diagram, or ground photographs of the plant (fig. 17.1). The larger the scale of aerial photographs, the easier will be the identification of industries. The use of stereographic pairs of aerial photographs is an obvious aid. Not only is visual acuity improved by the use of stereoscopy, but seeing the vertical configuration of the industrial units aids greatly in the identification process. Simple pocket stereoscopes permit the use of aerial photography as small as 1:20,000 for much industrial interpretation.

Applied Aspects of Aerial Photograph Interpretation of Industry

There are many ways in which aerial photograph interpretation can be applied to industrial operations or problems associated with industrial activity. One use is for inventories.

Inventories from Aerial Photographs

In those industries where raw materials or finished goods, such as coal, ores and other minerals, pulpwood, and lumber, are piled in the open, aerial photographs expedite the estimation of the quantity of materials. In some instances it has been demonstrated that estimates made from aerial photographs are not only less costly, but probably more accurate than those made on the ground. Obviously, neatly stacked piles of pulpwood, saw logs, or lumber are easy to measure, and accurate inventories can be made from large scale stereographic sets of aerial photographs. Reliable inventory work usually requires aerial photography at a larger scale than 1:1,000.

If, on the other hand, the stockpiles are mounds of varying shapes and sizes, accurate inventories from aerial photographs can still be made. Large stockpiles of coal or other minerals have to be inventoried periodically. It has been determined that if aerial photographs are used for the inventory, a savings of approximately 25 percent can be made as compared to the cost of a ground survey. Some practitioners of aerial photograph inventories claim their method is accurate to within 2 percent, whereas they claim the lowest cost ground surveys average only 15 percent accurate.

In both ground and aerial surveys of stockpiles, the most accurate ground and aerial photograph methods employ the same kinds of measurements. The stockpile is mapped in detail with closely spaced contour lines at one or two foot intervals. Smaller intervals are used near the top of piles where the area dimensions of the pile are smaller. After contouring, the area circumscribed by each contour is measured. Using the contour interval and the area, the volume of each layer can be determined. Tables for specific minerals are used to convert the cubic units into number of tons in each layer. Frequently, because of the settling of material in the pile, a compaction factor is used to give a more accurate estimate of weight. The total number of tons in the pile is obviously the sum of all the layers.

Similar kinds of measurements can be made of excavations, and the amount of material removed from an open pit mine over a period of time can be calculated. This type of data is of value to industry, and it could provide a low cost, accurate way for governments to monitor and tax mineral industries on a tonnage basis.

It is also very easy to calculate the storage capacity of tanks from stereographic sets of aerial photographs using the simplest measuring tools, such as height finders and scales. Measurements are made from the stereoscopic image of the height and width of cylindrical or spherical storage units, and standard formulas for determining their volumes are used to calculate capacities.

Aerial Photograph Interpretation of Industrial Water Pollution

There is presently a great concern with problems of water pollution, and industries are major contributors to that problem. One of the major objectives in dealing with water pollution is to find the sources of pollutants in surface waters.

Pollutants from industrial and other sources enter streams and lakes at discharge points which are called "outfalls." Pinpointing the location of outfalls can be effectively done on aerial photographs.[5] By use of 70 mm conventional color and CIR photography at scales of 1:3,000 to 1:9,500, twenty-six of twenty-seven major outfalls on the upper Cuyahoga River in northern Ohio were accurately identified in one study. Outfall identification was not easy because of the narrow, meandering course of the river and the dense vegetation along its banks. Conventional color film was found to be equal to CIR film in the identification of outfalls. The following signatures aided in the identification of stream pollution:[6]

1. Change in stream color
2. Discharge structures themselves
3. Possible surface drainage courses leading from buildings to the stream edge as revealed by topographic and vegetation characteristics
4. Higher reflectance due to bubbles and foam associated with a high velocity discharge
5. A dampening or smoothing effect on waves in the stream because of the greater density of discharge fluids

The first three detection mechanisms were the most important in determining pollution and outfalls on the upper Cuyahoga River. Actually, standard black and white panchromatic, minus blue aerial photography frequently gives the location of outfalls since they commonly image in a lighter tone than the body of water into which they flow.

In a subsequent study it was demonstrated that nine by nine inch color infrared photographs indicated the amount and kind of pollution in a stream.[7] The film was analyzed by color microdensitometer. Water samples were taken at the time of photography to check on the reliability of this method.

The unique definition qualities of color infrared film (Kodak Ektchrome Infrared Aero, type 8443) were used to determine the extent of sulfuric acid contamination of spoil banks in abandoned coal strip mines in the Appalachians.[8] Sulfur minerals associated with coal deposits upon exposure decompose to form sulfuric acid. The sulfuric acid poisons the ground, leaving it devoid of a vegetation cover for as long as fifty years. The problem extends beyond the immediate area of the spoil banks since the poisoned areas are subject to erosion resulting in contamination of downstream areas. The bare earth of the contaminated areas images in the color cyan on the color infrared positive.

It was demonstrated by a survey that a comprehensive statewide study of coal refuse sites can be made for the purposes of reclamation from high altitude color infrared photography at scales of 1:120,000.[9] In addition to locating almost 200 coal refuse pile areas and slurry ponds, it was found possible to assess the character of the site in terms of its areal extent, vegetation cover, and degree of soil pollution. The quality of the survey was checked using large scale photography (1:20,000), ground survey, and industrial area data. The small scale aerial photography measurements were found to be approximately 85 percent accurate. The survey was completed in ninety days at a much lower cost than is possible by any other method.

Nonphotographic Imagery in Industrial Analysis

The use of nonphotographic imagery in industrial analysis when compared to aerial photographs is, in general, in an early stage of development. This is particularly true for satellite imagery, but less so for data supplied by nonphotographic sensor systems carried aboard aircrafts. In the latter case airborne thermography has been widely applied to a variety of heat related industrial problems for almost a decade. The use of thermography in industrial analysis, however, is the major exception to what is primarily an area of experimental research.

Thermography as Applied to Industry

The first nonphotographic sensor systems made available for civilian use after radar were infrared scanners which operated in the near and middle 0.7-5.5 micron, and far, 8-13 micron, infrared bands. These units were commonly referred to as heat sensors, and eventually the term *thermography* was applied to this phase of remote sensing.

Essentially thermography measures the temperature of various surfaces from a distance. Aside from its airborne mode, thermography plays a significant role in medicine and energy conservation. It is an aid in the diagnosis of diseases such as cancer, deep seated infections, and circulatory problems as these conditions may reveal themselves by variations in temperatures over different parts of the body. Mobile ground scanners have proved their worth in measuring heat loss from buildings and heat-using equipment in industrial plants. The technique has already been commercialized with services provided to industry and home owners who are concerned about the proper insulation of their homes in the growing energy crunch. An example of thermographic scanned data and its utility when reproduced on film is shown in figures 17.12 and 17.13.

Figure 17.12 is a film reproduction of a thermal scan of a portion of the Connecticut River near Haddam, Connecticut showing a thermal discharge into the stream from a nuclear power plant. Figure 17.12A shows the heated water, white or light gray in tone, moving downstream to the sea during low tide. Figure 17.12B shows the heated water moving upstream during high tide and probably causing problems in the normal dissipation of heat from the river and thus potentially affecting the environmental character of the stream.

Figure 17.13 shows a simultaneous radiometric trace and thermal scanning technique designed to determine the range of temperatures in the stream near the discharge point. It is a more self-contained remote sensing technique than those relying on supplemental ground or surface temperature measurements which are then later corollated with the thermal scanned data. Using a thermal mapping system to scan the stream an airborne radiometer simultaneously measured water temperatures along the center of the scan line. The radiometer can detect temperature differences of 0.1° C and prints out a strip chart to match the output of scanned data and/or the film copy of that data. Simultaneous time markings in seconds are recorded on both the radiometer strip chart and the thermal imagery. This permits matching the radiometric temperature trace and the thermal imagery.

Since a relationship between the radiometric data and the scanned data can be established, an isothermal map of water temperatures in the stream can be compiled. For example, on the filmed thermal image, a given shade of gray can be matched with a relative temperature. At point A on the radiometer strip chart a temperature change of approximately 8° C is recorded, and on the filmed thermal image at A is the greatest contrast in tone. It should be noted that the

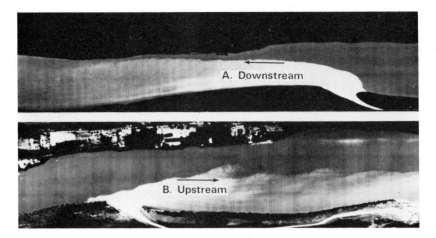

Figure 17.12. These two thermal images were made in September 1968 with an infrared scanner operating in the 8-14 micron band. They show the thermal plume in white and light gray from the Connecticut Yankee power plant near Haddam, Connecticut. With the ebb tide (A), the plume moves down the Connecticut River toward the sea. With high tide (B), the plume moves upstream, probably affecting the ecological balance of the river. (Courtesy of HRB-Singer, Inc., Energy and Resource Systems)

Figure 17.13. A simultaneous radiometric temperature measurement and thermal scan of the Connecticut Yankee power plant thermal plume. (See fig. 17.12) (Courtesy of HRB-Singer, Inc., Energy and Resource Systems)

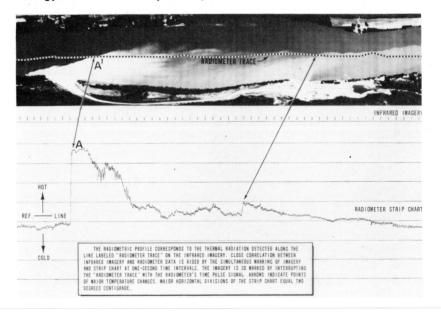

airborne radiometer measured relative temperature differences which in many environmental studies may be more important biologically than the absolute temperature.

If absolute and/or more detailed temperature data from thermal discharges into bodies of water are to be collected using remote sensing methods, along with data on the movement and dispersion of thermal plumes, a considerable effort is required. The amount of ground truth data to be acquired is directly proportional to the desired detail.[10]

Another example of thermal sensing as applied to electrical generating facilities and thermal pollution is seen in Plate 22. Large volumes of water are required to cool condensing units in thermal and nuclear power stations. This water reaches a high temperature and its discharge may pose serious environmental problems. The water can be cooled in a tower or confined cooling pond before being discharged into a stream, lake, or ocean. If a pond is used, the efficiency of the cooling pond must be constantly checked. One method is to use a grid of permanent monitoring stations scattered through the pond. Another method is to monitor temperatures from a boat. Both methods provide data which are confined to a limited number of points and interpolated for those areas not sampled.

By use of a calibrated airborne thermal scanner, a continuous areal readout of temperature data can be obtained for the entire pond in only a few minutes. The scanner used in this study was a quantitative infrared line scanner sensing within the 8-14 micron wave band from an altitude of 457 meters (1500 feet). At this altitude a number of passes had to be made over the pond; however, since atmospheric attenuation of infrared increases with altitude, a low flight was preferred. Ground truth temperatures obtained at selected points in the pond were used to convert and correct the relative temperature differences sensed by the scanner into absolute data which was eventually mapped at intervals of 1° F. The sensor can record temperatures to within 0.1 °C (0.18 °F).

Data collected were recorded on magnetic tape and a color image reproduced via the Digicolor film recording system. This film image uses colors ranging from white for warmest temperatures, 39° C (102° F) and over, through red, yellow, cyan, blue, and magenta to black, for temperatures of 28.9° C (84° F) or lower. The full range of 18° F is divided into six intervals of 3° F with each interval represented by one of the colors. If greater temperature detail is desired, the six colors, exclusive of white and black, can be assigned to different 1° F intervals to cover a temperature range of 6° F. The process can be repeated for each 6° F interval, and in this way successive images produced on film can be used to prepare an isothermal map in 1° F intervals (fig. 17.14).

Another example of the application of thermography to industrial problems is seen in figure 17.15. Figure 17.15A is a standard daytime oblique aerial photograph taken near Williamstown, Pennsylvania showing a culm bank, or coal refuse pile. It was suspected that there was a subsurface fire in the culm bank, but the exact location of the fire was not known. Figure 17.15B is a nighttime far infrared scanned image of the same coal refuse pile. It shows the location of at least two fires as shown by the light "hot spots" on the image. The same data can be recorded on tape and processed by computers and other equipment to provide even more precise information on temperatures and location. With this knowledge fire fighters can determine the source of the underground fires and attack them more efficiently.

With the energy crunch, unnecessary heat loss is of great concern. Airborne thermal scanners are particularly effective in pinpointing heat loss from buildings and industrial sites. Figure

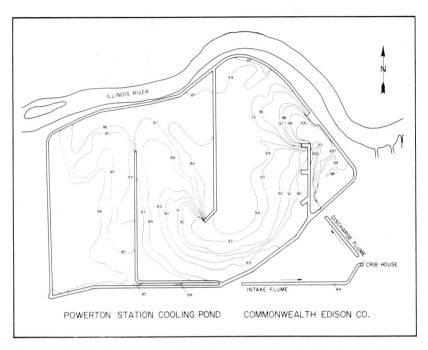

Figure 17.14. The map of 1° F isotherms was constructed from data compiled using the Digicolor film recording system. (Courtesy of Daedalus Enterprises, Inc.)

17.16 is an aerial infrared scanner image of the University of Iowa Campus in Iowa City taken on the night of November 22, 1976. The arrow shows a "hot spot" in what appears to be a parking lot. The "spot" is caused by the heat released from a 1.3 centimeter (1/2 inch) leak in an underground steam line. It was estimated that the heat loss from this break had cost the University up to $10,000 before it was repaired following its discovery on the thermal image. Other leaks were also discovered from this scan which cost about $6,000, which represented only 1/10 of 1 percent of the heating budget for the entire year.

An interesting thermal image is that in figure 17.17. It is a nighttime aerial image of the Atlantic-Richfield refinery south of Philadelphia, Pennsylvania. The location of intense heat producing units associated with refineries, such as cracking towers, are clearly shown by the white "hot spots." The sensor also singled out full and empty storage tanks. Those which are full are warmer and appear white or light gray, while those which are empty are colder and darker. This type of data could not be obtained from conventional aerial photography.

Satellite Imagery in Industrial Analysis

The application of satellite imagery in the study of industrial activity is limited, but of growing importance. The resolution of satellite sensors is usually not sufficiently detailed to provide the data desired for most industrial studies. There is some evidence that machine processed Landsat multispectral digitized data can be used to distinguish industrial complexes in

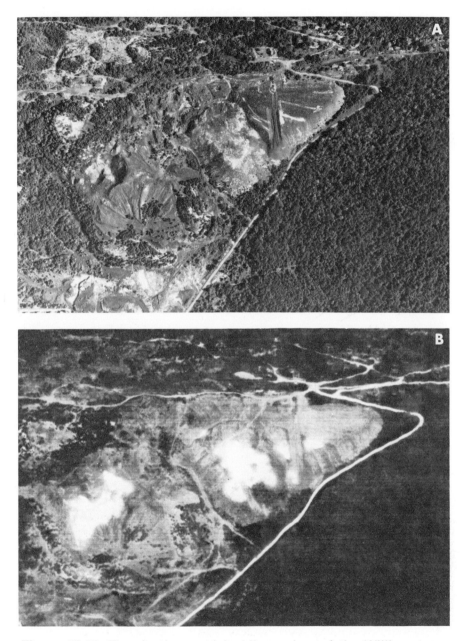

Figure 17.15. The daytime aerial oblique view of the Williamstown, Pennsylvania, culm bank (A) gives little or no evidence of the presence of a subsurface fire in the coal refuse. The infrared scanned image (B) taken in the evening shows high temperatures and the location of the fire at the light tone areas. (Courtesy of HRB-Singer, Inc., Energy and Resource Systems)

Figure 17.16. A large heat leak from a buried steam line is clearly shown by the light spot (arrows) on this night thermal image of the University of Iowa campus. Other heat losses from sub-surface heating lines can also be seen, as well as those from buildings. (Courtesy of Daedalus Enterprises, Inc.)

urban areas from other forms of land use, but only if the industrial areas are large. Roof tops of industrial buildings and roads provide the distinguishing spectral elements to the sensor.

Those industrial phenomena which cover large areas on the earth's surface, such as some of the extractive industries, can be imaged by satellite sensors. Strip mining is one of these industries. One study, using the Landsat 1, band 5 imagery of the Cumberland Plateau of Tennessee, was able to identify strip mines as light toned jagged lines on the positive film.[11] A negative image proved even more effective. High altitude photography was used as a check on the Landsat data and indicated a high level of reliability.

Other investigators working with Ohio data have applied the Landsat-CCT (computer compatible tape) mode to strip mining studies with considerable success.[12] The CCT permits the use of an automated data processing and mapping system which not only is faster but more accurate, detailed, and economic than that of visually examining frames of Landsat imagery for

Figure 17.17. An infrared image of a petroleum refinery. Processing units such as cracking towers can be located by the white "hot" spots. (Courtesy of HRB-Singer, Inc., Energy and Resource Systems)

evidences of strip mining. Computers are used to edit and refine the spectral measurements on the tape. Categories of land use pertinent to strip mining areas are developed from the spectral measurements after selecting "typical" areas with the aid of data obtained from aerial photographs and ground surveys. Not only are strip mine areas identified, but the stripped areas can also be grouped into subcategories, such as newly stripped, which are rough surfaced areas; smoothed but bare earth areas; and stripped areas now planted to grass. The data can be mapped in color at scales as large as 1:24,000. At this scale a single Landsat spectral signal which represents an area 57 by 79 meters (approximately 187 by 260 ft) on the ground is imaged by a rectangle 3.2 mm by 4.6 mm (0.125 by 0.18 inches) on the map. Figure 17.18 is a black and white copy of a color original which was produced using the system described above. The increase in stripped area in one year is obvious and can be precisely measured from the imagery.

One final example of the potential use of space satellite data as applied to industry is the tracking of a smoke plume from a large smelter near Sudbury, Ontario, 113 kilometers (70 miles) downwind.[13] The imagery used was Landsat 1 multispectral bands 4 and 5. Particulate concentrations in the plume can be determined, and the effects of fallout and changes in plant operation on the plume can be estimated. Such data is useful as it pertains to environmental decisions. Figure 17.19 clearly shows the plume.

It is expected that with time more uses and applications will be found for satellite and high altitude imagery in industrial analysis.

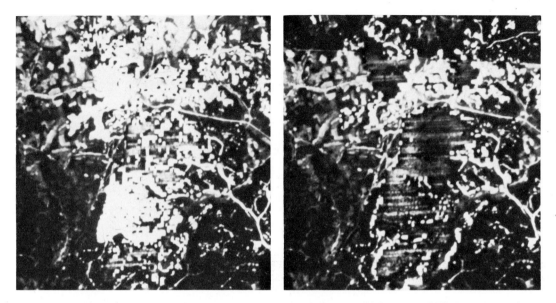

Figure 17.18. Stripped earth category mapped from 1972 and 1973 Landsat data superimposed on a NASA photograph obtained on September 1973. Approximate scale 1:40,000. Ohio Power Company mine in Muskingum County, Ohio. Between August 1972 and September 1973 the imagery indicates an additional 868 acres was stripped. (Courtesy of Bendix Aerospace Systems Division)

NOTES

1. C. Strandberg, "Water Pollution Analysis," in *Manual of Color Aerial Photography,* eds. J. Smith and A. Anson (Washington, D.C.: American Society of Photogrammetry, 1968).

2. J. Estes and L. Senger, "The Multispectral Concept as Applied to Marine Oil Spills," *Remote Sensing of Environment* 2 (1972):141-63.

3. T. Chisnell and G. Cole, " 'Industrial Components'—A Photo Interpretation Key," *Photogrammetric Engineering* 34 (1968):590-602.

4. For a more structured outline of recognition features for the major categories of industry, the reader is referred to Chisnell and Cole, p. 601.

5. K. Piech and J. Walker, "Outfall Inventory Using Air Photo Interpretation," *Photogrammetric Engineering* 38 (1972):907-14.

6. Ibid.

7. S. Klooster and J. Scherz, "Water Quality by Photographic Analysis," *Photogrammetric Engineering* 40 (1974):927-33.

8. Strandberg, "Water Pollution Analysis."

9. F. Wobber et al., "Coal Refuse Site Inventories," *Photogrammetric Engineering and Remote Sensing* 41 (1975):1163-71.

10. F. Scarpace, R. Madding, and T. Green III, "Scanning Thermal Plumes," *Photogrammetric Engineering and Remote Sensing* 41 (1975):1223-31. The authors demonstrate the difficulties and complexities of precise and detailed airborne thermal scanning in their study of thermal plumes in Lake Michigan.

11. J. Rehder, "The Uses of ERTS-I Imagery in the Analysis of Landscape Change," in *Remote Sensing of Earth Resources,* vol. 3, ed. F. Shahrokhi (Tullahoma, Tenn.: University of Tennessee Space Institute, 1974), pp. 573-86.

12. R. Rodgers, L. Reed, and W. Pettyjohn, *Automatic Mapping of Stripmine Operations from Spacecraft Data* (Ann Arbor, Mich.: Bendix Aerospace Systems Division, 1974).

13. K. Templemeyer and O. Ey, "Use of Remote Sensing to Study the Dispersion of Stack Plumes," in *Remote Sensing of Earth Resources,* vol. 3, ed. F. Shahrokhi (Tullahoma, Tenn.: University of Tennessee Space Institute, 1974), pp. 255-72.

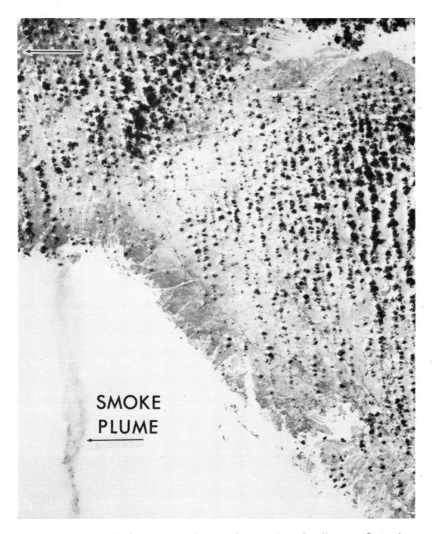

SMOKE
PLUME

Figure 17.19. The stack plume from the Sudbury, Ontario, refinery is clearly imaged as it moves off land and over Georgian Bay for a distance of 70 miles. The imagery is Landsat multi-spectral band 5, fall 1972. The refinery is located just off the upper left edge of the illustration. An arrow points toward its approximate location. (Courtesy of K. Tempelmeyer)

18

Application of Remote Sensing in Weather Analysis

Jack R. Villmow
Northern Illinois University

Introduction

U.S. ORBITS WEATHER SATELLITE;
IT TELEVISES EARTH AND STORMS;
NEW ERA IN METEOROLOGY SEEN

So read the headline in the *New York Times* on April 2, 1960. Tiros I—Television and **I**nfra**R**ed **O**bservation **S**atellite, with its television cameras 724 kilometers (450 miles) above the earth, sent back fairly clear images of storm systems and other cloud forms. Earlier, Vanguard II and Explorer VI had sent relatively crude cloud pictures, but Tiros I was hailed as the equivalent to meteorology in the twentieth century that the telescope had been to astronomy in the seventeenth century. A decade and a half later the new Synchronous Meteorological Satellite-2 (SMS-2) began transmitting. It was the western link in the National Environment Satellite Services (NESS) Geostationary Operational Environmental Satellite System (GOES). The first SMS-2 systems-operational picture, a full-disc 4 kilometer (2.5 mile) resolution visible image (subpoint 80°W), is shown in figure 18.1, and is dated 1835 GMT, 11 February 1975.[1]

Early satellites pictured clouds mainly; now "stationary" weather satellites orbiting 35,000 kilometers (22,000 miles) above the earth take pictures of the *same* scene at defined intervals. Probably more important is the remote sensing of electromagnetic radiation (temperature) of the surface of the earth and of the contents of the atmosphere. It is this sensing of invisible radiation which gives a measure of temperature, a relationship established in the nineteenth century by Stefan and Boltzmann; that is, the amount of radiation emitted varies with the fourth power of the absolute temperature. When the varying temperatures of different surfaces are studied, much can be learned about the heat economy of the earth-atmosphere system.

Figure 18.1. First systems-operational picture of the Geostationary Operational Environmental Satellite System (GOES), 11 February 1975. (NOAA)

Electromagnetic Radiation: Key to Remote Sensing in Weather Analysis

The Stefan-Boltzmann finding provides a good starting place for explaining how satellites far above the surface of the earth are able to distinguish among the various temperatures at the surface as well as within the sea of air surrounding the earth. In chapter 2 it was stated that the total radiating power of an object is proportional to the fourth power of its absolute temperature. Such objects are black bodies, or nearly so. Black bodies are nearly perfect absorbers and radiators at all wavelengths. The surface of the earth and the sun are in this category. The earth radiates, on the average, at 290 °K. The sun, on the other hand, radiates at approximately 6,000 °K, and, therefore, it sends out approximately 500,000 times as much radiant energy as the earth.

Kirchoff's principle states that the abilities of an object to absorb and radiate are equal. A good absorber, such as the earth, therefore, is also a good radiator, but certain gases in the atmosphere, such as carbon dioxide, water vapor, and, in particular, ozone are selective absorbers. Their maximum absorption is only in certain wavelengths of the total electromagnetic spectrum. The remainder essentially pass through these gases, or through atmospheric windows as they are known. When this energy escapes upward and away from the surface of the earth, whether day or night, it can be readily "tuned in" by radiometers in a satellite, being conveyed as valuable temperature information about the surface.

Liquid cloud droplets or ice particles in clouds, on the other hand, have their own absorptive powers. At night the cloud particles will radiate predominantly in nonvisible wavelengths which are often cooler than the earth itself. These wavelengths can be sensed by the satellite and labeled accordingly. During the daytime, however, clouds absorb some solar energy, but mostly they reflect visible solar energy at very short wavelengths which are picked up as visual images by satellites. Radiometers continue to sense the nonvisible radiation by clouds during the daytime as well as during the night.

Electromagnetic radiation can be best illustrated as a very long sliding scale (see fig. 2.1).[2] It can be seen that at one end are very short wavelengths of energy, much too short to be seen, such as X-rays and gamma rays. Slightly lower, but still considered short, are visible wavelengths of solar radiation which do not have as much energy as X-rays or gamma rays. Infrared rays have still less energy—this is where most of the radiant energy of the earth is found. It is not characterized by enough energy to be visible, but certainly possesses enough energy to be measured. Increasing the radiant energy of an object means that its temperature increases greatly and vice versa. Thus, the wavelengths of radiant energy, whether visible (sun or reflection of the sun from an object, such as a cloud top or snow field), or infrared (earth's surface, cloud droplet, ice particle in a cirrus cloud, or water vapor in the atmosphere) are the identifiers, or trade marks, of the object doing the radiating.

Some Practical Applications of Electromagnetic Radiation to Satellite Weather Analysis[3]

Visible Radiation

Consider now the two major sectors along the electromagnetic sliding scale: the short wavelength, high energy, visible, sunlight sector on the one hand, and the long wavelength, low energy, invisible, earth-atmosphere sector on the other. Television cameras record reflected radiation using a special spectral interval to reduce the effects of light scattering in the atmosphere. The brightness of the image picked up by the cameras is an indicator of the reflectivity or *albedo* of the earth and clouds (table 18.1). This is measured by degrees of whiteness on the resulting photographlike image. Clouds are the single most important image recorded by cameras on weather satellites, and their brightness is a function of such factors as:(1) angle of the sun (time of day, time of year), (2) angular position of the cloud in relation to the sun and the camera, and (3) the albedo of the cloud itself.

Cloud albedos, as given in table 18.1, are determined by (1) thickness, (2) distribution of size of water droplets and/or ice particles, (3) state of water in the cloud—ice, water, or both, and (4) the morphology of the upper cloud surface. Generally, water clouds are brighter than ice clouds, and thick clouds are more reflective than thin ones. The varying shades of white will also be indicative of cloud structure, helping to differentiate one cloud from another, or one cloud deck from another of differing altitude. The arrangement of the clouds in various

TABLE 18.1
Albedos

Cloud Types	Percent of Reflectance	
Cumulo-nimbus—large, thick	92	
Cumulo-nimbus—small	86	
Cirro-stratus—thick with lower clouds	74	
Cumulus and strato-cumulus—80% cover	69 ⎫	
Strato-cumulus—80% cover	68 ⎭	over land
Stratus—thick	64 ⎫	
Strato-cumulus—within cloud sheet	60 ⎬	over ocean
Stratus—thin	42 ⎭	
Selected Earth Features		
Fresh snow	75-95	
Sand—depending on vegetation	60 down to 17	
Snow—up to a week old on mountains	59	
Coniferous forest	12	
Lakes, oceans	7-9	

configurations is another apparent factor on weather satellite images: cellular or globular as opposed to linear, for example.

Invisible Radiation

Infrared radiation—which does not have enough radiation to be visible—is sensed by scanning radiometers which measure the longer wave, lower energy, invisible electromagnetic energy emitted by (1) clouds, (2) atmospheric gases, and (3) the land/water surfaces of the earth. Recalling that the earth radiates nearly as a black body, it is possible to convert its radiation imprint into temperatures. This is extremely important in terms of satellite remote sensing applications to weather analysis. Before the extremely complex interrelationships among earth-atmosphere-space-sun can be understood, a more refined type of bookkeeping strategy must be developed about the heat budget of the earth. Almost all atmospheric heat comes originally from the sun. About 20 percent of the incoming solar energy is absorbed directly by the atmospheric gases, dust, and clouds, and about 50 percent of the solar radiant energy succeeds in reaching the surface directly or by scattering. Most heating of the atmosphere, however, is done indirectly from the sun and directly by the surface of the earth. The earth, in other words, is the chief radiator for the great sea of air which covers the earth. It becomes imperative, therefore, to gain more understanding about the nature of the earth as a radiating body. Infrared radiation sensors on remote sensing satellites will help provide these understandings by generating a great number of digitized views of the radiating surface of the earth during both night and day.

The earth and its atmosphere normally radiate at absolute temperatures ranging from 200 °K to 300 °K (−73 °C to 27 °C) with wavelength intensity peaks well within the category of longer wavelengths of invisible electromagnetic energy. Scanning radiometers on the satellites measure this radiance and convert it to imagery on which the earth and clouds are shown in shades of gray. Dark areas are warm surfaces, while bright ones are cool. Relative cloud heights are also readable on the images. Higher clouds normally will be colder and appear brighter than low clouds. Low clouds are normally warmer because they are closer to the radiating surface of the earth. Low clouds, therefore, will be darker gray.

Global weather analysis is done in both the USSR and the United States by computer, providing information about relative cloud height based on brightness of clouds. Recalling the need to quantify heat exchanges at the surface of the earth, it is important to note that the maximum resolution of satellite images ranges from 7.4 kilometers (4.6 miles) along the subpoint of NOAA's ITOS to 15 kilometers (9.3 miles) along the track of the Meteor satellite. (NOAA: National Oceanic and Atmospheric Administration; ITOS: Improved Tiros Operational System) In addition, instrument error varies from satellite to satellite and from the warmer (shorter) to the colder (longer) end of infrared radiation from 2 °C to 8 °C, so refinements must be made before accurate sensing is possible.

Since the earth receives essentially all its short-wave energy during the day from the sun, it, in turn, radiates at a higher temperature during the day than at night. Infrared daytime pictures are significantly brighter than nighttime pictures because the earth-radiator is warmer then. The geographic knowledge of land/water heating and cooling capabilities leads to some interesting observations in infrared images: during the day the *relatively* cool water appears brighter than the warmer land surface; at night the *relatively* warm water appears darker than the cooler land surface.

Distinctions between radiating land surfaces and radiating clouds are based upon several factors, the most important of which are location and time of year. In low latitudes near sea level, clouds invariably are cool and, therefore, brighter than land surfaces. In high latitudes this is true only in summer. In winter the surface of the earth is often as cold as, and sometimes colder than, the clouds. Consider the case of temperature inversions in high latitude winters. Under such circumstances the clouds might be warmer, and hence darker, than the surface of the earth beneath them. If the surfaces of clouds and earth are similar in temperature, infrared images would disclose little brightness contrast between them.

The Two Radiations Compared

Some general observations can be made between conventional photographs and infrared imagery. The general outlines of land/water boundaries and of clouds appear similar on both images. Infrared sensors image water bodies in various shades of gray depending on the radiating temperature of the water surface. On conventional photographs water bodies are relatively dark except for occasional solar glare. Towering cumulo-nimbus appear bright (cold) on infrared images and bright (highly reflective: 92 percent) on images sensed in the visible portion of the spectrum (table 18.1). Cloud structure and heat distribution are revealed on infrared photography in shades of gray. On imagery made in the visible bands of the electromagnetic spectrum variations in brightness are essentially a measure of reflectivity; however, solar angle, time of day, and morphology of the cloud itself may affect the imagery. In summary, brightness on infrared imagery is a measure of heat (or cold), while on visible-band imagery it is basically a measure of albedo (or reflectivity).

The Application Sequence

The preceding examination of the role of electromagnetic radiation in remote sensing of the earth/atmosphere system, along with several practical applications of two very different means of registering elements and their heat budgets, provides a background for a series of case studies in meteorology. In each, interpretation will be guided by (1) observation, (2) analysis, and (3) potential forecasting. Detectors on a satellite are the remote sensors or observers. The images which are produced by these sensors must be analyzed in light of basic meteorological principles. Finally, extrapolation of the analysis to a position in time after the observation allows the understandings to bear fruit in forecasting weather.

Another technique in remote sensing is used along with satellite imagery in smaller scale weather analysis. This is radar imagery which is particularly valuable for short-range, local forecasting. Where appropriate, the role of imagery will be introduced and explained in the case studies.

The observation-analysis-forecasting sequence in the utilization of satellite imagery is simply one of the values in this utilization of remote sensing. Another one, which probably is less glamorous, but is of much greater significance, is the utilization of remote sensor imagery to analyze the heat budget of the earth and atmosphere. The science of meteorology is still operating with far too little data and far too little proven hypotheses to elevate it to the level of a mature predictive science. Satellite imagery will add to basic data and provide for the evolution of hypotheses to demonstrated truth. Only when the complete set of atmospheric equations are developed and proven can satellite imagery be fully appreciated, understood, and applied.

Case Studies

Background for Case Studies

Actual applications of weather satellite imagery to meteorological analysis is possible in a number of ways. The World Meteorological Organization, in *The Use of Satellite Pictures in Weather Analysis and Forestry,*[4] separates weather analysis in the extratropical (middle and high) latitudes from analysis in tropical latitudes. This is a clear indication of an awareness of the very different approach necessary in understanding low latitude weather as compared to more traditional avenues of understanding which can operate elsewhere. A recent publication by the American Society of Photogrammetry, uses an approach geared toward individual meteorological parameters—such as hurricanes, precipitation, winds, water vapor, and thunderstorms, followed by analyses of air pollution, heat budget, and the "hardware" involved in the sensing operation.[5] Synoptic and mesoscale analyses of weather phenomena accompanied by over forty slides has recently appeared in a modest-sized booklet published by the U.S. Government.[6] It includes, in addition to weather analysis, sound background on hardware systems and data interpretation.

The approach used here is based on the easy availability of both imagery and analysis in a widely respected journal of meteorology which also enjoys rather wide circulation in reference libraries throughout this country and Canada as well, the *Monthly Weather Review,* now published by the American Meteorological Society in Boston, Massachusetts. Before 1974 it was a publication of the U.S. Department of Commerce. Shortly after the first satellite images were received in 1960, this publication began a series entitled "Picture of the Month." This feature provides an on-going text in remote sensing techniques for weather analysis and a reference laboratory of source materials. The pictures for the period from June 1967 through December 1976 are listed in Appendix B. The case studies which follow are drawn heavily, but not exclusively, from this source.

Case Study 1: Typhoons and Hurricanes

One of the most clearly recognized weather map symbols is that which represents a hurricane or typhoon:$. The accuracy with which this symbol reflects reality is seen in figure 18.2. The image shows Hurricane Bernice on July 13, 1969. It began as a tropical disturbance on July 8; after two days it intensified into a tropical storm; then for 72 hours was a hurricane, before weakening into a tropical depression again on July 15. What is particularly significant about Bernice and its satellite identification is its origin over a thousand miles south-southeast of Los Angeles in an area with a sparse meteorological network. Early location of the evolving storm by satellite made it possible to send aircraft to the trouble spot to examine it in more detail and to track it carefully. Alerts were issued for the California coast as heavy surfs caused much erosion and endangered life. Note particularly that during the three days Bernice had hur-

Figure 18.2. Hurricane Bernice was a full-blown hurricane off the west coast of Mexico from July 12 to 14, 1969. (NOAA)

ricane status, the general geometry of its clouds remained true to its weather map symbolization. (The term hurricane is used when the storm originates east of the International Date Line, and typhoon when it originates west of the line.)

A good illustration of remote sensing by three different techniques was illustrated when hurricane Beulah appeared and spent most of its life in the Caribbean Sea and Gulf of Mexico. The satellite view in figure 18.3 provides the familiar arrangement of clouds within a hurricane. It is possible to estimate the wind speed in hurricanes or typhoons when the size of the storm, measured in degrees latitude, is checked against the curvature of the spiral cloud bands.

In addition to satellite imagery of hurricanes, these intense storms are monitored by weather reconnaissance planes and by ground radar. High altitude (18,288 meters or 60,000 feet) photographs from weather planes provide resolution of some details of hurricanes which satellite imagery cannot deliver. Radar units image the large water droplets in the storm. In the images from satellites, high altitude aircrafts, and ground radar units the hurricane cloud pattern is unmistakable.

Hardware in satellite technology has greatly improved in the past decade and a half. On September 22, 1975, the view of hurricane Eloise provided 1 kilometer (0.6 mile) resolution (fig. 18.4). Note how clearly the eye is identified on Eloise as it appeared shortly before landfall along the Florida panhandle. At this point it is well to compare the satellite image with the *Daily Weather Map* which was produced less than five hours earlier (fig. 18.5).

Figure 18.3. Satellite view of hurricane Beulah, centered at 95° W, 25° N on 19 September 1967. (National Satellite Center and EDS, Marine Branch)

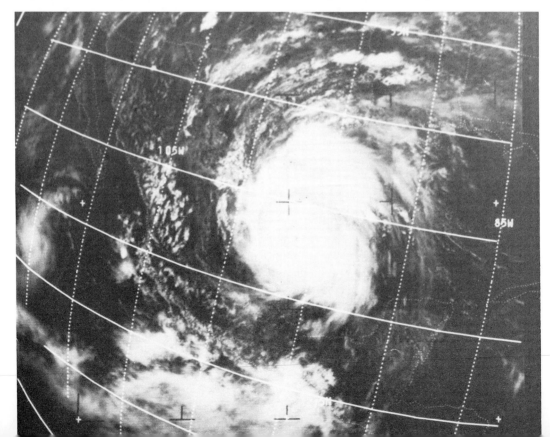

Figure 18.4. Hurricane Eloise off the Florida coast, 22 September 1975. (NOAA)

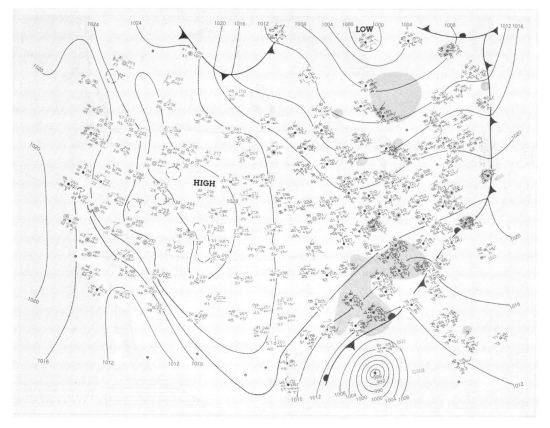

Figure 18.5. Surface weather map of Eloise in the Gulf of Mexico, 22 September 1975 at 7 A.M. EST. (From Daily Weather Map, Weekly Series, September 22, 1975, U.S. Department of Commerce)

Observe from the map that only one weather station in the vicinity of the storm reported no indication of either the extent or the character of·cloud cover, both of which are clearly seen on the satellite image. Application of a nomogram to the estimation of wind speeds, along with a careful scrutiny of the brightness during the daytime or gray shading on the nighttime infrared image, would reveal much to a meteorologist concerning the heights of clouds, the layering of clouds, and temperatures associated with the several cloud layers.

In addition to identifying the typical cloud pattern in most of the intense and potentially destructive tropical storms, a study and comparison of visual wavelength and infrared satellite images add considerably to understanding the character of these storms. For example, total area covered, average time to develop, average duration of the hurricane stage, clues to deterioration, and the geometry of the path are among the more important answers which remote sensing imagery provides. Along with radar interpretation which assists in precipitation estimates, it is now possible to accurately predict the origin of hurricanes as well as their trajectory, probable life span, and total meteorological impact.

Case Study 2: Snowfall—Good Surface Reflector.

Fresh snow has one of the highest albedos (table 18.1) of all natural surfaces found on the earth; only large, thick cumulus clouds have higher albedos. An extensive February storm system moving northward through the eastern portion of the United States provided an excellent opportunity to illustrate the usefulness of reflectivity of both clouds and snow. Figure 18.6A illustrates a severe winter storm centered just off the Carolina coast. It has associated with it distinctive masses of cumulus clouds over South Carolina and Georgia. Both lower clouds and an extensive higher layer of cirrus clouds were present over North Carolina and northward. Farther north in New England (1 on fig. 18.6A) there was only a high thin film of cirrus, and the darker ground below is visible through the clouds. Leeward of Lake Ontario snow showers can be discerned, while patches of ice are clearly visible on the Great Lakes, especially Lake Huron.

On the next day a massive high pressure system of Arctic air from Canada dominated most of eastern United States. In figure 18.6B, the major heavy snowfall, which broke records in many sections of Georgia and the Carolinas, appears in sharp contrast to the areas around it where no snow fell, or where it had fallen and melting occurred. Snow-covered areas north of the Ohio River also stand out, but less dramatically so, from their surroundings. At 2, high cirrus, oriented by the direction of jet flow, can be identified, and off the coast at 3 a narrow band of low continuous cloud cover is seen.

The major research significance of this precise identification of the vast snowcover lies in its application to heat budget analysis. Where this fresh snowcover lies, practically none of the incoming solar radiation is being used to heat the surface of the earth. Almost all of it is reflected back through a very dry air mass to space. Normally, the United States north of the Ohio River will experience this phenomenon during February, but the addition of a large area in the South profoundly affects the total heat budget of the continent.

An example of a highly localized meteorological phenomenon, lake effect snowfall, is illustrated in figure 18.7. Cloud bands from **A** to **B** in both satellite images are related to the 18 inch snowfall which occurred leeward of the eastern end of Lake Erie. The upper image shows the larger scale motion by means of cloud patterns. From point **C** northeastward a strong cold front can be identified by a dense and extensive band of clouds, behind which a large clearing is

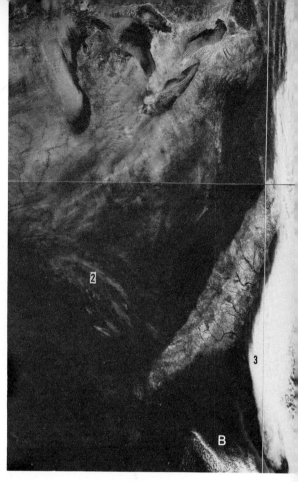

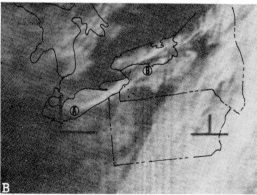

Figure 18.6. (A) Major storm centered off the east coast of the United States as seen by visual satellite imagery during the morning of 10 February 1973. (B) Snow-cover and snow showers over Great Lakes region—the following morning clear skies. (NOAA)

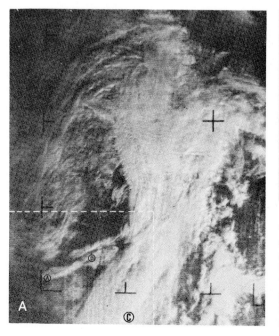

Figure 18.7. (A) Cloud bands associated with heavy lake snow A to B within arctic air mass; cold front extends from north-eastern New York state to southwestern Pennsylvania. Morning of 23 November 1970. (B) An enlargement of the larger photograph. (Both photographs from E. W. Ferguson, "Picture of the Month," *Monthly Weather Review,* vol. 99, p. 247)

easily seen. Although all of the Lake Erie region was behind the cold front and in the drier polar air mass, strong westerly winds were moving across the length of the relatively warm lake. These succeeded in evaporating considerable moisture and producing bands of clouds clearly identified on the lower image and resulting in heavy lake-effect snows to the lee of the water. Satellite imagery here provides a means of "tracking" the cloud bands and, along with surface radar, helps to estimate the amount of precipitation which will fall.

Case Study 3: Frontal Systems

The lake-induced snowfall, following the passage of a strong cold front and occurring within the cold polar outbreak illustrated in figure 18.7, leads logically to close examination of one of the more easily identifiable weather patterns by satellite imagery—the cold front itself. Figure 18.8 facilitates comparison between conventional imagery and infrared imagery of the same phenomenon. Since the infrared view is foreshortened, letters designating the same cloud elements are employed to make the two images more directly comparable. Cirrus clouds, the highest, coldest, least dense of all cloud forms, are more easily identified on infrared imagery than on other imagery, and provide essential data on altitude temperature and distribution of these clouds. **H** best illustrates this as it points out cirrus very distinctly on the infrared photograph, but only hazily on the other photograph because of the thinness of the cloud.

A gray form in the infrared image, however, is not less a distinct image, but rather an indication of the warmer temperatures of lower cloud forms such as those labeled **C. D** and **F** appear to be nearly equally bright (white) on the infrared image and, therefore, approximately the same temperature, and presumably the same altitude, but on the visual-band image **F** is less

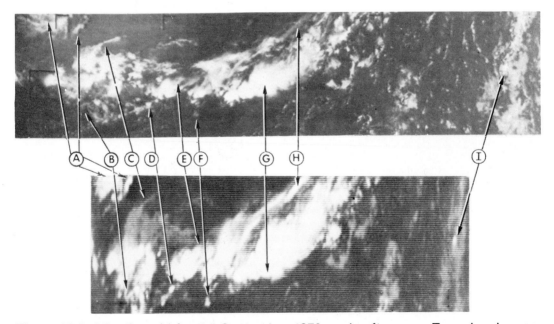

Figure 18.8. Atlantic cold front 4 September 1970, early afternoon. Top: visual-range image. Bottom: infrared image. (NOAA)

distinct than **D**. **D** is probably an actively growing convective cell, while **F** is, at the same time, a dying cumulo-nimbus. The growing cloud is reflecting much light and appears bright on the visual-band photograph, but the dying cloud has less and less surface from which reflection can occur and appears less bright. High cirrus clouds are more easily identified on infrared than on visual-band photographs, but convective clouds can be identified in terms of developmental stage when the two kinds of imagery are compared. Cloud elements **G** and **I** appear equally bright on both images, and the conclusion must be that they represent active convective groupings which are highly reflective (visual-band image) and have cold tops (infrared image).

Clearly the forecasting potential is greatly enhanced when knowledge about the age or stage of development of the cloud is known, as well as its altitude. A growing convective cluster means increased probability of severe weather; a dying group points toward stabilization of the weather picture. Successive images made at regular time intervals give a satisfactory clue regarding the developmental character of clouds.

One of the most critical problems arising in the forecasting procedure is timing. The rate of movement of a vigorous weather system can be observed and measured by satellite imagery in successive images. This is clearly shown in figure 18.9. The upper image shows the frontal

Figure 18.9. Successive views of a cold front, western North Atlantic Ocean. (A) Afternoon, 3 March 1971. (B) Morning, 4 March 1971. (NOAA)

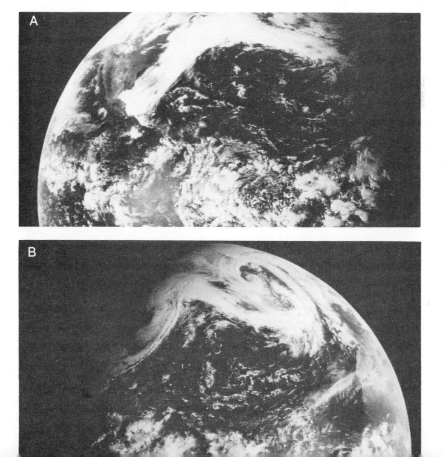

passage, along with a series of prefrontal cloud lines, advancing over the Florida peninsula in midafternoon. A vast area of cloud cover extends behind and to the northwest of the front. Early the next morning (lower image), the prefrontal convective cloud line stands out unusually sharply because the low morning sun casts shadows west of that line. On earlier images, too, both the prefrontal line and the front itself were sharply defined and very long, and it was possible to pinpoint the time of the frontal passage very accurately. The Miami, Florida, weather office had predicted the frontal passage for 2100 EST; it actually passed at 2048 EST!

On figure 18.9B note the scalloped, wave pattern extending out over the North Atlantic Ocean, with a wave crest off the New England coast. The broad cloud-covered warm front extends southeastward toward North Africa with the wide-open warm sector opening equatorward; the rapidly moving cold front is now out over the ocean. A true textbook model of an extratropical cyclone is beautifully seen in this scene by cloud imagery alone.

Occasionally an unusually severe Arctic outbreak, strongly assisted by meridional (north to south) flow aloft, will steer cold, dry air southward along the eastern edge of the Rocky Mountains, across eastern Mexico and the western Gulf of Mexico, and at right angles to the "narrows" of Mexico just west of Yucatan. At the Isthmus of Tehuantepec, a relatively deep cold air mass will spill over the low Sierra Madre Range and down its western slope onto the Pacific Ocean (fig. 18.10). Although the main body of cold air remains on the Atlantic Ocean side (**F**) of the Sierra Madre, the spillover sends a prefrontal squall line out over the Pacific Ocean as a

Figure 18.10. Arctic front invades the tropics, 3 February 1970. (NOAA)

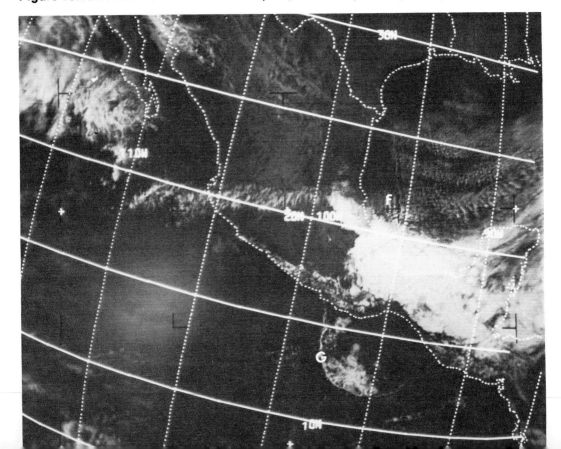

well-marked arch of clouds (**G**). Here, the simple geometry of the cloud form, reflecting the light of the sun, is visible proof of the heat budget adjustment of the earth.

The flow pattern which made the phenomenon shown in figure 18.10 possible is one of the necessary adjustments within the general circulation to assure that the tropics will not heat up excessively. Periodic expulsion of Arctic air with its later invasion of the tropics is achieved by the strong meridional flow just described and illustrated. Reexamine figure 18.9 and note there that the warm front has moved northward (poleward) from one day to the next. An expulsion of warm tropical air with its subsequent invasion of polar regions is illustrated in that sequence. These two kinds of exchanges are necessary to maintain the heat budget of the earth and can be catalogued by satellite imagery.

Case Study 4: Infrared Image of Ocean Surface

A very high resolution infrared color enhanced image of a part of the Gulf Stream-North Atlantic Drift at 2000 local time provides a dramatic illustration of a special technique of emphasizing the varying quantities of heat, and therefore, radiating power, of different parcels of water (plate 23). Color enhancement here has rendered the following equivalents in degrees Celsius:

Violet	20-23
Red	18-20
Dark orange	16-18
Orange	15-16
Light orange	13-15
Light blue	11-13
Blue	9-11
Dark blue	6-9
White	less than 6

Nearly two-thirds of the surface of the earth is covered with water, and the fluid character of water (as in the case of air) means that heat energy is transported in great quantities in the global exchange of heat energy. Color enhanced infrared images help to give a measure of that heat exchange; cyclonic and anticyclonic whirls of heat energy are distinguishable in several places. In the southeast corner cool cloud tops are shown in blue and white. The urban heat island—Philadelphia, Pennsylvania—is clearly identified immediately southeast of the cloud clusters in the upper left hand corner of this infrared image.

Some measure of the tremendous ability of water to transport heat energy is seen when it is pointed out that a 2.5 meter (8.2 feet) vertical section of water has the capacity to transport heat energy equivalent to a *total* vertical section of the atmosphere. In subtropical and middle latitudes approximately 25 percent of all meridional heat exchange is accomplished by water transport. Images such as the one in plate 23 can be carefully gridded and digitized to provide considerable insight into daily transports of heat energy by water to assist in the analysis of the heat budget economy of the earth.

Case Study 5: Radar Assists to Satellite Data

The surface radarscope is generally used in concert with other sensors. On it "echoes" from water droplets and larger snowflakes create pattern configurations similar at times to the visual and infrared patterns observed on satellite imagery. More commonly these patterns include: (1) cells—single convective echoes; (2) areas—related or similar cells readily associated geographically; (3) lines—related or similar echoes that form a pattern exhibiting a length-width ratio of 5:1 and a length of at least 30 nautical miles (35 statute miles); (4) hook echoes—the signature of tornadoes appearing as a hook extension from the rear lower quadrant of a thunderstorm echo; (5) wave patterns—exhibiting a simple sinusoidal geometry; (6) snow echoes—diffuse patches with some banding during severe storms; (7) fine lines—basically meteorological discontinuities, neither precipitation nor ground clutter, often leading edges of cold fronts or cold air outpourings ahead of thunderstorms.

Figure 18.11 shows one of the common echo patterns. It is a well-defined line associated with a pronounced cold front passage. The weather map for this day, figure 18.12, shows an unusually long cold front extending from James Bay to the central Gulf of Mexico, with a pronounced meridional orientation. The radar pattern shows a 16 kilometer (10 mile) wide, 258 kilometer (160 mile) long pattern just west of Pittsburgh, Pennsylvania, at 1430 GMT, only a short time after the *Daily Weather Map* depiction of the front located between Columbus, Ohio, and Pittsburgh.

Radar echo intensity codes provide a means by which a theoretical rainfall rate in centimeters (inches) can be "read" from the radar picture by means of a contouring system. The radar image is similar to satellite imagery of the same phenomenon (case study 3: frontal analysis). As a remote sensor it is likely to yield more specific information about severe weather

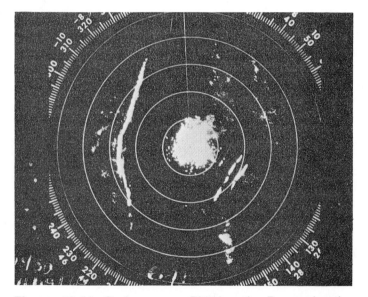

Figure 18.11. Radarscope, Pittsburgh, Pennsylvania. Range circles have a radius representing 46.4 kilometers (28.8 miles). Morning of 19 November 1969. (From P. R. Gullick and C. E. Goodall, "Picture of the Month," *Monthly Weather Review,* vol. 98, p. 510)

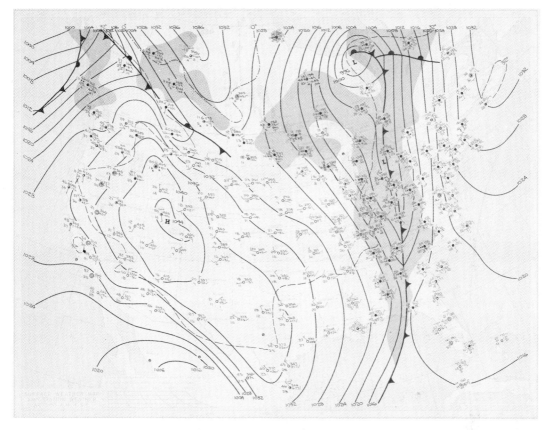

Figure 18.12. Surface weather map, 19 November 1969. (*Daily Weather Map, Weekly Series,* November 19, 1969, U.S. Department of Comnerce)

to a given forecasting center; however, it also can be used along with satellite imagery to fill in almost all details of a weather system.

Case Study 6: Atmospheric Dynamics

One of the greatest contributions of satellite imagery to weather analysis is its ability to convey a realistic picture of the motion of the atmosphere by means of the shape and/or pattern of cloud formations. It is by means of this motion of air that heat and moisture are carried from one place to another on the earth. Heat is not only in the air itself (sensible heat), but also locked up in the water vapor (latent heat)—the single most important gas in the atmosphere as far as weather is concerned. The air itself, of course, is invisible, so it is not possible to see sensible and latent heat transported from place to place. Clouds, however, are visible water droplets and ice particles, and their motion, as evidenced by shape and/or pattern, is an index of the movement of the air in which they are suspended. The next three examples, although they show clouds, actually illustrate the motion of heat energy and moisture from place to place.

In figure 18.13, a long "rope" of clouds extending from the Sea of Okhotsk (**A**) to mid-Pacific (**B**), has to its south and west a linear arrangement of cumulus clouds extending a thousand miles from the edge of the pack ice (**C**) to the mid-Pacific. Strong northwesterly airflows, both aloft and at the surface, off the cold Asiatic continent have succeeded in developing cloud lines parallel to the flow pattern. At **D** and east of the "rope" of clouds extending from **A** to **B** the air is lighter and from the northeast, suggesting that horizontal shear (change of direction in air flow), along with convergence of air, accounts for the regularity and extensiveness of the linear clouds.

Figure 18.13. Visual-range satellite image in the North Pacific region off the coast of Siberia, midday 16 February 1969. (NOAA)

Two patterns of air flow are easily described on figure 18.14. Although anticyclonically curved jet streams are often identified on satellite imagery by cirrus clouds, cyclonically curved jets are far less frequently identified because their clouds dissipate easily in a region of subsiding air associated with downward motion between ridge axis and trough axis aloft. The jet location on this figure, extending from **A** to **B,** is an exceptional opportunity to "see" the jet where it rarely can be seen. Note how well this cirrus cloud sheet relates to the upper air flow on the *Daily Weather Map* in figure 18.15. Where the height contours are closest together, the jet is best defined, and it is clear from the 500 millibar chart, approximately 4.8 kilometers (3 miles) above the surface of the earth, that the flow pattern, as evidenced by cirrus clouds, is a real one.

Mountain wave (fig. 18.14, **C-D**) clouds over northern California and Nevada represent the second pattern on the satellite image. This is also seen as a region of strong upper air flow

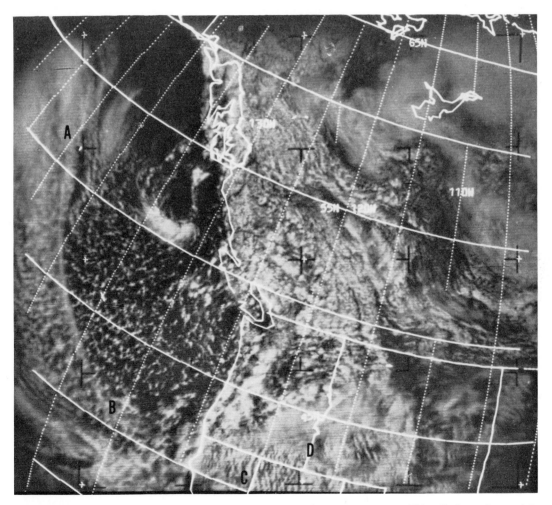

Figure 18.14. Visual-range satellite image along the west coast of North America, mid-day 25 April 1970. (NOAA)

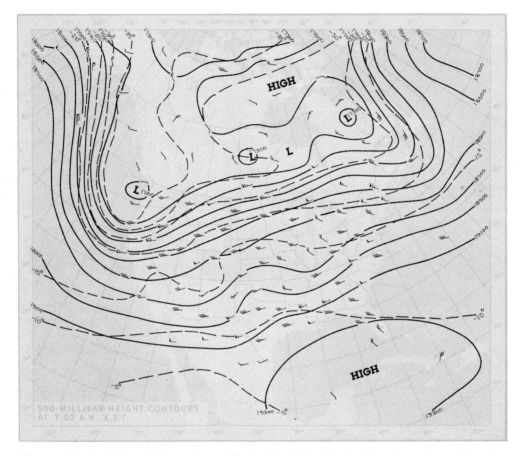

Figure 18.15. 500 millibar chart, 26 April 1970. (*Daily Weather Map, Weekly Series,* April 26, 1970, U.S. Department of Commerce)

on the 500 millibar chart. Strong winds crossing the north-south mountain barrier formed by the Cascade Range-Sierra Nevada mountains are caused to ripple, and clouds resulting from these ripples are distinctly banded in pattern.

A somewhat similar pattern is seen on figure 18.16. This pattern covered large parts of Arkansas and Louisiana shortly after violent thunderstorm and tornado activity had dominated weather in large areas to the northwest. The arrangement of clouds is probably the result of gravity waves moving southeast away from violent thunderstorms and. tornadoes in the southern Great Plains. It is almost as if a pebble had been dropped into the fluid air and ripple waves were moving concentrically outward from the point of disturbance.

Another invisible quality of the atmosphere is made visible by satellite imagery in figure 18.17. Here, infrared imagery reveals a relatively warm high layer of clouds at (**A**). These are barely visible on the visual-band image because they are so thin. The lower cloud deck which extends westward over Nashville, Tennessee, (circled dot) is colder as measured on the imfrared

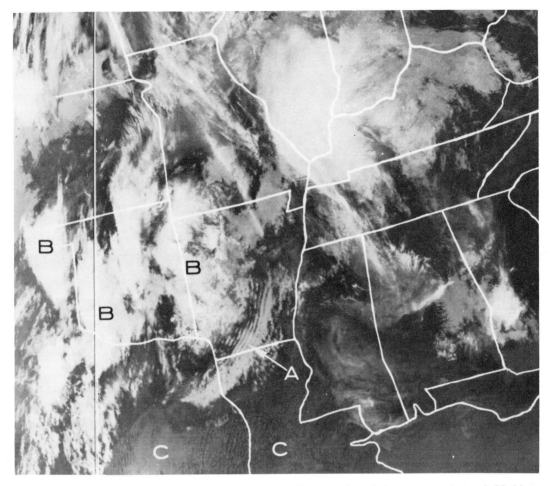

Figure 18.16. Visual-range satellite image Arkansas/Louisiana, morning of 23 May 1973. (NOAA)

image. Rawinsonde (radio balloon instrument) measurements confirmed that near Nashville there was a marked temperature inversion, with the warm air layer concentrated between 850-915 millibars. The darker shading in the infrared image, therefore, probably represents the clouds at that altitude, and the lighter imaged cloud sheet extending eastward from Nashville is, in all probability, not only colder but also nearer the surface of the earth, probably at the 950 millibar level or even lower.

A significant, and apparently necessary, ingredient in the weather recipe for tornadoes is a dry air intrusion at low or middle levels in the troposphere directed toward a low level moist tongue of air. It represents the means by which convective instability is released, and highly destructive storms result. A clue to the probability of such severe weather, therefore, can be given by a "void" in the satellite imagery in a region of otherwise widespread cloud cover.

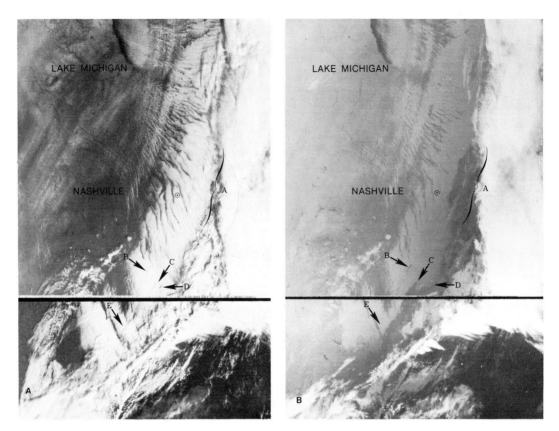

Figure 18.17. (A) Visual-range satellite image centered on Nashville, Tennessee, mid-morning 6 December 1972. (B) Infrared satellite image of the same place and at the same time. (Courtesy of NOAA, National Environmental Satellite Service)

Figure 18.18 shows a dry slot (D) projecting from southwestern Missouri. In three hours this dry tongue projected rapidly northeastward from near the Arkansas-Missouri line to the Missouri-Illinois border. Fourteen tornadoes were reported within a few hours of this penetration by the tongue of dry air. The significance of this satellite guide to severe weather is well worth noting.

Present/Future

The preceding case study ended on a predictive note. If the dry tongue intrusion model works, an immense step forward in more reliable tornado forecasting will have been made at a time when tornadoes are apparently greater in number and more destructive than ever before. This is only one example of remote sensing in weather analysis which is as future-oriented as it is presently developed. After less than two decades, meteorologists are only beginning to see some of the potentialities for forecasting weather by analyzing remote sensing images. They have yet to develop a systematic analysis procedure and are still at the "Isn't it amazing . . ."

Figure 18.18. Visual-range satellite image of central United States, midday 18 January 1973. "D" is dry slot. (Courtesy of NOAA, National Environmental Satellite Service)

stage of looking at images. It is important for meteorologists to apply their growing understandings of remote sensing imagery as soon as possible to increase their forecasting skills.

Some of the new and exciting possibilities for using remote sensing include: (1) developing a body of data leading to a global heat budget; (2) sensing and measuring all atmospheric pollutants; (3) assisting in the maturing of cloud physics theory; and (4) forecasting weather at several levels. Examination of satellite and radar imagery shows meteorologists exactly what the weather at the surface of the earth is like everywhere. When the whole story is known, they can start to test hypotheses and carefully construct a complete set of forecasting rules which will work. Then and only then will weather analysis and forecasting be a true science.

NOTES

1. *Weatherwise* 28, no. 4 (1975):cover and A151.
2. Richard A. Anthes et al., *The Atmosphere* Columbus: Merrill, (1975), p. 22.
3. Ralph K. Anderson et al., *The Use of Satellite Pictures in Weather Analysis and Forecasting,* Technical Note No. 124 (Geneva: World Meteorological Organization, 1973). Some of the illustrations in this section are adapted from chapter 2 of this reference.
4. Ibid.
5. "Weather and Climate: Measurement and Analysis," chap. 21 in *Manual of Remote Sensing,* ed.-in-chief Robert G. Reeves (Falls Church, Va.: American Society of Photogrammetry, 1975).
6. Jimmie D. Johnson et al., *Environmental Satellites: Systems, Data Interpretation and Applications* (Washington, D.C.: U.S. Department of Commerce, National Oceanic and Atmospheric Administration, National Environmental Satellite Service, May 1975).

19

Application of Remote Sensing in Cartography

Robert S. Weiner
University of Connecticut

Introduction

Manual Versus Computer Cartography

As the end of the twentieth century nears, one of the major topics which tends to be discussed by cartographers is the question of manual cartography as opposed to computer-generated or other machine-produced cartography. In this discussion two variable factors generally rise to the surface: time and cost-effectiveness. In terms of time, there is no question that conputer-generated cartography is quicker, assuming that the programer has the ability to produce a workable program which will be accepted by the computer without too much difficulty. Computers, therefore, excell over human cartographers in terms of the amount of time needed to produce a map once the program has been accepted.

In considering the factor of cost-effectiveness, however, the situation is much different. From the standpoint of cost, it can be assumed that the total expenditure for the person-hours to manually produce a map would be approximately $300, even when the map is to be in color. This is far less than the cost of a similar map which was computer-generated.

For large academic institutions, large computers are becoming standard equipment. For small or medium-sized institutions, the availability of such equipment for use in a cartography class cannot be taken for granted. There is the possibility, however, that the latter institutions might have access to a computer on a time-sharing basis, and that they might even have one of several types of plotters which are available to produce maps. Because of comparative costs, however, academic cartographers probably will continue to use manual methods.

In the professional field of cartography, there is no data available concerning the number of cartographers employed in relation to the size of the employing organization. For a small map company, without the resources or facilities to produce

The author wishes to express his thanks to Professor Bruce L. LaRose who read this chapter and made valuable comments and suggestions. The author also wishes to thank former students of his in geography/cartography at Briarcliff College for permitting the use of their graphics in this chapter.

computer-generated maps, there is no alternative to manually-produced maps. Even in some large cartographic agencies of the federal government, where computer facilities are not a major problem, many maps are produced manually.

From observation, it can be assumed that the bulk of professional cartographers today are involved to some degree in manually-produced maps. This chapter is written from the point of view that most maps which will be produced in the near future will be done using manual techniques.

Thematic Maps

One of the many problems cartographers face as they go about their professional tasks is that of securing up-to-date, accurate data for the purpose of compiling a thematic map. Robinson defines thematic maps as those which ". . . specifically communicate geographical concepts such as the distribution of densities, relative magnitudes, gradients, spatial relationships, and all myriad interrelationships and aspects among the distributional characteristics of the earth's phenomena."[1] These maps also are referred to as "special purpose maps." Raisz has called the nonquantitative maps "chorochromatic maps" if color-patches are used to show the distribution of certain types of data.

With the evolution of the Earth Resources Observation Systems (EROS) program, created in 1966, at least a partial solution to the problem of acquiring up-to-date, accurate data for thematic maps has been accomplished. With the wide variety of imagery presently available, it is possible for the geographer/cartographer to produce large-, medium-, and small-scale thematic maps of physical and/or cultural phenomena which appear spatially distributed over the surface of the earth. Lindgren has stated, ". . . it is for the first time economically possible to acquire land-use data quickly and efficiently over large areas. This is an indispensable capacity if we are ever to control urban growth and protect our land resources."[2] Lindgren also stresses the importance of Landsat to the developing countries: ". . . the value of ERTS [Landsat] to the countries of the developing world is inestimable. In some parts of these nations the ERTS photos themselves are better than any existing map."[3]

Types of Imagery Used to Produce Thematic Maps

Virtually all types of remote sensing imagery can be used to produce thematic maps. The basic types of imagery from which maps of various scales can be produced include: high altitude color or color infrared (CIR) aerial photography; side looking airborne radar (SLAR) imagery; black and white vertical aerial photography; multispectral (MSS) imagery, particularly false-color composite in bands 4, 5, and 7; and thermal infrared imagery (TIR). The type of imagery which is used depends upon a number of factors: (1) the availability of the imagery based upon cost, quality of the imagery desired, and the season of the year desired; (2) the final scale of the product; and (3) the ability of the cartographer to analyze and interpret the imagery. It is possible, therefore, to produce a high-quality thematic map from all forms of remote sensing imagery involving such themes as land use/land cover; inventories of agricultural resources; changes in the evolution of an urban renewal project; regional geomorphology/hydrology; and thermal pollution in a river or lake. For beginners it would be wise to obtain various kinds of imagery of an environment with which they are familiar.[4]

Scales for Thematic Map Reproduction

The scale of the imagery has always been a problem for the cartographer. Most imagery is produced and available at scales of 1:5,000,000, 1:1,000,000, or 1:500,000. Maps produced directly from this imagery on a 1:1 basis would be considered small- or medium-scale maps, according to the standards set by the American Society of Photogrammetry.[5] These scales tend to leave the cartographer who wishes specific details and very distinct differentiation of spatial patterns at a loss because of a lack of detailed data found at these scales. Many authorities in the field of remote sensing have indicated that the technical aspects of producing imagery at a much larger scale would be costly, but not impossible. In addition to the problem of cost there is the problem involving international relations with nations which do not wish to have their land areas "photographed" at large detailed scales.

To overcome the problem of scale, the cartographer can either be very selective about the type of imagery used, or employ some method whereby the scale of imagery can be enlarged before the map is compiled. For example, plate 24 shows that the details of a complex urban area can be extremely well defined in a CIR aerial photograph taken from a high altitude by NASA aircraft, such as the U-2 or RB57F. High altitude CIR imagery can be enlarged and enhanced to provide additional detail for the cartographer (plate 7). The map in figure 19.1 was drawn from Skylab infrared imagery.

For the enhancement of Landsat imagery there is at least one piece of optical equipment on the market today the primary objective of which is to reproduce imagery from space so that a large scale map can be drafted from it. According to the manufacturer, the magnification

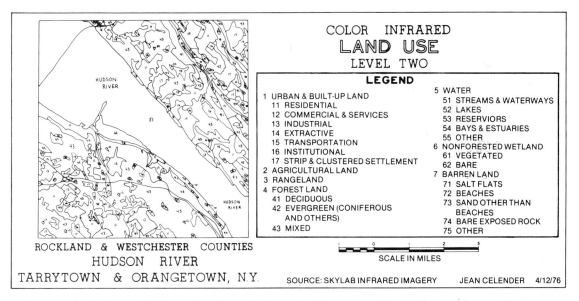

Figure 19.1. A preliminary land use interpretation map of middle and lower Hudson Valley, N.Y. using Skylab infrared imagery. (Cartographer and image interpreter, Jean Celender)

specifications of this equipment can be as great as 14 times. Using imagery on a scale of 1:1,000,000, the final compiled map would be approximately 1:70,000.

Eyton and Kuether, on the other hand, have suggested a simpler piece of optical equipment which, while not producing a thematic map of the quality of that produced by commercial equipment, is of acceptable quality for learning experiences and some professional mapping.[6] In terms of cost, this simpler piece of equipment, which Eyton and Kuether call a "Throwback Projector" (fig. 19.2), can be put together by a good carpenter with materials costing less than $50, assuming that most departments have access to a slide and/or opaque projector as part of their standard audiovisual equipment.

The writer and his students have been able to produce original maps using the Throwback Projector at scales of approximately 1:422,000 (1 cm represents 4.2 kilometers, or 1 inch represents 6.7 miles). In figure 19.1 observe the amount of detail that was obtained using a Level II Land Use Classification system.[7] It is possible to obtain an even larger scale map using this method if the projector is repositioned, but then the high-quality resolution of imagery is sacrificed.

To make a large-scale thematic map using high-altitude CIR imagery is a relatively simple procedure. The first step involves making the base map upon which the final data will be placed. Base maps can be prepared from a U.S. Geological Survey quadrangle (7 1/2 minute series, usually at a scale of 1:24,000). If a smaller or larger base map is needed, it can be obtained by using a proportional square grid of any desired size as an overlay to the topographic

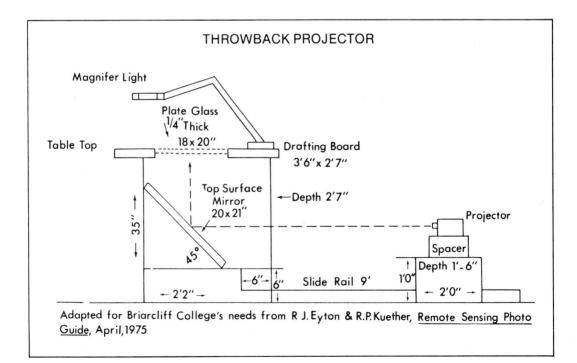

Figure 19.2.

map. A similar grid, with squares of the desired proportion, placed beneath the map sheet on a light table will make it possible to grid transfer the political boundaries with their proportional dimensions.

The grid map may be placed in an opaque projector or copied onto a 35 mm slide which can be inserted into a 35 mm projector. The 35 mm or opaque projector can be moved along the slide rail until the desired scale is obtained on the base map. The details of the thematic map can then be compiled for the final map.

Overcoming Difficulties of Interpreting and Analyzing Imagery

False-Color Composite

With a false-color composite image for comparison, a cartographer can begin to make interpretations of imagery. The ways in which physical, cultural, and biotic features of the land appear on high altitude CIR and Landsat color composites have been discussed in preceding chapters. These can be referred to if readers are in doubt about the ways in which particular features are displayed on this type of imagery. Experience with false-color imagery combined with ground truth studies soon will make a new remote sensing analyst able to interpret the various shades or hues of a particular image into recognition and interpretation of the target.

Identi-map

Figure 19.3 represents a preliminary draft of a thematic map made from SLAR imagery. LaRose coined the term Identi-map to represent the beginning research stage for this type of thematic map. His use of the term was inspired by an article on interpretation of radar imagery by Bryan.[8] Various parts of the imagery are enclosed in boxes and identified by letters or numbers. The letters or numbers are then used to identify various objects or signatures by comparing the imagery with a USGS topographic quadrangle of the area and by actual field observations.

To make a SLAR Identi-map the imagery must first be oriented to a topographic map. The cartographer must realize that just as there is scale distortion along the periphery of an aerial photograph, there is also scale distortion at the edges of SLAR imagery. Once the imagery has been properly oriented, a base map of the coverage area is drawn. Specific features, physical and/or cultural, can be enclosed in boxes with trailers leading into the map margin where the features are identified by numbers or letters. In the legend of the map the letters or numbers are listed along with the features they represent.

Simulation of Computer Decision Making

LaRose also has developed an exercise to give the beginning student experience in decision-making procedures similar to what a computer goes through in deciding the dominant false color in small areas of the image, known as pixels. A picture element, or pixel, is defined as "the numerical value of the energy radiated by an instantaneous field of view in each of the multispectral sensor bands."[9] Squares on such a map give the impression that they represent pixels, and some students have referred to their maps as "pixel maps." The size of the squares on the graph paper which is used and the scale of the imagery, however, preclude the possibility that these are actually pixels. LaRose, using computer language, has suggested that these maps be referred to as "Redigitized Maps." The purpose of using computers with Landsat MSS im-

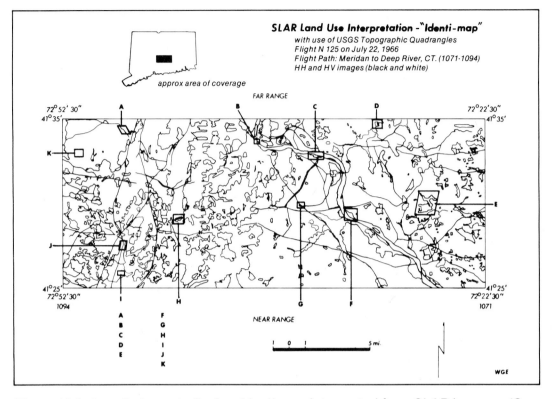

Figure 19.3. A preliminary draft of an Identi-map interpreted from SLAR imagery. (Cartographer and interpreter, Wayne G. Eddleman)

agery is to make geometric corrections of the original imagery through a technique known as "precision processing."[10]

To produce a "redigitized" decision-making surface, a sheet of graph paper is placed over a false-color composite. As the cartographer views each square of the graph paper, a decision is made as to which color dominates in the square. The square is either colored accordingly or a letter is used to represent a specific color, as a computer printout would do once the computer has decided the dominant color in the square.

Ground Truth

Another project which is of help in the development of a trained image interpreter/analyst is a ground truthing project of an area with which the trainee is familiar, such as the local area. This is done by going into the field with a topographic map of the area and a corresponding Landsat color composite. In the field the various colors in the imagery which were reflected by different signatures or targets can be identified and mapped.

There are problems involved in such a process, however, and the problems must be overcome. Kitchel and Weiner have discussed some of the problems which they experienced in a ground truth project in a tropical region.[11] Among them were problems of cloud cover in the imagery, the many scales involved with more than one type of imagery, and the physical inaccessibility of many of the areas to be investigated. Another problem was the time lag between the date of the image and the date of the ground truth project. This was particularly a problem when ground truthing agriculture because of frequent changes in tropical agricultural patterns. Lougeay has turned his attention to the problems of remote sensing ground truth in middle latitude areas.[12]

Map Compilation

For the cartographer compiling a first thematic map from remote sensing imagery—in this case the map of a state, a possible starting point might be to use black and white ERTS imagery or parts of the mosaic of the United States in band 5 (fig. 1.1). This will result in a relatively small-scale map of a state of about 1:4,000,000 to about 1:3,000,000, the scale depending upon the state chosen and the size of the format upon which the map is to be produced. This imagery permits only a broad regional interpretation of the topic to be depicted, such as the general geomorphology and hydrology patterns of the state.

To make a state or regional physical map using the ERTS-1 mosaic of the United States, band 5, a base map is again needed. When using a state format a good source of the base map can be one of those published in the *1970 Census of Population* volume 1. If one wishes to broaden the area of coverage, an economic region or subregion might be selected. For example, Bogue and Beale divide the United States into 13 major economic regions and 118 economic subregions, using counties as the building blocks of these regions.[13]

The ERTS-1 mosaic of the United States does not have the political boundaries on the imagery; however, an atlas is available which has satellite photographs with the state boundary lines printed on the photography.[14] This atlas is helpful for locating geomorphic features. On the base map, the major geomorphic features are outlined, and if desired, an overlay of hydrology can be added. The latter will add to the detail of the final product.

As more experience is gained, the cartographer can finally turn to a large-scale, more detailed map and use a Level II Land Use/Land Cover classification. Using the Throwback Projector and a selection of 35 mm slides of the false-color composites of the United States, it is possible to produce a detailed Land Use/Land Cover map with a Level II Classification at a scale of about 1:444,000 (1 cm represents 4.44 kilometers, or 1 inch represents 7 miles). These maps can be prepared for photographic reproduction in color. For this a separate flap or overlay must be made for each color, filling in the specific color patches on each overlay with solid black ink. Black and white line negatives (orthochromatic film) are made from which the final map can be produced by color-proofing techniques, or as offset plates for printing the map on paper.

The Future of Remote Sensing in Relationship to Cartography

While the procedures which have been described in this chapter represent a start in the types of thematic maps which can be compiled from Landsat imagery, as well as from high altitude color and color infrared (CIR) photography, they are only a beginning. Other thematic maps which have been developed include such topics as geology, particularly mineral and petroleum deposits; environmental themes; water resources, including floor damage; and range, marine, forest, and agricultural resources. NASA reported in *1973 Significant Accomplishnents in Remote Sensing,* ''A cartographic investigation in progress involves Skylab

S-190A and S-190B imagery to demonstrate the feasibility of topographic mapping of remote or inaccessible areas.''[15] Such maps are needed in the decision-making process of the world's policy makers. It is the cartographers' responsibility to prepare these maps.

In terms of economic applications, a midwestern power utility cooperative is considering the use of remote sensing imagery to make a land use/land cover map of a territory 103,600 square kilometers (40,000 square miles) in area. The map would aid the company in deciding on the routes which it will use for electrical transmission lines.

The uses of remote sensing in aiding the development of the lifestyles of people everywhere are in the early stages of development. As more people come to realize the value of remote sensing, and the part it can play in making a better life, there will be a major need for persons trained in all applications of remote sensing, including cartographic uses of remote sensor imagery.

Opinions of the value of remote sensing range from over-elation to a waste of the taxpayers money. Both extremes are probably far from reality. It is certain that imagery from space does not represent the panacea for the problems of the world. On the other hand, the argument that there should be a reordering of priorities and a redirecting of the use of the taxpayers' money for solving immediate problems is shortsighted. Without a continuing space program of the United States, there would be no progress in the field of remote sensing with its increasing contributions to knowledge. Perhaps the point of view that the money spent on the remote sensing program represents a major investment in the future of all peoples is the more realistic approach to remote sensing and the entire space program.

NOTES

1. Arthur H. Robinson and Randall D. Sale, *Elements of Cartography,* 3d ed. (New York: John Wiley & Sons, 1969), pp. 10-11.
2. D. T. Lindgren, "Land Use," *Photography from Space to Help Solve Problems on Earth,* NASA Goddard Space Flight Center (Washington, D.C.: U.S. Government Printing Office, 1974), p. 9.
3. Ibid.
4. John B. Rehder, "Landsat Imagery: Pictures of Your Place from Space," *Journal of Geography* 75 (1976):354-59.
5. Robert G. Reeves, ed., *Manual of Remote Sensing* (Falls Church, Va.: American Society of Photogrammetry, 1975), p. 2091.
6. J. Ronald Eyton and Richard P. Kuether, *Remote Sensing Photo Guide,* preliminary draft (Urbana-Champaign: University of Illinois, Department of Geography, April 1975), p. C-4.
7. Larry Alexander et al., *Remote Sensing: Environmental and Geotechnical Applications, the State of the Art,* Engineering Bull. no. 45 (Los Angeles: Dames and Moore, 1974), p. 22.
8. M. Leonard Bryan, "Interpretation of an Urban Scene Using Multichannel Radar Imagery," *Remote Sensing of Environment* 4 (1975):49-66.
9. Ralph Bernstein and George C. Stierhoff, "Precision Processing of Earth Image Data," *American Scientist* 64 (Sept.-Oct., 1976):504.
10. Ralph Bernstein, "Results of Precision Processing (Scene Correction) of ERTS-1 Images Using Digital Image Processing Techniques," ERTS-1 Symposium, NASA Goddard Space Flight Center, IBM Corporation, Gaithersburg, Md., March 1973.
11. Alice Kitchel and Robert S. Weiner, "Ground Truthing Problems in Puerto Rico," paper given at the Association of American Geographers, Remote Sensing Workshop, Briarcliff College, April 1976.
12. Ray Lougeay, "Remote Sensing Ground Truth Problems and Considerations," prepared for the Association of American Geographers, short course on remote sensing, Seattle, April 1974.
13. Donald J. Bogue and Calvin L. Beale, *Economic Areas of the United States* (New York: Free Press of Glencoe, 1961).
14. James R. Grady, executive producer, *Photo-Atlas of the United States* (Pasadena, Calif.: Ward Ritchie Press, 1975).
15. NASA, *1973 Significant Accomplishments in Remote Sensing,* Earth Observations Division, Science and Applications Directorate (Houston: Johnson Space Center, November 1974), p. 64.

20

Remote Sensing of Earth Resources: Summary and Look Ahead

Merrill K. Ridd

University of Utah

Ruth I. Whitman

National Aeronautics and Space Administration

Background

To forecast the future of a technology as dynamic as remote sensing is risky business. If the next decade produces advancements in science as remarkable as those of the past decade, any view of the future can be only a guess. Perhaps the best way to introduce a look into the future is to summarize the past.

As has been mentioned, the term "remote sensing" is only about a decade old, but its art and science are ancient. As stated in chapter 1, the human eye is the original remote sensor, in the visual mode. Even the technology of remote sensing, the recording of distant objects by a mechanical device, goes back over a hundred years. Table 20.1 sketches some highlights in the evolution of the technology of remote sensing of the earth's surface features.

Airborne photography of the earth has been noted as early as 1858, some twenty years after the invention of photography. Following a few military applications in World War I, operational programs in aerial photography were underway by the 1930s, largely for agricultural purposes and for topographic mapping, with budding interest in geological and resource applications.

Although remote sensing by use of conventional photography from aircraft remains the most widespread mode of remote sensing even today, two developments have occurred in the post-World War II era that have revolutionized and accelerated a thousand-fold (in terms of data) the process of sensing the earth's surface from above. Those two are (1) the rise of nonphotographic sensors, and (2) the advent of orbiting spaceborne vehicles. A third development, the emergence of high speed computers to handle the flood of data, has been an essential part of the process as well. Actually the gestation period for nonphotographic sensing has been quite long. Radar was introduced as early as 1903, while its use for earth surface mapping developed only in the past twenty years. Other nonphotographic sensors, such as thermal infrared and passive microwave sensing, are likewise recent as mapping devices.

TABLE 20.1
Selected Historical Highlights in Remote Sensing of the Earth

1839	Photography invented (Daguerre and Niepce)
1858	Photograph from a balloon (Tournachon)
1862	Multispectral imaging introduced (Sutton)
1903	Photography from carrier pigeon (Neubrouner)
1903	Early form of radar introduced (Hulsmeyer)
1907	Photograph from a rocket (Maul)
1909	Photograph from an airplane (Wilbur Wright)
1930s	Widespread operational aerial photography
1940s	Operational use of CIR film
1950s	Sidelooking Airborne Radar (SLAR) develops
1959	First television photograph of earth from space (Explorer 6)
1960	Television InfraRed Observation System (TIROS) weather satellite
1961	First photography from manned suborbital spacecraft (Mercury 3)
1962	First photography from manned orbital spacecraft (Mercury 6, 7, 8: Glenn, Carpenter, Shirra)
1968	First photography from an extraorbital satellite (Apollo 8: Borman, Lovell, Anders)
1969	First satellite photography in distinct spectral bands (Apollo 9: McDivitt, Scott, Schweikart)
1972	First Earth Resources Technology Satellite (ERTS-1 or Landsat 1)
1973	Skylab, manned orbiting space station
1975	Second Earth Resources Technology Satellite (Landsat 2)
1978	Third Earth Resources Technology Satellite (Landsat 3)

The greatest single event in the recent mushrooming of remote sensing of the earth occurred with the launching of the Earth Resources Technology Satellite (ERTS), subsequently renamed Landsat, providing its 18-day frequency of electronically sensed data of nearly all the earth if continuously engaged. Never before had there been such a breadth of coverage, let alone the potential of an 18-day recurrence. Between January 1975 and January 1976, Landsat 2, along with Landsat 1, provided a 9-day imaging interval. The versatility of the four-band electronically sensed data coupled with the frequency of the data sets has sent thousands of earth and resource scientists scurrying about still discovering its utility and application, and inventing ways of handling the masses of data. The intervening Skylab experiment established the functional complementarity of return-film photography and electronic multispectral imaging.

The new science of remote sensing is coming of age. The expanding army of researchers swells each year with specialists from dozens of disciplines dealing with hardware, software, and a growing list of applications. Agencies in agriculture and resource management are finding new applications at federal, state, and local levels. Several states have established comprehensive and systematic programs in remote sensing to improve the data base for resource and environmental decision making. A growing number of industries is engaged in remote sensing applications, especially those in agriculture, forestry, or, the largest of all commercial users, the mineral industry.

Universities have seized the opportunities as well—serving two main functions: to lead with the vanguard of researchers in hardware, software, and applications development, and to provide the training for people moving into the job market, as well as retraining for agency people already on the job. Training typically consists of a combination of skills that interrelate to varying degrees the following five specialties: remote sensing, aerial photograph interpretation, cartography, quantitative techniques, and computer use. The number and quality of courses offered in all these skills areas has multiplied in the past decade. In remote sensing, aside from aerial photograph use, the number of courses offered in various disciplines with varying emphases, approaches, and applications is increasing in colleges and universities. In geography alone, the number of schools offering remote sensing exceeds one hundred, with a number of departments offering three, four, or five courses, from introductory surveys to graduate research seminars.

Naturally, with all this investment of manpower and money in colleges, universities, agencies, and industry, the opportunity for equipment makers is growing rapidly and competition is keen—for the simplest items to the most sophisticated electronic/optical advancements. Herein lies one of the hazards. The equipment buyer, whether college, university, agency, or industry, is faced with a complex and rapidly changing array of alternatives. The question of add-on capabilities, intercompatabilities, and the ever-present threat of obsolescence is a serious dilemma where budgets are limited. There will be no easy answer to the problem. Users can only attempt to keep current with the trends in the technology, anticipate the changing format of available data (photographic and digital), examine their own evolving needs and applications, and then shop for the most effective combination of present and future equipment, given their budget constraints.

Remote Sensing Data

In the future, as at the present time, the form of available data and the means of interpreting it will be fundamental issues. Persons interested in entering remote sensing as an instructional or research objective need to consider a number of related factors regarding the changing nature of the data and the use of it.

Data Format

Currently, remotely sensed data are made available to the user in two basic formats: film products and digital data. Most of the frontier research work is carried on in the latter to improve the data gathering, data processing, and analysis for particular applications. As for Landsat systems, the one which will continue to provide the bulk of our earth resource remotely sensed data from space for the next decade or so will continue to be digital, as telemetered from the satellites' electronic scanners. Any photographic products, therefore, will be at least one generation away from the original digital data.

The advantage of digital data is two-fold: (1) it is the *original* data, carrying all the information that electronics can bring back from what the sensor detects, and (2) it is far more *versatile* than the photograph in that it is a quantitative expression of the scene (selected components of the scene), which can be manipulated and compared, composited, and displayed in endless ways through computer line printers, computer plotters, cathode ray tubes photographs, and so on. The principal disadvantages at the present time are cost and the requirement that the analyst have access to and be trained in the use of relatively sophisticated equipment. Many colleges and universities, however, now have adequate computer hardware to accomnodate the NASA-produced computer compatible tapes, and software programs are available from a variety of sources. With access to a skilled computer technician to adapt the program(s) to the campus computer, the investigator or teacher can do sophisticated automatic mapping from digital remotely sensed data. Commercial suppliers are continually improving equipment for data handling and display of data in analog form.

The advantages of the photographic format until now have been largely (1) for display purposes, as it has more nearly represented the familiar view of the scene, and (2) for photograph interpretation purposes, as the least expensive and simplest way of getting into the remote sensing business. Anyone with conventional aerial photograph interpretation skills can apply the same techniques of visual pattern recognition to Landsat images and derive a great deal of information for teaching or research purposes. The user may get by with no equipment, and the cost of photographic imagery is inexpensive.

The relative advantage of the realistic, familiar photograph is diminishing as digital processing improves. Through recent developments, the present digital format from Landsat, when suitably processed and displayed, retains the comfortable appearance of the photographic product plus all the flexibility and information content inherent in digital data. Compare figures 20.1 and 20.2. The former is a computer reconstituted picture and the latter is a conventional photographic image. The reason many photographic products are still in use is principally a function of cost and convenience. It is cheaper to reproduce an image by photographic processes than to reconstruct the image with the computer, and obtaining the photograph is still more convenient to the general user.

Figure 20.1. Special reconstituted image of New Orleans area from CCT.

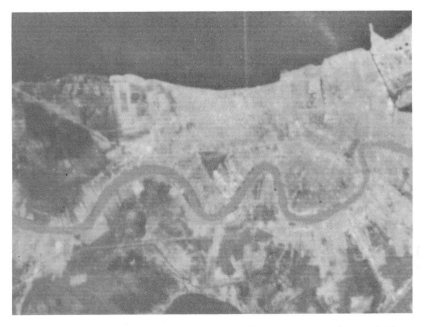

Figure 20.2. Conventional photographic image product of New Orleans area. (EROS Data Center)

Innovative spinoffs from space-related remote sensing and other developments continue at many levels. In the more advanced centers of remote sensing research and development, the analyst utilizes an electron beam recorder or laser beam recorder to extract digital data from computer compatable tapes in order to produce photographic products in black and white or color. As the instantaneous field of view (IFOV) narrows with subsequent sensors and through careful use of the dynamic range in processing, the resolution and reality of digital products will continue to improve. At a level closer to the general investigator, many aids are becoming available at relatively low cost, making an increasing variety of products and displays accessible. Such electronic/optical/photographic devices include minicomputers, digitizers, density slicers, edge enhancers, cathode ray tubes, instant development cameras, and other locally controlled, interactive fixtures. Displays may be made available at various scales in various forms—such as, line printer, computer plotter, cathode ray tube image, and photographic copy—for use in the classroom or laboratory. Through the use of digital processing investigators may select output products ranging from the realism of familiar photography through symbolic characters or patterns that single out thematic elements of the environment, such as urban features, forest types, oil slicks, or thermal pollution patterns.

Data Flow

The technology for data gathering through continuously orbiting satellites with their multiband scanners has advanced far ahead of our interpretive and application methodologies. Even in the pioneering Landsat system, a single scene with its approximately 7.8 million pixels generate, with its four bands, over 30 million data bits to be sorted out and understood. This was repeated every nine days, with the two satellites, if activated each time, providing 40 over-passes in a year, or 1.2 billion elements of data for each scene (185 × 185 kilometers) each year. From Landsat 3, another 2 billion data elements will be available on the scene each year.

For many purposes that refinement will be welcome and lead to solutions of resource inventories and management problems. There is an inherent danger, however, of falling into the trap of assuming more detail and more data are innately better. There will continue to be many applications where our present level of data, for example, from Landsat 2 and 3, are more useful and easier to manipulate than the two-fold or hundred-fold increase in data bits that their successors could bring.

Landsat transmitters are not activated at all times. They are governed by ground control according to meteorological conditions and user needs. Generally they are engaged over the United States, except when known widespread cloud cover exists. Elsewhere in the world the data is transmitted only in response to requests by users or by existing foreign ground stations.

Through an open space policy, the United States government has agreed to make all Landsat data available to any requesting nation. In support of a policy of open dissemination of Landsat data, the agreement signed between the United States and any country having a ground station makes the Landsat data available at the cost associated with data handling.

Worldwide data will continue to pour out to interested individuals and institutions with increasing volume if the satellite program continues. Users will have the opportunity and obligation to select information from a great mass of imagery and tapes containing visual, multispectral scanner, thermal infrared, radar, passive microwave, and other data. As part of a community of users at various levels, each one will need to focus continually not only upon how to manipulate the mass of data, but upon the quality and meaning of the data for individual specific purposes. In other words, success as users will depend upon (1) the ability to handle masses of data through the computer, and (2) the ability to select and find meaning in the data. As in all inquiry, this reflects back to the nature of the individual problem.

Remote Sensing System

Developments in remote sensing might be outlined in terms of the three major components of the total system: the scene, the sensor system, and the processing system.

The *scene* is the portion of the earth in "view" of the sensor. It is the place where the problem is conceived and conceptualized. The "observable" characteristics of the scene that may be sought by the sensor are spatial, spectral, and temporal. The *spatial* characteristics are the geometric/cartographic features that allow us to determine position. The *spectral* characteristics are those that allow detection by electromagnetic (EM) energy to identify the quantity and possibly the quality of elements of the scene. The *temporal* characteristics are those dynamic characteristics which typify the earth surface environments as the quality, quantity, or position change through time. While we have little control over the scene, the more we learn of its characteristics the more we find there is to learn, and the better we can utilize remote sensing to learn it. Thus, part of our future application of remote sensing depends on our advancing knowledge of the scene through other senses and methods.

The *sensor system* is the data gathering device. It consists of the platform (aircraft or spacecraft), the sensor proper, and the data transmission component. It is designed to pick up the spectral characteristics of the scene at an appropriate scale resolution (spatial) and with appropriate frequency (temporal) to serve the needs of the problem. The dilemma is that there are limitless problems perceived by countless investigators. The designers, therefore, attempt to create a combination of sensors mounted on a variety of platforms with varying frequencies of over-flight in order to satisfy a broad range of user problems. Individual users are dependent on the system designer to make accessible the information-bearing dimensions of the scene of interest to them. As succeeding sections of the chapter are read, the potential user may judge whether the forthcoming sensor systems seem to be designed to serve interpretation needs better than present sensors.

One of the areas for investigation is the interrelationship of multistage (varying altitudes from low to high aircraft and spacecraft) and multisensor (photographic, MSS, microwave, etc.) data. Other factors being constant, the higher the altitude the less the detail. On the other hand, the higher the altitude, with a given angle of view, the broader the area of coverage of the scene and hence the less expensive in money and time. Variations on this generalization can be introduced in other optical sensors (e.g., telephoto lenses) or nonoptical sensors (e.g., improving resolution in scanners). Each investigator must resolve the trade-off values according to the parameters of the problem and the changing nature of available products. In many cases a combination of stages and sensors may be optimal. The old adage that ''if a little is good a lot is better'' does not necessarily apply to data. The pitfall of improperly processing unneeded data may become an increasing menace.

The *processing system* may be considered to include the entire range of activities from receiving the raw data from the spacecraft or aircraft to the manipulation and display of data in the hands of the user. The first part of this stream of processes is beyond control of the user, administered by NASA or some other agency or firm. The user has a basic choice here to opt for computer photographlike products or photographs, depending on his/her data handling capability, the nature of the job, and time constraints.

Prospects for the future are bright in this area as data handling facilities are advancing rapidly. Global communication systems are emerging to transmit data with the speed of light, and agencies administering processing equipment, both photographic and digital, are striving to improve delivery time on orders. At present the lag may range from two weeks to two months. Efforts are underway to close the gap to days so that near ''real time'' observation may be obtained directly by the user. To benefit from this forthcoming advantage, users need to be improving internal data handling procedures through well adapted processing algorithms and implementation procedures. Again, it is the nature of the job or objective and the equipment available to the user that dictates.

An Operational Landsat System?

Should Landsat, now an experimental system, become a more permanent operational system? Communication satellites and weather satellites, after a fairly lengthy period of development, have established remarkably sound records as operational systems. Satellite-based earth resource sensor systems, in the few years of experimental trials, have demonstrated dramatically their utility across a broad range of earth resource and environmental investigations and applications. Among the demonstrated uses of Landsat, agricultural applications probably hold the greatest potential in terms of forecasting crop production, of assessing crop damage caused by drought, pests, and storms, and for observing seasonal rangeland conditions. Applications related to mineral exploration, water resources, flooding, land use, environmental monitoring, and mapping have all been demonstrated. Yet Landsat is still designated as experimental. This experimental designation has important implications, two of which are: (1) there is no governmental commitment to continue to launch Landsat satellites, and (2) the instrumentation, and even the satellite parameters, can be changed prior to launch.

With the substantial financial and time commitments made by agencies, institutions, industry, and thousands of individuals, coupled with Landsat's demonstrated value in resource management, the case for an operational Landsat-type system is mounting. The advantages of such a system are evident. The major need is to be assured of a continuation of data obtained by the satellite system in a format that is predictable in order for the users to justify investments in equipment, manpower, and programs for use of its imagery. A back-up satellite should be ready in the event of failure of one already in orbit because it takes at least three years to build a satellite and prepare it for flight. Users also require rapid retrieval of data. Time lag in obtaining data has become a matter of concern and discussion among users, but efforts are underway by NASA and industry to approach a "real time" data delivery system.

Some of the salient issues impinging on the decision of an operational Landsat-type system are considered in the Staff Report to the Senate Committee on Aeronautical and Space Sciences.[1] The fourteen issues discussed therein are listed below:

Issue 1.   Should a commitment be made that assures the availability of Landsat-type data in the 1980s?

Issue 2.   Should improvements in the quality of Landsat data and services be initiated at this time?

Issue 3.   Should government ownership and operation of the various Landsat system elements be curtailed?

Issue 4.   How can industry initiatives in Landsat activities be accelerated?

Issue 5.   How can user knowledge of an involvement with Landsat capabilities be accelerated?

Issue 6.   How can the U.S. avoid undue international sensitivities to a U.S. owned and operated Landsat system?

Issue 7.   What assurances are there that the general taxpaying public will have a reasonably equal opportunity to benefit from this system?

Issue 8.   Over the long term how can a loss of U.S. technological leadership be avoided?

Issue 9.   Can the interfaces between the system's segments be better defined in order to avoid unnecessary overlap and duplication of effort?

Issue 10.   Pertinent to present and future Landsat issues, how can a properly integrated decision-making process be provided within the federal government?

Issue 11.   Should the Landsat system evolve to a system of individual satellites each dedicated to a specific purpose or should it evolve to a consolidated system of multipurpose satellites?

Issue 12.   Should the worldwide network of remote terminals obtain their data directly from the satellite or should data be routed to them from a central data facility?

Issue 13.   Are there alternate sources of data acquisition that might be more cost effective than Landsat?

Issue 14. How should data from all sources be integrated into a total global information system?

Answers to the above questions, derived by design or default, will determine the course of remote sensing for the coming decades.

The Near Future

Additional earth resource satellites are in preparation stages for launching by the United States within the next few years as an extension of the earth resources remote sensing activity.

Landsat C/3

The third satellite in the Landsat series, Landsat 3, was launched in 1978. Prior to launch it was known as Landsat C, so the term Landsat C/3 is being used here. Launch had been delayed from 1977 because the multispectral scanner (MSS) on Landsats 1 and 2 performed long after design life of one year, thus providing longer availability of MSS Data.

The sensor package on Landsat C/3 varies from that of its forerunners in two significant ways: (1) it has a fifth band on MSS, and (2) it has a return Beam Vidicon (RBV) sensor with a 40 meter (131 feet) Instantaneous Field of View (IFOV).

The additional band on the MSS is in the 10.5-12.5 micrometers (thermal infrared) region. The Instantaneous Field of View (IFOV), or spatial resolution, of this band is 240 meters (787 feet). This provides first high resolution thermal data taken simultaneously with visible and near IR data from space. (The other four bands will continue to have an IFOV of 80 meters (263 feet), as in Landsats 1 and 2.) The new thermal data should be helpful in crop condition determination, thermal pollution monitoring in water, and other monitoring activities where temperature difference is an important parameter.

The Return Beam Vidicon (RBV) on Landsat C/3 consists of two panchromatic instruments with adjacent swaths which have IROV's of 40 meters. This means that there will be four RBV images for each Landsat frame. These four higher resolution images are registered to the standard MSS image to test the value of higher resolution in Landsat data analysis.

Probably the most important change resulting from Landsat C/3 will occur on the ground. The major problem in the present Landsat system has been in data delivery to users. Before the first launch of Landsat, it was thought that the primary analysis technique would be conventional photointerpretation procedures, i.e., that the analyst would look at the photographlike image (produced by NASA or another agency from the data tapes through a computer program) and proceed to analyze the scene using the basic procedures developed by the photograph analysts over the preceding twenty years. A decision therefore, was made by the U.S. Geological Survey and NASA that the basic product from the Landsat would be photographlike images. These images would be available to the public, primarily from the EROS Data Center, Sioux Falls, South Dakota.

Digital processing techniques, however, utilizing MSS data tapes directly, developed and spread quickly. Catalysts for this innovation were many, but notably a group at the Laboratory for the Application of Remote Sensing (LARS) at Purdue University and a group at the University of Michigan, now the independent Environmental Institute of Michigan (ERIM). The analysis was accomplished in the digital domain prior to the formation of the information image.

NASA and USGS are converting from what was basically an image producing system to a digital system to meet new demands. With Landsat C/3 the basic data product produced at the NASA Goddard Space Flight Center will be a digital tape. EROS Data Center, as the primary data distribution center, will produce images that have been geometrically corrected (corrected to agree with a map projection) or will copy radiometrically and geometrically corrected tapes.

Heat Capacity Mapping Mission

An item of major interest to scientists, which cannot be investigated with the thermal data from Landsat C/3, concerns the heat holding capacity of different earth materials. Scientists have shown that temperature extremes are most pronounced at about 2:00 P.M. and just before sunrise. Between 9:30 A.M. and 10:30 A.M., the local time of the Landsat overpass, the temperatures of most of the earth's constituents are nearly the same. Landsat, therefore, is poorly suited for heat capacity detection. As this is being written, thermal mapping missions are being planned. The Heat Capacity Mapping Mission will be flown to determine how the characteristic ways in which bodies of a given material change temperature during the diurnal cycle. Each day, over certain portions of the United States, heat data will be recorded just prior to the 2:00 P.M. and just prior to the 2:00 A.M. equator crossing times. It is postulated that the different heat holding capacities will help identify earth crust constituents from space.

Water is one of the major heat holding constituents on the earth, so it is possible that soil moisture and crop condition can be determined from this data. The IFOV for this satellite is 500 meters (1,640 feet) so this data must be applied to relatively large scale phenomena. For example, soil moisture in small fields cannot be determined using this data.

Landsat D

With the launch of Landsat C/3, the next major launch of a Landsat is planned for 1981. A new multispectral scanner, called the thematic mapper (TM), is to be installed on Landsat D. Even though the MSS on Landsat C/3 incorporates a new thermal channel, TM truly represents the second generation multispectral scanner. This scanner will have six bands (the possibility of a seventh band is actively being studied) which will be more optimally spaced for digital analysis. Figures 20.3 and 20.4 show the kind of spectral plots that helped scientists select the spectral bands for the thematic mapper. Typical vegetation reflectance curves do not give any impression of the large variability of vegetation signatures as a function of growth stage (time of year) or of vigor. The signatures of different kinds of vegetation change so much that a single unique signature for each vegetation type cannot be found. For that reason, scientists looked at the vegetation reflection signature and the solar irradiance curve reflection and absorption zones, which fortunately correspond to the atmospheric windows shown in figure 20.4. Table 20.2 compares specifications of the four Landsat scanners.

The second advanced feature of the thematic mapper is in the area of improved radiometric sensitivity. Radiometric sensitivity is often measured in terms of the Noise Equivalent Change in Reflectance. It is well known that it is impossible to make a "perfect" measurement. It is necessary, therefore, to make a measurement which follows the changing signal caused by changing reflectance rather than the noise that is induced by the measuring device. The thematic mapper is designed to introduce less noise than the existing multispectral scanner. It would be very rare for a single scene imaged by Landsats 1, 2, and C/3 to contain all the data levels in any channel because those levels have to cover the entire range of reflectances

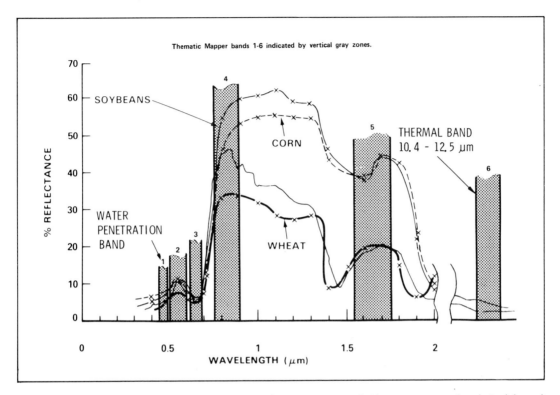

Figure 20.3. Typical reflectance curves from representative crops and related band widths of thematic mapper.

from dark to bright surfaces and from the different sun angles that occur in the summer and the winter. This total change in reflected energy which the detector can sense is called the dynamic range of the sensor. The TM, on the other hand, will be able to transmit more data, so the dynamic range of the sensor will be divided into additional levels. This should allow the analyst working in the digital domain to separate items whose spectral response is too close to be separated with the present MSS systems.

The third major, and possibly most important, improvement in the thematic mapper will be in spatial resolution. The IFOV of the thematic mapper is 30 meters (98 feet). This means that the majority of the signal comes from .2 acres on the ground instead of the 1.1 acre unit as in the present Landsats. To do a machine classification, best results are obtained when a field contains approximately 20 pixels. With an IFOV of 30 meters, this can occur in much smaller fields. TM, therefore, should allow analysis of smaller fields than the present MSS.

The thematic mapper to be used on Landsat D is more complicated than the multispectral scanners now being used, and the improvement "does not come for free." The major cost will be in increased data rates. Each scene is going to contain an order of magnitude of ten times more data than is provided by the MSS of Landsats 1, 2, and C/3. One way of holding down the amount of data is to limit the amount of time the TM is turned on. To accomplish this, the ground handling system for Landsat D is being sized to handle only 50 scenes a day.

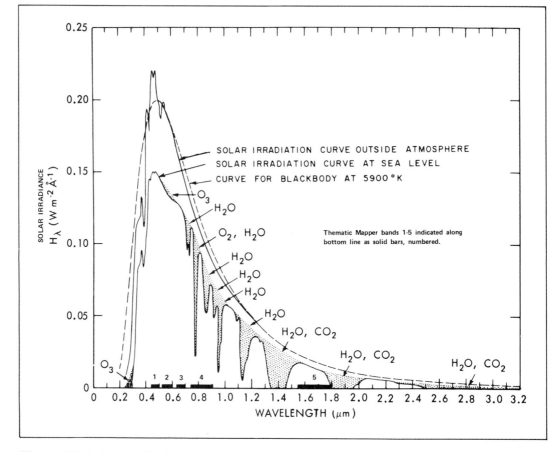

Figure 20.4. Atmospheric windows in the solar spectrum and related band widths of Thematic Mapper.

As previously stated, one of the major problems in the "operational" use of the data has been in delivery of the data products. Landsat D, like Landsat C/3 will be a digital system, meaning the primary data storage medium will be digital. Another feature the TM will have is that there will no longer be tape recorders on board the spacecraft. The 18-day repeat coverage has only been a reality in the United States, Canada, Brazil, and parts of Europe where direct readout ground stations exist—plus in the future a few countries where ground stations are under construction or planned. When Landsat D is launched, data taken over the other areas of the world will no longer be recorded on tape and then played back when the satellite is in sight of a ground station; it will be relayed to White Sands, New Mexico, through the Tracking Data Relay Satellite System (TDRSS) being built for NASA. Data will be sent to NASA Goddard Space Flight Center for preprocessing via a Domsat (Domestic Satellite) link (fig. 20.5). It can then be sent to users the same way.

TABLE 20.2

Landsat Scanner Specifications

	I and II	C	D^a
	Multispectral Scanner		Thematic Mapper
	(MSS)	(MSS)	(TM)
Scanner Characteristics			
Spectral bands (micrometers)			
Blue			0.45-0.5
Green	0.5-0.6	0.5-0.6	0.52-0.6
Red	0.6-0.7	0.6-0.7	0.63-0.69
Near IR	0.7-0.8	0.7-0.8	
Near IR			0.76-0.90
Near IR	0.8-1.1	0.8-1.1	
Mid-IR			1.55-1.75
Thermal IR		10.4-12.6	10.4-12.5
Instantaneous field of view[b]		78 meters (1-4)	30 meters
Thermal IR		234 meters	120 meters
Quantizing levels:		64 (6 bits)	256 (8 bits)
Sampling frequency:		1.4 samples/IFOV	1.0 samples/IFOV
Data rate:		15 MB/S	100 MB/S
Scene:		185 × 185 KM	185 × 185 KM
Mission Parameters			
Orbit (APOGEE), Sun Synchronous		914 KM	705 KM
Local Time at descending node		9:30 AM	9:30 AM
(equatorial crossing)			
Coverage cycle duration (2 satellites)		9 days	16 days
Image and Processing Characteristics			
Information per scene		$2(10^8)$ bits	$2(10^9)$ bits
Total scenes		500/day[c] (all land masses and coastal areas)	50/day[d]

[a]The MSS instrument will also be flown on Landsat D if money is provided.

[b]Sometimes referred to as spatial resolution.

[c]Potential

[d]Processed by U.S. Master Data Process at Goddard Space Flight Center (augmented by foreign ground station data acquisition).

Seasat A

Another satellite launch watched for by the earth resources remote sensing community is that of Seasat A, planned for 1978. While this mission was designed as an ocean monitoring system, it has as one of its instruments the first active microwave imaging radar flown from space. This is an L band (1.14 ghz) radar with a 20 degree incidence angle. Relatively little of the data will be taken over land, but the excitement concerning this data is high.

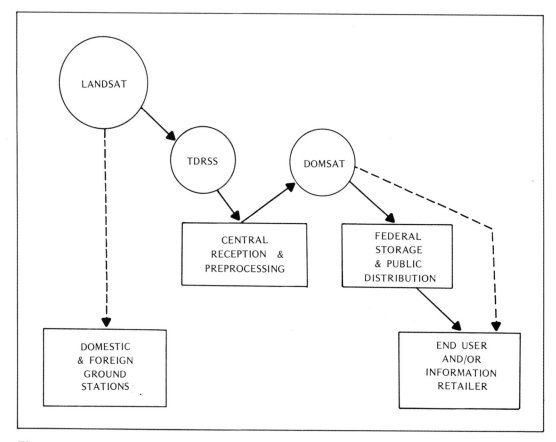

Figure 20.5. Landsat Follow-on Baseline Data Distribution System.

Dreams

Beyond these approved flight systems are the dreams of the scientists (fig. 20.6). NASA is looking forward to the availability of the Space Shuttle. Using the Space Shuttle, the scientists of the world can test instruments from space altitudes—they can determine in careful, scientifically controlled experiments whether the improved information is worth the increased cost. They can determine from space whether this band or that band is better, or whether this look angle improves the information or degrades it beyond usefulness. NASA is particularly interested in the opportunity to test instruments—such as radars with L, C, or X bands and with various polarization and incidence angles—and then bring them back, modify them, and fly them again prior to committing them to expensive free flyers, that is, to unmanned satellites. Unmanned satellites with their associated costs and complexity would be reserved for proven instruments.

One of the first active microwave instruments that will be flown on the shuttle is SIR-A (Shuttle Imaging Radar A). This will be an L band active microwave instrument like the one flown on Seasat-A, but the angle of incidence will be approximately 50 degrees from nadir. This

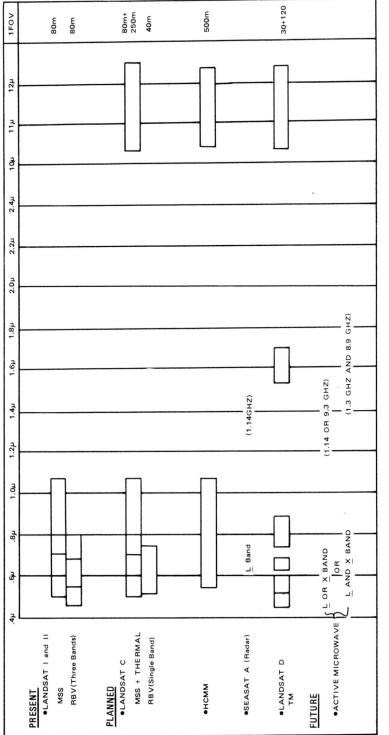

Figure 20.6. Space sensors—now (January 1978) and in the future.

will enhance terrain features (shadows) that are important for geological applications. It is intended to evaluate the applicability of this data in various disciplines. SIR-A is scheduled to be flown first on an orbital test flight (OTF) mission in July 1979. This mission will be primarily to test the shuttle and the L band instrument will be flown "piggyback." Other active and passive microwave instruments are being planned.

Beyond the Space Shuttle is a whole world of potential geostationary satellite applications. If permanent viewing of the same location in the world is wanted for earth resources requirements, geostationary satellites—similar to the weather satellites—could be put into orbit. The instruments and techniques of the future will be continuing refinements of those now in use or planned. Most important, however, will be the answers to the question: where do scientists go from here? What is needed is a new, creative way of thinking—one that does not use this new source of information to do things in the old ways, but rather one that matches the dynamic quality of the information source. These and other ideas—or dreams—continue to be studied by scientists in NASA, in universities, and in industry.

NOTES

1. *An Analysis of the Future Landsat Effort,* Staff Report prepared for the Committee on Aeronautical and Space Sciences, U.S. Senate (Washington, D.C.: U.S. Government Printing Office, 1976), p. 9.

Appendix A:
Sources of Remote Sensing Imagery

1. Manned spacecraft photography, Skylab photography, all Landsat standard products, computer enhanced Landsat images, NASA aircraft photography, U.S. Department of Interior photography, black and white aerial mapping photography, and Apollo and Gemini photography can be purchased from:

EROS Data Center
Sioux Falls, South Dakota 57198

The EROS Data Center provides geographic computer searches for Landsat, Skylab, NASA aircraft, and aerial mapping photography at no charge. To obtain this information, send the latitude and longitude of a point, or the latitudes and longitudes of the corners of a selected area to the EROS Data Center. If the geographic coordinates of a point or area are unknown, a geographic computer search will be made from other information, such as geographic names, locations, or a map. When a computer search is requested such additional information as time of year, minimum imagery quality, maximum cloud cover acceptable, and type of product (black and white, color, or color infrared) should be forwarded to the EROS Data Center.

Imagery which is achieved at the EROS Center may be reviewed at any one of the EROS Data Reference Files. These are located at:

Alaska
Public Inquiries Office
U.S. Geological Survey
108 Skyline Building
508 Second Street
Anchorage, Alaska 99501

California
Public Inquiries Office
U.S. Geological Survey
Room 7638, Federal Building
300 North Los Angeles Street
Los Angeles, California 90012

Hawaii
University of Hawaii
Department of Geography
Room 313C, Physical Science Building
Honolulu, Hawaii 96825

Massachusetts
U.S. Geological Survey
5th Floor
80 Broad Street
Boston, Massachusetts 02110

Missouri
Topographic Office
U.S. Geological Survey
900 Pine Street
Rolla, Missouri 65401

New York
Water Resources Division
U.S. Geological Survey
Room 343, Post Office and Court House
    Building
Albany, New York, 12201

Ohio
Water Resources Division
975 West Third Street
Columbus, Ohio 43212

Oregon
Bureau of Land Management
729 NE Oregon Street
Portland, Oregon 97208

Tennessee
Maps and Survey Branch
Tennessee Valley Authority
20 Haney Building
311 Broad Street
Chattanooga, Tennessee 37401

Washington
U.S. Geological Survey
Public Inquiries Office
Room 678 U.S. Court House
West 920 Riverside Avenue
Spokane, Washington 99201

The EROS Data Center is part of the National Cartographic Information Center (NCIC). The NCIC provides information concerning cartographic data, aerial photography, and space imagery. NCIC offices are located at:

National Center, Stop 507
12201 Sunrise Valley Drive
Reston, Virginia 22092

Air Photo Sales
U.S. Geological Survey
Federal Center, Building 25
Denver, Colorado 80225

Map and Air Photo Sales
U.S. Geological Survey
345 Middlefield Road
Menlo Park, California 94025

2. Information concerning Landsat and Skylab imagery, as well as data from Tiros and Nimbus weather satellites, ATS geostationary satellite imagery, and full-disc and section images of the earth from ESSA, NOAA, and SMS geosynchronous satellites is available at:

NOAA Satellite Data Services
Information Services Division
National Climatic Center
World Weather Building
Suitland, Maryland 20233

3. Landsat imagery can be inspected prior to ordering at numerous NDAA Browse Files. These are located at:

University of Alaska
Artic Environmental Information and Data
    Center
142 East Third Avenue
Anchorage, Alaska 99501

Inter-American Tropical Tuna Commission
Scripps Institute of Oceanography
Post Office Box 109
LaJolla, California 92037

National Geophysical and Solar Terrestrial
  Data Center
Solid Earth Data Service Branch
Boulder, Colorado 80302

National Oceanographic Data Center
Environmental Data Service
2001 Wisconsin Avenue
Washington, D.C. 20235

Atlantic Oceanographic and Meteorological
  Laboratories
15 Rickenbacker Causeway, Virginia Key
Miami, Florida 33149

National Weather Service, Pacific Region
Bethel-Pauaha Building, WFP 3
1149 Bethel Street
Honolulu, Hawaii 96811

National Ocean Survey—C3415
Building No. 1, Room 526
6001 Executive Boulevard
Rockville, Maryland 20852

Atmospheric Sciences Library—D821
Gramax Building, Room 526
8060—13th Street
Silver Spring, Maryland 20910

National Environmental Satellite Service
Environmental Sciences Group
Suitland, Maryland 20233

Northeast Fisheries Center
Post Office Box 6
Woods Hole, Massachusetts 02543

Lake Survey Center—CLx13
630 Federal Building and U.S. Courthouse
Detroit, Michigan 48226

National Weather Service, Central Region
601 East 12th Street
Kansas City, Missouri 64106

National Weather Service, Eastern Region
585 Stewart Avenue
Garden City, New York 11530

National Climatic Center
Federal Building
Asheville, North Carolina 28801

National Severe Storms Lab
1313 Halley Circle
Norman, Oklahoma 73069

Remote Sensing Center
Texas A and M University
College Station, Texas 77843

National Weather Service, Southern Region
819 Taylor Street
Fort Worth, Texas 76102

National Weather Service, Western Region
125 South State Street
Salt Lake City, Utah 84111

Atlantic Marine Center—CAM02
439 West York Street
Norfolk, Virginia 23510

Northwest Marine Fisheries Center
2725 Montlake Boulevard East
Seattle, Washington 98112

University of Wisconsin
Office of Sea Grant
610 North Walnut Street
Madison, Wisconsin 53705

4. Landsat photo mosaic maps of the conterminous United States and Alaska can be purchased from:

Aerial Photography Field Office
U.S. Department of Agriculture
Agricultural Stabilization and Conservation
222 West 2300 South
P.O. Box 30010
Salt Lake City, Utah 84125

5. Skylab, Gemini, Apollo, and Apollo-Soyuz photographs, along with a Landsat mosaic of New Mexico (black and white (1:1,000,000) can be purchased from:

Technology Application Center
The University of New Mexico
Albuquerque, New Mexico 87131

Audio-visual slide programs of space photographs can be purchased from:

Audio-Visual Institute
6839 Guadalupe Trail N.W.
Albuquerque, New Mexico 87107

Annotated slide sets of space photographs from Skylab, Gemini, and Apollo can be purchased from:

Pilot Rock, Inc.
P.O. Box 470
Arcata, California 95521

Professor Noel Ring has supplied the following information on the availability of Skylab urban area color photography.

Akron, OH (SL3-83-156), Albuquerque, NM (SL4-52-005), Allentown Bethlehem-Easton, PA, NJ (SL3-88-070), Altoona, PA (SL3-83-161), Appleton, WI (SL3-83-141), Atlantic City, NJ (SL3-40-125), Augusta, GA (SL4-90-047), , Austin, TX (SL4-94-123);

Bakersfield, CA (SL2-04-127), Baltimore, MD (SL3-83-166), Baton Rouge, LA (SL3-88-040), Battle Creek, MI (SL2-16-158), Bay City-Miland, MI (SL3-88-230), Beaumont-Port Arthur, TX (SL4-136-3474), Binghamton, NY (SL3-46-234), Birmingham, AL (SL4-70-274), Boise, ID (SL3-22-099), Boston-Lowell-Brockton, MA (SL3-46-305), Bridgeport, CT (SL3-88-275), Buffalo-Niagara Falls, NY (SL3-128-3996), Burlington, VT (SL3-40-038);

Canton, OH (SL3-22-188), Cedar Rapids, IA (SL3-46-264), Champaign-Urbana, IL (SL3-81-336), Charleston, WV (SL3-28-002), Charlotte, NC (SL3-88-353), Chattanooga, TN (SL4-52-065), Chicago-Gary, IL, IN (SL3-88-222), Cincinnati-Hamilton, OH (SL3-40-308), Cleveland, OH (SL3-83-155), Colorado Springs, CO (SL3-83-039), Columbia, SC (SL3-88-147), Columbus, GA (SL4-20-277), Columbus, OH (SL3-40-310);

Dallas-Fort Worth, TX (SL3-22-116, SL3-40-238), Danbury, CT (SL3-88-275), Davenport-Rock Island, IL, IA (SL2-10-251), Dayton-Springfield, OH (SL3-28-018), Daytona Beach, FL (SL4-70-282), Dayton-Springfield, OH (SL3-28-018), Daytona Beach, FL (SL4-70-282), Decatur, IL (SL2-81-334), Denver CO (SL3-22-106), Des Moines, IA (SL2-10-246), Detroit-Ann Arbor-Pontiac, MI (SL3-88-151), Duluth-Superior, MN, WI (SL2-5-384), Durham, NC (SL3-88-150);

Elmira-Corning, NY (SL4-58-318), El Paso, TX (SL4-92-024), Erie, PA (SL3-116-1947), Eugene, OR (SL2-5-464), Evansville, IN (SL4-52-061), Fall River, MA (SL3-122-2597), Fitchburg-Leominster, MA (SL3-46-305), Flint, MI (SL3-88-230), Fort Wayne, IN (SL3-28-017), Fresno, CA (SL3-40-193), Galveston, TX (SL2-103-1003), Gastonia, NC (SL3-40-115), Grand Rapids, MI (SL3-88-226), Green Bay, WI (SL3-83-142), Greensboro, NC (SL3-40-116);

Harrisburg, PA (SL3-88-067), Hartford-New Britain, CT (SL3-88-276), Honolulu, HA (SL3-128-3010, Houston, TX (SL3-124-2719), Huntington-Ashland, KY, WV (SL3-28-021), Huntsville, AL (SL3-46-016), Jackson, MI (SL3-22-184), Jackson, MS (SL4-92-295), Jacksonville, FL (SL4-193-7169), Johnstown, PA (SL3-84-160), Kalamazoo, MI (SL2-16-158), Kansas City, MO-KS (SL4-92-056), Kingsport, TN (SL3-88-053), Knoxville, TN (SL3-88-050);

Lafayette, IN (SL4-64-296), Lafayette, LA (SL4-94-181), Lake Charles, LA (SL4-A4-203), Lancaster, PA (SL3-46-030), Lansing, MI (SL3-83-149), Las Vegas, NV (SL3-28-059), Lexington, KY (SL3-40-308), Lincoln, NB (SL2-10-134), Little Rock, AR (SL4-64-386), Los Angeles-Long Beach, CA (SL4-142-4541), Louisville, KY (SL2-81-343), Lubbock, TX (SL4-90-091), Lynchburg, VA (SL3-28-026);

Macon, GA (SL3-88-144), Madison, WI (SL4-76-008), Manchester, NH (SL3-40-305), Mansfield, OH (SL3-121-2423), Melbourne, FL (SL4-70-283), Memphis, TN (SL3-87-045), Miami-Fort Lauderdale, FL (SL4-64-102), Milwaukee, WI (SL4-76-010), Minneapolis, St. Paul, MN (SL3-2-009), Mobile, AL (SL4-92-300), Modesto, CA (SL4-92-339), Monterey, CA (SL2-04-020), Montgomery, AL (SL4-70-275), Muskegon, MI (SL3-83-146), Nashua, NH (SL3-46-305), Nashville, TN (SL4-52-063), New Bedford, MA (SL3-40-131), Newburgh, NY (SL3-88-274), New Haven, CT (SL3-88-276), New London, CT (SL3-88-277), New Orleans, LA (SL3-46-276), Newport News-Hampton, VA (SL3-40-121), New York-Newark-Paterson, NY, NJ (SL3-46-301), Norfolk-Portsmouth, VA (SL3-28-029);

Oceanside, CA (SL4-142-4543), Ogden, UT (SL3-88-014), Oklahoma City, OK (SL4-64-282), Omaha, NB (SL2-81-175), Orlando, FL (SL2-16-280), Pensacola, FL (SL4-92-301), Peoria, IL (SL3-88-220), Petersburg, VA (SL3-28-027), Philadelphia-Camden, PA, NJ (SL3-116-1953), Phoenix, AZ (SL3-86-011), Pittsburgh, PA (SL3-83-159), Pittsfield, MA (SL3-46-303), Portland, ME (SL3-46-307), Portland, OR (SL3-128-3012), Poughkeepsie, NY (SL3-88-274), Providence, RI (SL3-46-304), Provo, UT (SL3-83-301), Pueblo, CO (SL3-83-040);

Racine, WI (SL4-76-010), Raleigh, NC (SL3-88-150), Reading, PA (SL3-88-068), Reno, NV (SL3-84-005), Richmond, VA (SL3-28-027), Roanoke, VA (SL3-28-025), Rochester, NY (SL3-40-036), Rockford, IL (SL4-76-008), Sacramento, CA (SL3-40-139), Saginaw, MI (SL3-88-230), St. Louis, MO (SL2-10-144), St. Petersburg, FL (SL4-64-026), Salem, OR (SL3-83-291), Salt Lake City, UT (SL3-22-322), San Antonio, TX (SL4-203-7773), San Bernadino-Riverside, CA (SL4-142-2542), San Diego, CA (SL3-40-329), San Francisco-Oakland-San Jose, CA (SL4-76-071), Santa Barbara, CA (SL4-142-4538), Santa Cruz, CA (SL2-04-119), Santa Rosa, CA (SL4-92-335);

Sarasota, FL (SL4-64-026), Savannah, GA (SL4-90-050), Scranton-Wilkes Barre, PA (SL3-46-032), Seattle-Tacoma, WA (SL2-5-458), Shreveport, LA (SL4-52-184), South Bend, IN (SL4-76-012), Spartanburg, SC (SL3-40-114), Spokane, WA (SL3-122-2515), Springfield, IL (SL2-10-254), Springfield, MA (SL3-46-304), Steubenville-Weirton, OH, WV (SL3-22-190), Stockton, CA (SL4-92-338), Syracuse, NY (SL3-88-269);

Tampa, FL (SL4-64-026), Terre Haute, IN (SL3-81-338), Toledo, OH (SL3-22-185), Topeka, KS (SL4-92-053), Tucson, AZ (SL2-04-203), Tulsa, OK (SL3-83-206), Tuscaloosa, AL (SL4-70-274), Utica-Rome, NY (SL3-40-037), Ventura, CA (SL4-149-4538), Waco, TX (SL3-83-291), Washington, DC (SL3-83-166), Waterbury, CT (SL3-88-276), Waterloo, IA (SL2-10-248), West Palm Beach, FL (SL4-64-103), Wichita, KS (SL4-A4-342), Wichita Falls, TX (SL4-64-280), Winston-Salem, NC (SL3-40-116), Worcester, MA (SL3-46-304), York, PA (SL3-88-066), Youngstown, OH (SL3-83-157); and

San Juan, PR (SL4-90-058), Ponce, PR (SL4-81-240), and Mayaguez, PR (SL4-81-240).

6. Satellite images of weather systems may be purchased from:

U.S. Department of Commerce
National Oceanic and Atmospheric Administration
National Environmental Satellite Service
Washington, D.C. 20233

National Environmental Satellite Service
Satellite Field Services Station
601 East 12th Street, Room 1724D
Kansas City, Missouri 64106

16 mm time-lapse motion pictures of clouds imaged by satellite over the western hemisphere may be purchased from:

California Institute of Earth, Planetary and
    Life Sciences
12208 Northeast 137th Place
Kirkland, Washington 98033

7. Aerial photographs may be ordered from:

Map Information Office
U.S. Department of the Interior
Geological Survey
Washington, D.C. 20240

U.S. Forest Service
Department of Agriculture
Washington D.C. 20250

Western Distribution Center
U.S. Geological Survey
Federal Center
Denver, Colorado 80225

Tennessee Valley Authority
Maps and Survey Branch
311 Broad Street
Chattanooga, Tennessee 37401

Eastern Distribution Center
U.S. Geological Survey
1200 South Eads Street
Arlington, Virginia 22202

U.S. Coast and Geodetic Survey
Department of Commerce, ESSA
Washington Service Center
Rockville, Maryland 20852

Western Laboratory (for areas west of the
    Mississippi River)
Aerial Photography Division
U.S. Department of Agriculture
Agricultural Stabilization and Conservation
    Service
2505 Parley's Way
Salt Lake City, Utah 84109

Mark Hurd, Inc.
345 Pennsylvania Avenue, S.
Minneapolis, Minnesota 55426

Eastern Laboratory (for areas east of the
    Mississippi River)
Aerial Photography Division
U.S. Department of Agriculture
Agricultural Stabilization and Conservation
    Service
45 South French Broad Avenue
Asheville, North Carolina 28802

Abrams Aerial Surveys
124 North Larch Street
Lansing, Michigan 48823

U.S. Department of Agriculture
Soil Conservation Service
Federal Center Building
East-West Hyway and Belcrest Road
Hyattsville, Maryland 20781

Stratex Instrument Company
Box 27677
Los Angeles, California 90027

8. Professor Gary Whiteford supplied the following sources from which Canadian space imagery may be obtained:

Integrated Satellite Information Services, Ltd.
P.O. Box 1630
Prince Albert, Saskatchewan, Canada S6V 5T2

National Air Photo Library
615 Booth Street
Ottawa, Ontario, Canada K1A OE9

Addresses of Aerial Photograph Libraries in Canada for oblique and vertical aerial photographs are:

Alberta
Director Technical Division
Alberta Lands and Forest
National Resources Bldg.
Room 325
109th Street and 99th Avenue
Edmonton, Alberta T5K 1H4

British Columbia
Director of Surveys and Mapping Branch
Department of Lands, Forests and Water
    Resources
Victoria, B.C. V8V 1X5

Manitoba
Director of Surveys
Department of Mines, Resources and En-
    vironmental Management
1007 Century Street
Winnipeg, Manitoba R3H 0W4

Maritimes
New Brunswick, Nova Scotia, Prince Edward
    Island
Mr. Neale Lefler
Maritime Resource Management Service
Box 310
Amherst, Nova Scotia B4H 325

Newfoundland
Department of Forestry and Agriculture
Building 810
Pleasantville
St. John's, Newfoundland A1A 1P9

Ontario
Photo Library
Administrative Services Branch
Whitney Block, Room 3501
Queen's Park
Toronto, Ontario M5S 2C6

Quebec
Ministere des Terres et Forets
Service de la Cartographie
1995 Quest, Boul. Charest
Quebec, Quebec G1N 4H9

Saskatchewan
Lands and Surveys Branch
1260-8th Avenue
Regina, Saskatchewan S4R 1C9

Appendix B:
Pictures of the Month

Appendix C pertains specifically to the material presented in Chapter 18: Application of Remote Sensing in Weather Analysis.

The first fifty "Pictures of the Month" are presented in *Photographs from Meteorological Satellites,* U.S. Navy Weather Research Facility, Norfolk, Virginia, June 1967. It covers the period from January 1963 to May 1967. After May 1967 the following "Pictures of the Month" are available in the *Monthly Weather Review,* providing excellent illustrations, sound analysis, and a clear state of the art of the application of remote sensing in weather analysis. Through December 1973, *Monthly Weather Review* was published by the U.S. Department of Commerce, Washington, D.C.; since January 1974 it has been published by the American Meteorological Society, 45 Beacon Street, Boston, Massachusetts 02108.

449

Appendix C:
Selected References

Aldrich, R. C. "Detecting Disturbances in a Forest Environment." *Photogrammetric Engineering and Remote Sensing* 41 (1975):39-48.

Aldrich, S. A.; Aldrich, F. T.; and Rudd, R. D. "An Effort to Identify the Canadian Forest-Tundra Ecotone Signature on Weather Satellite Imagery." *Remote Sensing of Environment* 2 (1971):9-20.

Alexander, R. H. "Geographic Data from Space." *Professional Geographer* 16 (1964):4-5.

———. "Central Atlantic Regional Ecological Test Site." In *4th Annual Earth Resources Program Review,* pp. 72-1 to 72-9. Houston: Manned Spacecraft Center, 1972.

American Society of Photogrammetry. *Manual of Photographic Interpretation.* Washington, D.C.: American Society of Photogrammetry, 1960.

———. *Manual of Photogrammetry.* 3d ed. vols. 1 and 2. Menasha, Wis.: George Banta Co., 1966.

———. Photointerpretation Subcommittee 1. "Basic Matter and Energy Relationships Involved in Remote Reconnaissance." *Photogrammetric Engineering* 29 (1963):761-99.

Anderson, J. R.; Hardy, E. E.; and Roach, J. T. *A Land-Use Classification System for Use with Remote Sensor Data.* U.S. Geological Survey Circular 671. Washington, D.C.: U.S. Government Printing Office, 1972.

Ashley, M.D. "Seasonal Vegetation Differences from ERTS Imagery." *Photogrammetric Engineering and Remote Sensing* 41 (1975):713-19.

Avery, T. E. *Interpretation of Aerial Photographs.* 3d ed. Minneapolis: Burgess Publishing Co., 1977.

Barrett, E. C. *Climatology from Satellites.* London: Methuen and Co., 1974.

Barrett, E. C., and Curtis, L. F. eds. *Environmental Remote Sensing: Applications and Achievements.* London: Edward Arnold, 1974.

Bastuscheck, C. P. "Ground Temperature and Thermal Infrared." *Photogrammetric Engineering* 36 (1970):1064-72.

Bauer, M. E. *Crop Identification and Area Estimation Over Large Geographic Areas Using Landsat MSS Data.* West Lafayette, Ind.: Laboratory for Applications of Remote Sensing, Purdue University, 1977.

Benson, A. S., et al., "Ground Data Collection and Use." *Photogrammetric Engineering* 37 (1971);1159-66.

Bernstein, R. "Digital Image Processing of Earth Observation Sensor Data." *IBM Journal of Research and Development* 20 (1976):40-57.

Bernstein, R., and Stierhoff, G. "Precision Processing of Earth Data." *American Scientist* 64 (1976):500-8.

Bishop, B. C. "Landsat Looks at Hometown Earth." *National Geographic Magazine* 150 (July 1976):140-47.

Bodechtel, J., and Gierloff-Emden, H. G. *The Earth from Space.* New York: Arco Publishing Co., 1974.

Bradford, W. R.; Lake, D. B.; and Plaster, M. "Microwave Radar." In *Handbook of Remote Sensing Techniques,* edited by W. R. Bradford. Orpington, Kent, England: Technology Reports Centre, 1973, pp. 173-216.

Brooner, W. G., and Simonett, D. S. "Crop Discrimination with Color Infrared Photography: A Study in Douglas County, Kansas." *Remote Sensing of Environment* 2 (1971):21-35.

Bryan, M. L. "Interpretation of an Urban Scene Using Multi-Channel Radar Imagery." *Remote Sensing of Environment* 4 (1975):49-66.

Canada Centre for Remote Sensing. *Earth Resources Technology Satellite Data User's Handbook.* Ottawa, Ont.: Department of Energy, Mines and Resources, 1973.

Carneggie, D. M.; Pettinger, L. R.; Hay, C. M.; and Daus, S. J.; Colwell, R. N.; et al. "Analysis of Earth Resources in the Phoenix, Arizona Area." In *Analysis of Earth Resources on Apollo 9 Photography.* Berkeley: University of California, 1969.

Chevrier, E. D., and Aitkens, D. F. W. *Topographic Map and Air Photo Interpretation.* Toronto: Macmillan Co. of Canada, 1970.

Chisnell, T., and Cole, G. " 'Industrial Components'—A Photo Interpretation Key on Industry." *Photogrammetric Engineering* 24 (1968):590-602.

Ciesla, W. F. "Insect Damage from High Altitude Color-IR Photos." *Photogrammetric Engineering* 40 (1974):683-90.

Cihlar, J., and Ulaby, F. T. *Microwave Remote Sensing of Soil Water Content.* Technical Report 264-6. Lawrence: University of Kansas Center for Research, 1975.

Coiner, J. C. *SLAR Image Interpretation Keys for Geographic Analysis.* Technical Report 177-19. Lawrence: University of Kansas Center for Research, 1972.

Coiner, J. C., and Morain, S. A. "Image Interpretation Keys to Support Analysis of SLAR Imagery," *Proceedings of the American Society of Photogrammetry Fall 1971 Convention.* Falls Church, Va.: American Society of Photogrammetry, 1971, 393-412.

Colwell, R. N. "Some Uses of Infrared Aerial Photography in the Management of Wildlife Areas." *Photogrammetric Engineering* 26 (1960):774-85.

———. "Spectrometric Considerations Involved in Making Rural Land Use Studies with Aerial Photography." *Photogrammetria* 20 (1965):15-30.

———. "Uses and Limitations of Multispectral Remote Sensing.' In *Proceedings of the Fourth Symposium on Remote Sensing of the Environment,* pp. 78-100. Ann Arbor: University of Michigan, Institute of Science and Technology, 1966. Also reprinted in the Bobbs-Merrill Reprint Series in Geography, no. G-39. Indianapolis: Bobbs Merrill Co.

———. *Manual of Multiband Photography.* NASA Grant No. 05-003-080. Washington, D.C.: National Aeronautics and Space Administration, 1968.

———. "Remote Sensing as an Aid to the Management of Earth Resources." *American Scientist* 60, no. 2 (1973):175-83.

Conway, D., and Holz, R. K. "The Use of Near-Infrared Photography in the Analysis of Surface Morphology of an Argentine Alluvial Floodplain." *Remote Sensing of Environment* 2 (1973):235-42.

Craib, K. B. "Synthetic Aperture SLAR Systems and Their Application for Regional Resources Analysis." In *Remote Sensing of Earth Resources,* vol. 1, pp. 152-78. Edited by F. Shahrokhi. Tullahoma, Tenn.: University of Tennessee Space Institute, 1972.

Davis, S. M. *Skylab: Earth Resources Experiment.* Purdue Research Foundation, 1976. One of the "Fundamentals of Remote Sensing" Minicourse Series produced by LARS/Purdue, it is available through the Continuing Education Administration, 116 Stewart Center, Purdue University, West Lafayette, Indiana 47907.

Denny, C. S., et al. *A Descriptive Catalog of Selected Aerial Photographs of Geologic Features in the United States.* U.S. Geological Survey Professional Paper no. 590. Washington, D.C.:U.S. Government Printing Office, 1968.

Duda, R., and Pecrot, A. J. *The Applications of ERTS Imagery to the FAO/Unesco Soil Map of the World.* Rome, Italy: Food and Agriculture Organization of the United Nations, 1977.

Eastman Kodak Company. *Applied Infrared Photography.* Kodak Publication no. M-28. Rochester, N.Y., 1968.

Edelson, E. "Gaudy New Way to See Things You Cannot See." *Smithsonian* 4, no. 5 (1973):22-29.

Estes, J. E. "Some Geographic Applications of Aerial Infrared Imagery." *Annals of the Association of American Geographers* 56 (1966):673-82.

Estes, J. E., and Senger, L. W. "The Multispectral Concept as Applied to Marine Oil Spills." *Remote Sensing of Environment* 2 (1972):141-63.

Estes, J. E., and Senger, L. W., eds. *Remote Sensing Techniques for Environmental Analysis.* Santa Barbara, Calif.: Hamilton Publishing Co., 1974.

Eyre, L. A. "High Altitude Color Photos." *Photogrammetric Engineering* 37 (1971):1149-53.

Eyton, J. R., and Kuether, R. P. *Remote Sensing Photoguide.* Urbana: Department of Geography, University of Illinois, 1975.

Garnier, B. J. "The Observation of Topographic Variations in Surface Radiative Temperatures by Remote Sensing of Environment." In *Proceedings of the Seventh Symposium of Remote Sensing of Environment,* pp. 495-99. Ann Arbor, Mich.: Institute of Science and Technology, 1971.

Garofalo, D., and Wobber, T. "Experimental Application of Remote Sensing to Solid Waste Planning and Management." In *Proceedings of the American Society of Photogrammetry 39th Annual Meeting,* pp. 210-32. 1973.

Gausman, H. W. "Leaf Reflectance of Near Infrared." *Photogrammetric Engineering* 40 (1974):183-91.

Goldman, C. R., et al. "Limnological Studies and Remote Sensing of the Upper Truckee River Sediment Plume in Lake Tahoe, California-Nevada." *Remote Sensing of Environment* 3 (1974):49-67.

Goodyear Aerospace Corporation. *Basic Concepts of Synthetic Aperture Radar.* Litchfield Park, Ariz.: Goodyear Aerospace Corporation, 1971.

Gut, D., and Hohle, J. "High Altitude Photography—Aspects and Results." In *Proceedings of the American Society of Photogrammetry 43rd Annual Meeting,* pp. 422-42. 1977.

Haefner, J. "Airphoto Interpretation of Rural Land Use in Western Europe." *Photogrammetria* 22 (1967):143-52.

Hardy, E., et al. "Enhancement and Evaluation of Skylab Photography for Potential Land Use Inventories." Report under NASA Contract No. NAS9-13364 (1975), Department of Natural Resources, NYS College of Agriculture, Cornell University, Ithaca, N.Y. 14853.

Hardy, N. E. *Interpretation of Side Looking Airborne Radar Vegetation Patterns: Yellowstone Park.* Technical Report 177-24. Lawrence: University of Kansas Center for Research, 1972.

Harper, D. *Eye in the Sky, Introduction to Remote Sensing.* Montreal, Quebec: Multiscience Publications, 1976.

Hay, C. M. "Agricultural Inventory Techniques with Orbital and High-Altitude Imagery." *Photogrammetric Engineering* 40 (1974):1283-93.

Henderson, F. M. "Radar for Small-Scale Land-Use Mapping." *Photogrammetric Engineering and Remote Sensing* 41 (1975):307-19.

Higgs, G. K., and Sullivan, M. C. *Image Interpretation for a Multi-level Land Use Classification System.* East Lansing: Michigan State University, Remote Sensing Project, 1973.

Holz, R. K. *The Surveillant Science: Remote Sensing of the Environment.* Boston: Houghton Mifflin Co., 1973.

Hooper, J. O., and Seybold, J. B. "Digital Processing of Microwave Radiometric Images." In *Remote Sensing of Earth Resources,* vol. 3, pp. 173-92. Edited by F. Shahrokhi. Tullahoma, Tenn.: University of Tennessee Space Institute, 1974.

Howard, W. A., and Kracht, J. B. "An Assessment of the Usefulness of Small-Scale Photographic Imagery for Acquiring Land Use Information Necessary to the Urban Planning Function." Technical Paper no. 71-2. Denver, Colo.: University of Denver, Department of Geography, 1971.

Hudson, W. D.; Armstrong, R. J.; and Meyer, W. L. *Identifying and Mapping Forest Resources from Small-Scale Color-Infrared Airphotos.* Research Report 304. East Lansing: Michigan State University, Agricultural Experiment Station, 1976.

Itek Corporation. *Skylab Multispectral Photography.* A colorful brochure available from Itek Optical Systems, 10 Maguire Road, Lexington, Mass. 02173.

Johnson, G. E. "Cover Type Identification Capabilities of Remote Multispectral Sensing Techniques." In *International Geography 1972,* pp. 973-75. Published for the 22nd International Geographical Congress. Toronto: University of Toronto Press, 1972.

Johnson, P. L., ed. *Remote Sensing in Ecology.* Athens, Ga.: University of Georgia Press, 1969.

Kanemasu, E. T. "Seasonal Canopy Reflectance Patterns of Wheat, Sorghum, and Soybean." *Remote Sensing of Environment* 3 (1974):43-47.

Klemas, et al. "Inventory of Delaware's Wetlands." *Photogrammetric Engineering* 40 (1974):433-40.

Knipling, E. B. "Leaf Reflectance and Image Formation of Color Infrared Film." In *Remote Sensing of Ecology,* pp. 17-24. Athens: University of Georgia Press, 1969.

Kohn, C. F. "The Use of Aerial Photographs in the Analysis of Rural Settlements." *Photogrammetric Engineering* 17 (1951):759-71.

Kroeck, D. *Everyone's Space Handbook.* Arcata, Calif.: Pilot Rock, 1976.

Lattman, H., and Roy, R. G. *Aerial Photographs in Field Geography.* New York: Holt, Rinehart, and Winston, 1965.

Leamer, R. W.; Weber, D. A.; and Wiegand, C. L. "Pattern Recognition of Soils and Crops from Space." *Photogrammetric Engineering and Remote Sensing* 41 (1975):471-78.

Leartek Corporation. *Resources Management Surveys.* Boulder, Colo.: Itek Corporation, Publications Department, 1975.

Lewis, A. J., *Geomorphic Evaluation of Radar Imagery of Southeastern Panama and Northwestern Colombia.* Technical Report 133-18. Lawrence: University of Kansas Center for Research, 1971.

Lewis, A. J., ed. "Geoscience Applications of Imagery Radar Systems." Special issue: *Remote Sensing of the Electromagnetic Spectrum (RSEMS)* 3 (July 1976).

Lewis, A. J.; MacDonald, H. C.; and Simonett, D. S. "Detection of Linear Cultural Features with Multipolarized Radar Imagery." In *Proceedings of the Sixth International Symposium on Remote Sensing of Environment,* pp. 879-93. Ann Arbor: University of Michigan, 1969.

Lillesand, T. M. *Fundamentals of Electromagnetic Remote Sensing.* Syracuse, N.Y.: State University of New York, College of Science and Forestry, 1976.

Limperis, T. "Target and Background Signature Study." In *Proceedings of the Third Symposium on Remote Sensing of Environment,* pp. 423-33. Ann Arbor: University of Michigan, Institute of Science and Technology, 1964.

Lindgren, D. T. "Dwelling Unit Estimation with Color-IR Photos." *Photogrammetric Engineering* 37 (1971):373.

Lintz, J., Jr., and Simonett, D. S. *Remote Sensing of Environment.* Reading, Mass.: Addison-Wesley, 1977.

Lougeay, R. "Patterns of Surface Temperature in the Alpine/Periglacial Environment as Determined by Radiometric Measurements." In *Icefield Ranges Research Project Scientific Results,* vol. 3, pp. 163-76. New York: American Geographical Society and Arctic Institute of North America, 1972.

———. "Thermal Contrasts Between Ice-cored Detrital Surfaces." In *International Geography 1972,* pp.

159-61. Published for the 22nd International Geographical Congress. Toronto: University of Toronto Press, 1972.

Lowman, P. D., Jr. *Space Panorama.* Zurich, Switzerland: Weltflugbild Reinhold A. Muller, 1968.

———. *Apollo 9 Multispectral Photography: Geologic Analysis.* Publication X-644-69-423. Greenbelt, Md.: NASA Goddard Space Flight Center, 1969.

Maul, G. A., and Gordon, H. R. "On the Use of the Earth Resources Technology Satellite (Landsat-1) in Optical Oceanography." *Remote Sensing of Environment* 4 (1975):95-128.

Maul, G. A., and Hansen, D. V. "An Observation of the Gulf Stream Surface Front Structure by Ship, Aircraft, and Satellite." *Remote Sensing of Environment* 2 (1972):109-16.

McCoy, R. M., and Metivier, E. D. "House Density vs Socioeconomic Conditions." *Photogrammetric Engineering* 39 (1973):43.

Mercanti, E. P. "ERTS-1 Teaching Us a New Way to See." *Astronautics and Aeronautics* (September 1973):33-63.

———. "Widening ERTS Applications." *Astronautics and Aeronautics* (May 1974):28-39.

Minshull, R. M. *Human Geography from the Air.* London: Macmillan & Co., 1968.

———. *Landforms from the Air.* London: Macmillan & Co., St. Martin's Press, 1969.

Monkhouse, F. J. *Landscape from the Air.* London: Cambridge University Press, 1965.

Morain, S. A., and Campbell, J. B. "Radar Theory Applied to Reconnaissance Soil Surveys." *Soil Science Society of America Proceedings* 38 (1974):818-26.

Morain, S. A., and Simonett, D. S. "K-Band Radar in Vegetation Mapping." *Photogrammetric Engineering and Remote Sensing* 42 (1976):730-40.

Munn, L. C.; McClellan, J. B.; and Philpotts, L. E. "Air Photo Interpretation and Rural Land Use Mapping in Canada." *Photogrammetria* 21 (1966):65-77.

Nalepka, R. F.; Colwell, J.; and Rice, D. P. *Wheat Productivity Estimates Using Landsat Data.* Ann Arbor, Mich.: Environmental Research Institute of Michigan, November 1976.

National Aeronautics and Space Administration. *Data Users Handbook.* Document No. 71S4249. Greenbelt, Md.: Goddard Space Flight Center, 1972.

———. *Ecological Surveys from Space.* NASA SP-230. Washington, D.C., 1970.

———. *Landsat Data Users Handbook.* Document No. 76SD4258. Greenbelt, Md.: Goddard Space Flight Center, 1976.

———. *Observing Earth from Skylab.* NF-56/1-75. Washington, D.C.: U.S. Government Printing Office (Stock no. 033-000-00627-6). A special issue of *NASA Facts* which provides both a very basic introduction to remote sensing and a well-written and colorfully illustrated review of EREP Programs.

———. *Photography from Space to Help Solve Problems on Earth.* Washington, D.C.: U.S. Government Printing Office, 1974.

———. *Proceedings of the NASA Earth Resources Survey Symposium,* vol. 1-A *Agriculture-Environment.* NASA TM X-58168, JSC-09930. Houston: Johnson Space Center, 1975.

———. *Skylab Earth Resources Data Catalog.* JSC 09016. Washington, D.C.: U.S. Government Printing Office (Stock no. 3300-00586). This should be included as a basic reference in all remote sensing libraries.

———. *Space Shuttle.* Washington, D.C.: U.S. Government Printing Office (Stock no. 671-202/2447), 1975. A Johnson Space Center publication illustrating the role of this vehicle in servicing future Skylab-type space stations.

National Research Council. *Remote Sensing with Special Reference to Agriculture and Forestry.* Washington, D.C.: National Academy of Sciences, 1970.

Nunnally, N. R. *Introduction to Remote Sensing: The Physics of Electromagnetic Radiation.* Washington, D.C.: Association of American Geographers, U.S. Geological Survey Contract no. 14-08-001-10921, 1969.

Olson, C. E., Jr. "Accuracy of Land Use Interpretation from Infrared Imagery in the 4.5 to 5.5 micron band." *Annals of the Association of American Geographers* 57 (1967):382-88.

Ordway, F. I. *Pictorial Guide to Planet Earth.* New York: Crowell Co., 1975.

Peplies, R. W. "Land Use and Regional Analysis." *Earth Resources Aircraft Program Status Review,* vol. 1. Houston: Manned Spacecraft Center, 1968.

Philpotts, L. E., and Wallen, V. R. "Infrared Color for Crop Identification." *Photogrammetric Engineering* 35 (1969):1116-25.

Piech, K., and Walker, J. "Outfall Inventory Using Air Photo Interpretation." *Photogrammetric Engineering* 38 (1972):907-14.

Rehder, J. B. "The Uses of ERTS-1 Imagery in the Analysis of Landscape Change." In *Remote Sensing of Earth Resources,* vol. 3, pp. 573-86. Edited by F. Shahrokhi. Tullahoma, Tenn.: University of Tennessee Space Institute, 1974.

———. "Landsat Imagery: Pictures of Your Place from Space." *Journal of Geography* 75 (1976):354-59.

Reeves, R. C., ed.-in-chief. *Manual of Remote Sensing,* vols. 1 and 2. Falls Church, Va.: American Society of Photogrammetry, 1975.

Richason, B. F., Jr. *Atlas of Cultural Features.* Northbrook, Ill.: Hubbard Press, 1972.

Richason, B. F. III, and Enslin, W. *Upper Kalamazoo Watershed Land Cover Inventory.* East Lansing: Michigan State University, Remote Sensing Project, 1973.

Rudd, R. D. *Remote Sensing: A Better View.* North Scituate, Mass.: Duxbury Press, 1974.

Schanda, E. "Passive Microwave Sensing." In *Remote Sensing for Environmental Sciences,* pp. 187-256. Ecological Studies 18. Edited by E. Schanda. New York: Springer-Verlag, 1976.

Scherz, J. P., and Stevens, A. R. *An Introduction to Remote Sensing for Environmental Monitoring.* Madison: University of Wisconsin, Institute for Environmental Studies, 1970.

Schut, G. H., and Van Wijk, M. C. "The Determination of Tree Heights from Parallax Measurements." *Canadian Surveyor* 19 (1965):415-27.

Schwarz, D. E., and Caspall, F. "The Use of Radar in the Discrimination and Identification of Agricultural Land Use." In *Proceedings of the Fifth Symposium on Remote Sensing of Environment,* pp. 233-47. Ann Arbor: University of Michigan, 1968.

Short, N. M., et al. *Mission to Earth: Landsat Views of the World.* Washington, D.C.: NASA, Scientific and Technical Information Office, 1976.

Simonett, D. S. "Land Evaluation Studies with Remote Sensors in the Infrared and Radar Regions." In *Land Evaluation,* edited by G. A. Stewart. Sydney: Macmillan Co. of Australia, 1968.

Smith, T. J., ed. *Manual of Color Aerial Photography.* Falls Church, Va.: American Society of Photogrammetry, 1968.

Society of Photo-Optical Instrumentation Engineers. *Scanners and Imagery Systems for Earth Observation. Proceedings of the Society of Photo-Optical Instrumentation Engineers* 51 (1974). San Diego, Calif.

Specht, M. R.; Needler, D.; and Fritz, N. L. "New Color Film for Water Penetration Photography." *Photogrammetric Engineering* 39 (1973):359-69.

Spurr, S. H. *Photogrammetry and Photo-Interpretation.* 2d ed. New York: Ronald Press Co., 1960.

Steiner, D. "Use of Air Photographs for Interpreting and Mapping Rural Land Use in the United States." *Photogrammetria* 20 (1965):65-80.

Stephens, A. "Application of High-Altitude Color Infrared Photography to Land Use Mapping in the Tennessee River Watershed. In *Proceedings of the American Society of Photogrammetry 39th Annual Meeting,* pp. 201-9. 1973.

Stephens, P. "Comparison of Color, Color Infrared, and Panchromatic Photography." *Photogrammetric Engineering and Remote Sensing* 42 (1976):1273-78.

Stone, K. H. "A Geographer's Strength: The Multiple-Scale Approach." *Journal of Geography* 81 (1972):354-62.

Stranberg, C. H. *Aerial Discovery Manual.* New York: John Wiley & Sons, 1967.

Suits, G. H., and Safir, G. R. "Verification of a Reflectance Model for Mature Corn with Applications to Corn Blight Detection." *Remote Sensing of Environment* 2 (1972):183-92.

Sullivan, M. C., and Schar, S. W. *Users Guide to High Altitude Imagery of Michigan.* Project for the Use of Remote Sensing in Land Use Policy Formulation. East Lansing: Michigan State University, 1973.

Sully, G. B. *Aerial Photo Interpretation.* Scarborough, Ont.: Bellhaven House, 1969.

Tarkington, R. G., and Soren, A. L. "Color and False-Color Films for Aerial Photography." *Photogrammetric Engineering* 29 (1963):88-95.

Thrower, N. J. W. "Land Use in the Southwestern United States from Gemini and Apollo Imagery." *Annals of the Association of American Geographers* 60 (1970):Map Supplement no. 12.

U.S. Geological Survey. *Availability of Earth Resources Data.* USGS INF-74-30. Washington, D.C.: U.S. Government Printing Office (Stock no. 1976-211-345/44). An exceptionally useful booklet for locating state and regional repositories of Skylab photography, as well as of other remote sensing materials, laboratories, and general assistance.

———. *The EROS Data Center.* 211-345/15. Washington, D.C.: U.S. Government Printing Office, 1975.

———. EROS Program. *Studying Earth from Space.* Washington, D.C.: U.S. Government Printing Office (2401-2060), 1972.

Waite, W. P., and MacDonald, H. C. "Snowfield mapping with k-band radar." *Remote Sensing of Environment* 1 (1970):143-50.

Weaver, K. F. "Remote Sensing: New Eyes to See the World." *National Geographic Magazine* 135 (1969):47-73.

Wenderoth, S., and Yost, E., with Kalia, R., and Anderson, R. *Multispectral Photography for Earth Resources.* 2d printing. Greenvale, N.Y.: Remote Sensing Information Center, 1974.

Wiegand, C. L.; Torline, R. J.; and Gautreaux, M. R. "Landsat Agricultural Land-Use Survey." *Photogrammetric Engineering and Remote Sensing* 43 (1977):207-16.

Williams, R. S., Jr., and Carter, W. D., eds. *ERTS-1: A New Window on Our Planet.* U.S. Geological Survey Professional Paper 929. Washington, D.C.: U.S. Government Printing Office, 1976.

Wobber, F., et al. "Coal Refuse Site Inventories." *Photogrammetric Engineering and Remote Sensing* 41 (1975):1163-71.

Wong, K. W.; Thornburn, T. H.; and Khoury, M. A. "Automatic Soil Identification from Remote Sensing Data." *Photogrammetric Engineering and Remote Sensing* 43 (1977):73-80.

Wray, J. R. "A Remote Sensing System for Detecting Gross Land Use Change in Metropolitan Areas." In *International Geography 1972,* pp. 986-87. Published for the 22nd International Geographical Congress. Toronto: University of Toronto Press, 1972.

Appendix D

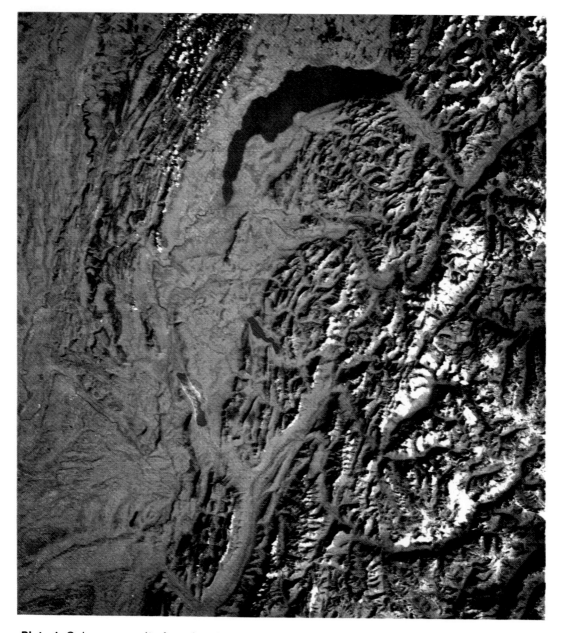

Plate 1. Color composite from Landsat imagery. The color photographlike print was produced by combining bands 4, 5, and 7 from the Landsat Multispectral Scanner system. This scene includes an area of eastern France and western Switzerland. Lake Geneva appears near the top of the image. The scene was imaged on October 9, 1972. The center of the image is 45°58′ north latitude; 6°11′ east longitude. At the time the image was made the elevation of the sun was 34°, and the azimuth of the sun along the horizon was 153°. The time of the imagery was 9:55 A.M. GMT. Vegetation appears as red in the scene. (Image provided by EROS Data Center)

Plate 2. Low oblique of downtown Fredericton, New Brunswick. (Photograph courtesy of Maritime Resource Management Service, Council of Maritime Premiers, Amherst, Nova Scotia)

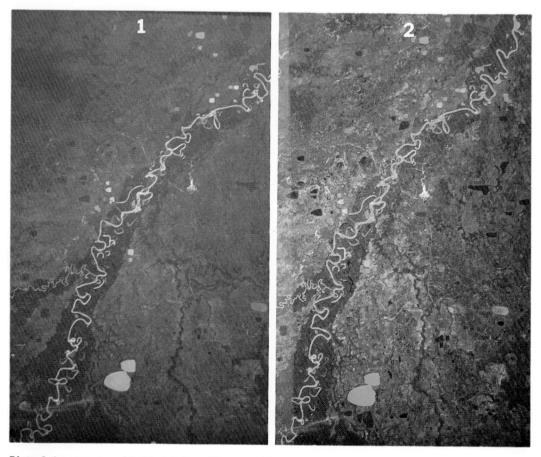

Plate 3. Images 1 and 2: Skylab 3 multispectral imagery of a portion of South America. (Source: NASA)

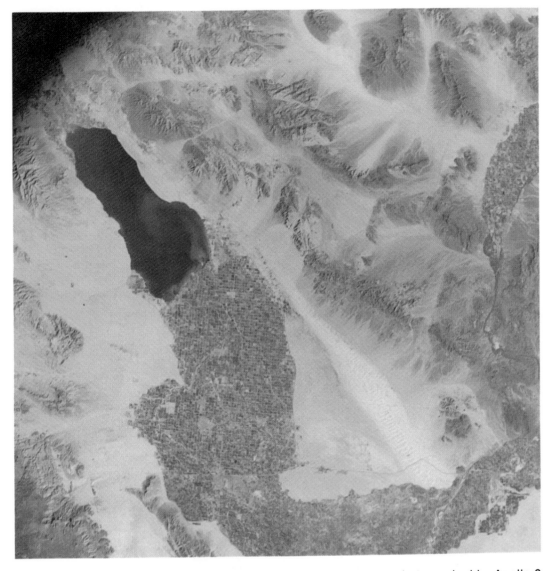

Plate 4. The Imperial Valley and the Salton Sea in California were photographed by Apollo 9 astronauts on their March 8-12, 1969 mission. The photograph was made on color infrared film in a Hasselblad camera. The wavelength sensitivity of the film, which was exposed through a Wratten No. 15 filter, was 510-900 nanometers. The photograph was made at an altitude of 208 kilometers (129 miles). Refer to an atlas and locate the following features on the photograph: Salton Sea; Mexicali, Mexico; Yuma, Arizona; All-American Canal; Coachella Canal; Chocolate Mountains; Superstition Mountain; Brawley; and Calipatria. (NASA photograph supplied by Technology Application Center)

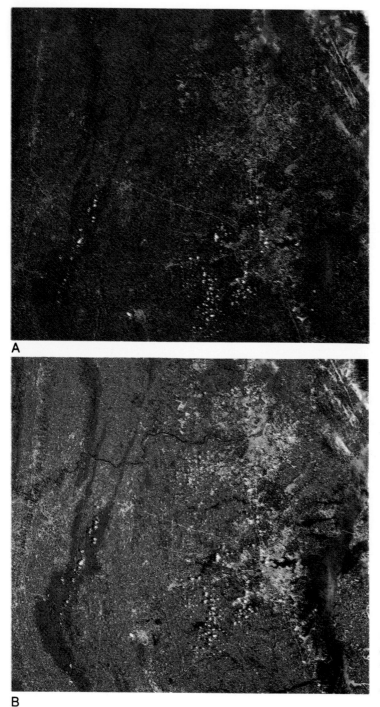

A

B

Plate 5. Washington, D.C. and Baltimore, Maryland, area imaged from Skylab by 190-A multispectral camera. (Source: NASA)

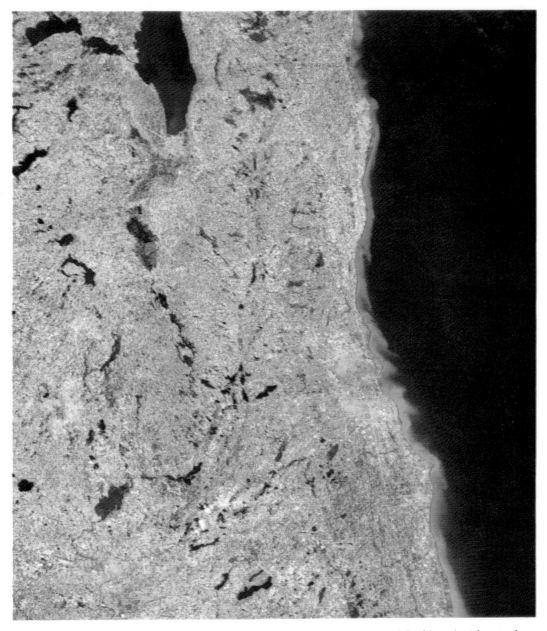

Plate 6. This color composite was prepared from MSS bands 4, 5, and 7 of Landsat imaged on April 29, 1976. The format center is 43°16′ N, 88°15′ W. The elevation of the sun was 52°, and the azimuth of the sun along the horizon was 129°. (EROS Data Center)

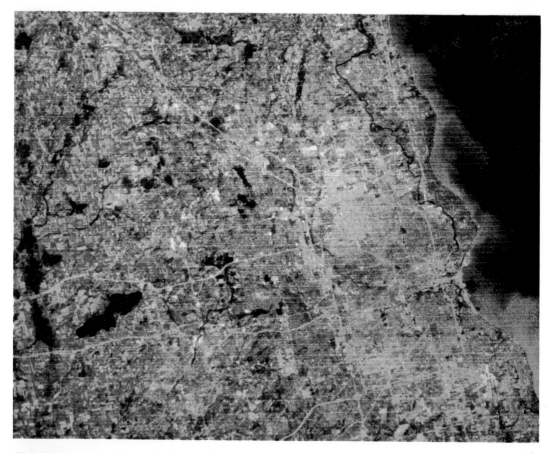

Plate 7. A small portion of a Landsat color composite can be enhanced and enlarged to provide the detail shown in this scene. Milwaukee is the large urban area on the right. Details of the Port of Milwaukee can be seen. Mitchell Field airport appears distinctly in the lower right-hand corner. This enlarged enhancement was prepared from the image on plate 6. (Enhancement and print by B. F. Richason, Jr.)

Plate 8. Cibachrome print of Diazo enhanced Landsat image of southeastern Wisconsin. Magenta colored areas are cities. Milwaukee occupies the eastern part of the image, and Waukesha can be located in the lower left quadrant. Green areas are forests or fields of crops. White and tan areas are bare fields or fields from which crops have been removed. Light blue areas indicate wet soils. Dark blue areas and brown areas are industry or commercial land. Some apartment areas show up in dark blue. (Enhancement and print by B. F. Richason, Jr.)

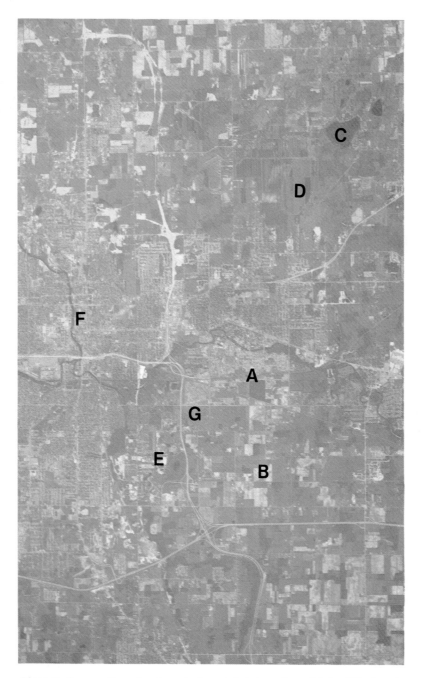

Plate 9. Conventional color photograph taken from high altitude of Lansing, Michigan, on June 10, 1972. The original scale was 1:120,000. See chapter 10 for explanation and identification of features. (NASA photograph supplied by EROS Data Center)

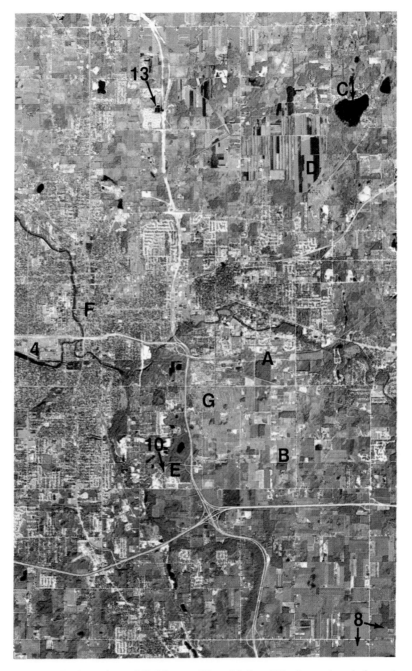

Plate 10. Lansing, Michigan. The high altitude color infrared photograph was made on September 15, 1972, at an original scale of 1:120,000. See chapter 10 for explanation and identification of features. (NASA photograph supplied by EROS Data Center)

Plate 11. High altitude color infrared photograph at an original scale of 1:120,000. The scene is near Lansing, Michigan, imaged by an RC-8 camera on June 10, 1972. (NASA photograph supplied by EROS Data Center)

Plate 12. High altitude color infrared photograph at an original scale of 1:60,000. The scene is near Lansing, Michigan, imaged by a Zeiss camera on June 10, 1972. (NASA photograph supplied by EROS Data Center)

Plate 13. East Lansing, Michigan. Original scale 1:60,000. September 15, 1972. Refer to chapter 10, figure 10.3 for explanation of features. (NASA photograph supplied by EROS Data Center)

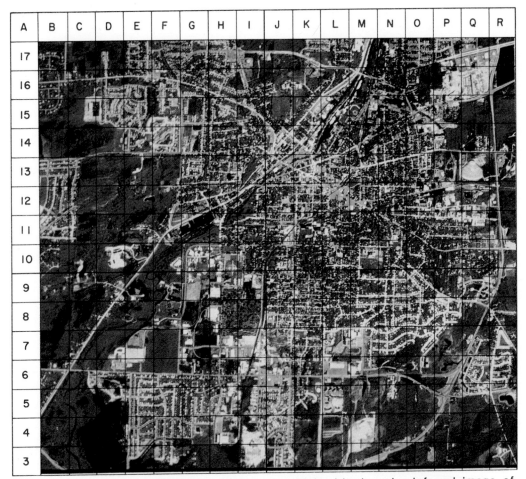

Plate 14. Enlarged and enhanced portion of a high altitude color infrared image of Waukesha, Wisconsin, made on July 31, 1974. The following features are identified:

D-15, D-16, E-15, and E-16	New subdivision (others can also be identified)
D-10	Subdivision under construction
E-9 and F-9	Oxbows along Fox River
D-6, D-7, F-6, and F-7	Large food chain warehouse
G-11, M-15, and N-15	Metal fabricating industries
Left part of M-15	Park and baseball diamonds
J-13, K-13, L-13, and L-12	Central business district
K-11 and L-11	Carroll College
M-11	Catholic high school
M-7	Public high school
Left part of L-7	Elementary school
L-6	Shopping center
H-8 and L-8	Cemetery
G and H, 6 to 10	Industrial park
Q-13	Drive-in theater
H-9 and H-10	Recreational area (four baseball diamonds)

(Original photograph: EROS Data Center. Enlargement and enhancement by B. F. Richason, Jr.)

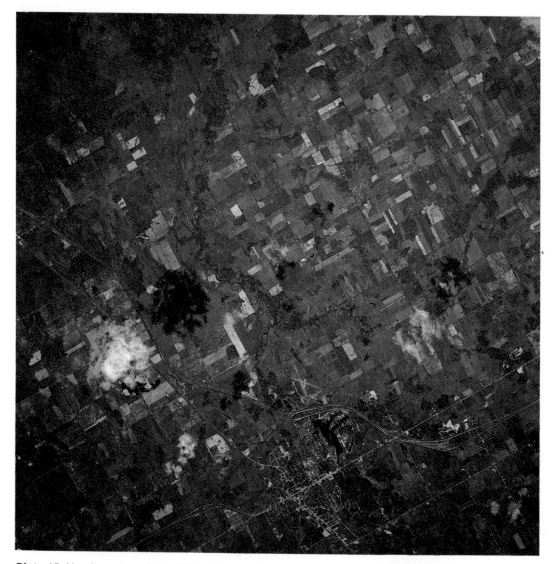

Plate 15. Northwestern Ohio. High altitude CIR photograph, mission 103, site 167, frame 161, at an original scale of 1:60,000. Photographed on September 10, 1969. (NASA photograph supplied by EROS Data Center)

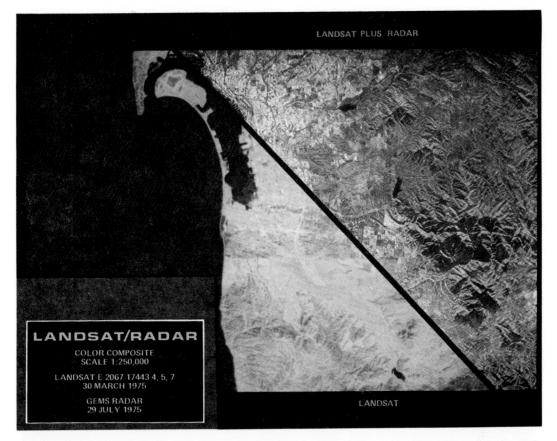

Plate 16. Recent research conducted by Goodyear Aerospace in cooperation with the U.S. Geological Survey EROS Data Center demonstrates the manner in which radar data can complement and enhance information collected via other sensors. Landsat multispectral scanner imagery of the San Diego, California, area is shown before and after combination with radar imagery. Note that the radar imagery provides unique emphasis of terrain relief and certain cultural features while the Landsat data provides unique data on vegetation type and condition. (Courtesy of Goodyear Aerospace and Aero Service)

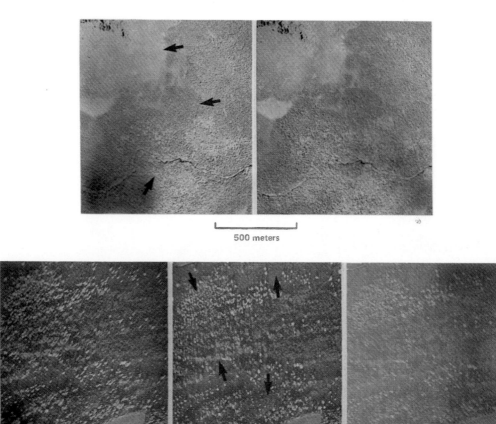

500 meters

300 meters

Plate 17.
Top. Color infrared pair over northeastern New Brunswick, June 13, showing fresh, spring foliage with maximum color variation. Vegetation includes deciduous trees bordering stream; conifers, center; and stringbog muskeg at lower right. (Canada Centre for Remote Sensing)

Bottom. Color and black and white stereo triplet printed from color negatives, St. John River floodplain, New Brunswick, November 24, showing remarkable vegetation color for such a late date. Bright green trees are spruce and white pine, yellowish trees in bog are tamarack, and bronze trees are deciduous species. Pasture is still green but bog has turned brown. (From research grant to W. H. Hilborn)

Plate 18. The interface of urban, suburban, and rural landscapes is apparent from the remotely sensed perspective. Observe the destruction between new housing areas and forest land in the left-hand portion of the photograph. Also, the boundary between the city and rural areas is evident at the top of this medium, altitude aerial photograph. (Remote Sensing Laboratory, Mississippi State University, NASA grant NGL-25-001-054, Application of Remote Sensing to State and Regional Problems)

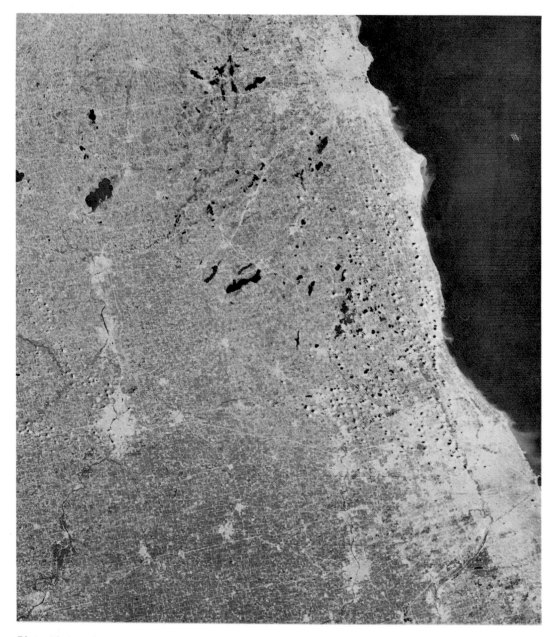

Plate 19. Landsat color composite image of northeastern Illinois and southeastern Wisconsin. The scene was imaged on August 9, 1972. The tan to white areas are cities. Straight white to tan lines are highways. Vegetation appears red. (EROS Data Center)

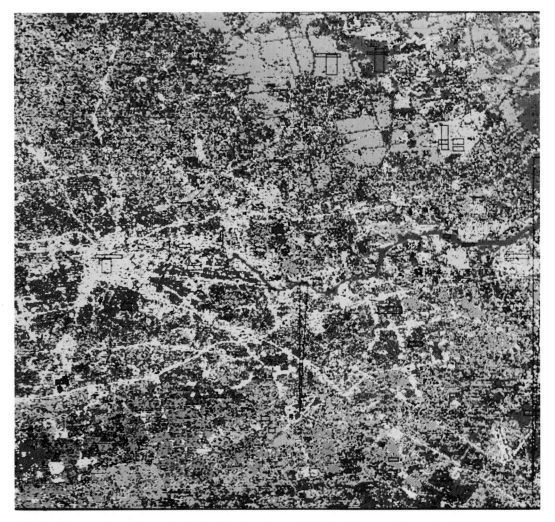

Plate 20. Supervised Computer Classification Map. (NASA, Johnson Space Center)

Plate 21. Color coded land use classification of Milwaukee County, Wisconsin, derived from machine processing of Landsat multispectral data (August 9, 1972). The classification results were displayed on a cathode ray tube digital display which was photographed. Commerce/industry appears as lavender; older housing as red; new housing as yellow; older upper-income areas are tan; grassy areas are light green; wooded areas are dark green; water is blue; clouds are white; and shadow areas are black. (Laboratory for Applications of Remote Sensing/Purdue University)

Plate 22. The circulation pattern of the Powertone Cooling Pond in Illinois is immediately apparent in this color-coded thermal mosaic. In addition, the highly efficient design of the pond is dramatically displayed here in six colors. Each color represents a 3° F thermal range commencing with red and continuing through magenta. The pond thus exhibits a temperature drop of 18° F from the point of discharge (white) to the intake point (black). Digicolor color-coded imagery of Powerton Cooling Pond, the result of a study performed for Sargent & Lundy Engineers and Commonwealth Edison of Illinois by Daedalus Enterprises, Inc. (Courtesy of Daedalus Enterprises)

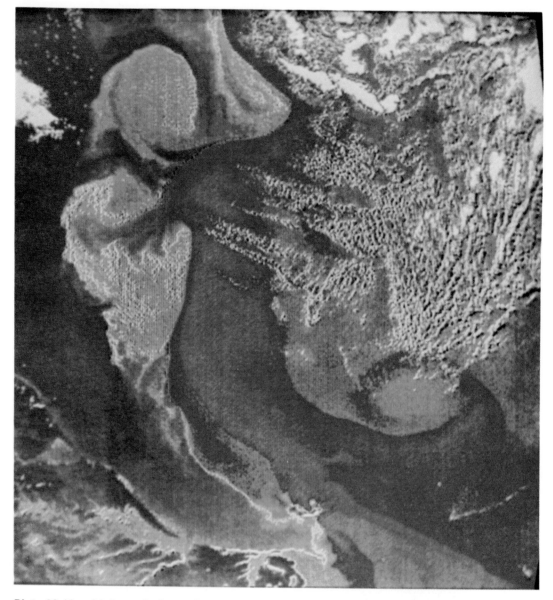

Plate 23. Very high resolution radiometer color enhanced image of the Gulf Stream. The scene was imaged on April 28, 1974. (Used with permission of the American Meteorological Society)

Plate 24. A high altitude color infrared (CIR) aerial photograph of a section of Boise, Idaho. (Courtesy of Roger A. Zanarini of Upland Industries, Inc., Omaha, Neb.)

Notes on Contributors

John E. Estes is Associate Professor, Department of Geography, University of California at Santa Barbara. He received his Ph.D. in geography from the University of California at Los Angeles in 1969. He has been involved in the field of remote sensing since 1963, contributing through federal government, private industry, and university employment. He has published widely in the field of remote sensing. From among his more than 70 publications, Dr. Estes is the author-editor of *Remote Sensing Techniques for Environmental Analysis* and coauthor-editor of "Fundamentals of Image Interpretation" in the *Manual of Remote Sensing.* He has been a principal or coprincipal investigator for contracts with the National Aeronautics and Space Administration, U.S. Coast Guard, U.S. Forest Service, Department of the Navy, and the State of California. He has served as director of remote sensing workshops for the AAG and the International Geographical Union, and is chairman of the AAG Commission on Geographic Applications of Remote Sensing.

L. Alan Eyre is Senior Lecturer in the Department of Geography, University of the West Indies, Kingston, Jamaica. He received the B.A. degree from the University of London, West Indies, and his Ph.D. in geography from the University of Maryland. His special fields of interest and research are population geography, regional geography of Latin America, Central America, and the Caribbean, and thematic cartography. He is the author of several books and numerous articles in geography and remote sensing journals.

James John Flannery, Associate Professor of Geography, University of Wisconsin-Milwaukee, was educated at the University of Wisconsin-Madison, receiving a Ph.D. in 1956. He has held full-time appointments at the Universities of Kansas, Pennsylvania, and Wisconsin-Milwaukee and temporary affiliations with Wisconsin-Madison, Temple, Missouri, Immaculata (Pa.), and Harvard. Major areas of interest are thematic cartography, remote sensing, and geography education, with approximately thirty years teaching experience in cartography and aerial photograph interpretation-remote sensing. Research interests are psychophysics as applied to map design, improved instruction in map reading, and maps and remote sensing imagery in urban area analysis. Dr. Flannery was the consulting editor on geography for D.C. Heath Social Studies series on the *Eastern Hemisphere.* He has participated in or directed a number of geography-education institutes for teachers.

Guernsey has held the rank of Professor of Geography at Indiana State University since 1960, and since 1968, has served as Director of the River Basin Research Center at the university. He received his Ph.D. in geography from Northwestern University in 1953, and in 1970 had a post-doctorate year at Florida State University. He has authored and coauthored six college textbooks, the latest being the revised edition (1976) of *Principles of Physical Geography*. Dr. Guernsey has also written 45 professional articles for 15 scientific journals, as well as 16 research monographs. He has served as technical advisor to the AMAX Coal Company, Argonne National Laboratory, the Wabash Valley Interstate Commission, the Vigo County Area Planning Department, and the State Planning Services Agency of Indiana.

Floyd M. Henderson is Assistant Professor of Geography at the State University of New York at Albany. He received his M.A. and Ph.D. degrees from the University of Kansas. His research and teaching interests center on remote sensing of environment with emphasis on radar imagery, land use analysis, applied remote sensing, and technology transfer. Dr. Henderson has served as a staff member at AAG remote sensing workshops.

Gary K. Higgs is an Assistant Professor of Geography at Mississippi State University where he is engaged in research and teaching in Remote Sensing Urban Geography, and Location Theory. He received his Ph.D. from the University of Illinois and has been engaged in research at the Center for Applied Urban Research at the University of Nebraska in Omaha. He has served on several state land use planning councils and commissions and has several publications in the field of urban remote sensing and human action spaces and activities in urban areas. He is a member of the AAG and NCGE remote sensing committees. Dr. Higgs has served as director and staff member for AAG and NCGE remote sensing workshops.

William H. Hilborn is professor of forestry at the University of New Brunswick, Canada. He earned the B.Sc.F. at the University of Toronto, the M.A. degree in geography at the University of Western Ontario. He has taught aerial photograph interpretation and remote sensing for twenty-five years. He has been engaged in a variety of research and consulting programs for both government agencies and industry. He has published articles on forestry in several Canadian and American scientific and professional journals. Currently, he is working on an instructor's manual in aerial photograph interpretation for the New Brunswick Land Surveyors Association. He has served as a staff member in remote sensing workshops for the NCGE and AAG in New York City and Toronto.

Ray Lougeay is an Assistant Professor of Geography and Director of the Environmental Studies Program at the State University College of Arts and Sciences at Geneseo, New York. His fields of specialty include climatology and remote sensing. Dr. Lougeay has spent many field research seasons in alpine areas of Alaska, the Yukon, and Europe. After receiving his doctorate at the University of Michigan in 1971, his research has involved the study of surficial climatic energy balances applied to the interpretation of thermal remote sensing. He has served as a member of the Remote Sensing Committee of the NCGE and has been a staff member of remote sensing workshops sponsored by the NCGE and the AAG.

Paul W. Mausel (B.A., M.A.—University of Minnesota; Ph.D.—University of North Carolina) is a professor of geography and Director of the Indiana State University Remote Sensing Laboratory (ISURSL). During the past ten years he has conducted research and published extensively in geography, soil science, and remote sensing journals. Machine processing of multispectral data as it applies to environmental research has been the focus of recent professional activities. As director of the ISURSL, he has participated actively in environmental contract research for EPA, U.S. Forest Service, NSF, various state and local regional planning agencies, and universities. Currently Dr. Mausel is engaged in establishing new remote sensing programs in environmental analysis for both college students and in-service scientists.

James W. Merchant, Jr., received his M.A. in geography from the University of Kansas in 1973. He is presently a graduate student in the Department of Geography and is a member of the research staff of the University of Kansas Applied Remote Sensing Program. His interests include environmental geography, applied remote sensing, water resources management, and cartography.

Victor C. Miller, with B.A., M.A., and Ph.D. degrees from Columbia University, has served as professional photogeologist/consultant in western United States and Canada. He has also been a professor of photogeology and geomorphology at the University of Libya, Tripoli (chairman), C. W. Post College (chairman), and is currently at Indiana State University. He has authored various papers in journals, as well as the McGraw-Hill text, *Photogeology,* (1961) which is still used worldwide.

Stanley A. Morain received his Ph.D. in Geography from the University of Kansas, Lawrence in 1970. He is now the Director of the Technology Application Center and Associate Professor of Geography at the University of New Mexico. His experience in remote sensing lies in the study of spectral properties and inventory procedures for soils, vegetation, and agriculture, and includes published works that have used multispectral, Landsat, radar, and thermal scanner data. Dr. Morain has worked as an international consultant and educator for the UN Food and Agriculture Organization, U.S. Agency for International Development, the USGS Office of International Geology, and the Asia Foundation. He has published more than 30 articles, maps, and other documents in a variety of professional journals.

James R. O'Malley (Ph.D., University of Tennessee) has been active in the field of remote sensing since the early 1970s. He has authored or coauthored a number of articles on the remote sensing of the cultural landscape. His major research interest is in the field of rural geography, particularly the detection and change of these landscapes. Dr. O'Malley is presently with the Department of Geography at West Georgia State College, Carrollton.

Rex M. Peterson is Remote Sensing Coordinator and Associate Professor, Remote Sensing Center, Conservation and Survey Division, University of Nebraska-Lincoln. He holds a Ph.D. from the University of Michigan and has done postdoctoral work in geography and remote sensing at the University of Kansas. His research interests cover a broad range in remote sensing.

John B. Rehder, with an M.A. (1965) and a Ph.D. (1971) from Louisiana State University, is an Associate Professor of Geography at the University of Tennessee. His teaching and research specialties include cultural geography, rural settlement geography, and remote sensing, with numerous publications in his fields of interest. His current research activities are concentrated in remote sensing. He is a member of both the NCGE and AAG remote sensing committees. During the summer of 1975 he was technical advisor for the Agency for International Development on the remote sensing of land use in Bolivia. During the summer of 1977 he was a member of the NASA-ASEE (American Society of Electrical Engineers) Engineering Design Team at NASA Langley Research Center. Currently, he is directing a NASA project on "The Verification of ERTS Data in the Geographic Analysis of Wetlands in Western Tennessee."

Benjamin F. Richason, Jr., is Professor of Geography and Chairman of the Geography Department at Carroll College, Waukesha, Wisconsin. He received the B.A. and M.A. degrees in geography from Indiana University and the Ph.D. in geography from the University of Nebraska. Dr. Richason was president of the National Council for Geographic Education in 1969, and in 1976 was recipient of the NCGE George J. Miller Distinguished Service Award. He developed and implemented the Audio-Visual-Tutorial method of instruction in introductory physical geography in 1966, and later expanded the AVT method to classes in geography of soils, aerial photograph interpretation, and remote sensing. He has published several books and numerous articles in professional and educational journals, along with a complete college-level AVT course in physical geography. During World War II he served with the 1st Photo Reconnaissance Squadron of the Army Air Force in the South Pacific. He has taught courses in cartography and aerial photograph interpretation since 1952, and remote sensing since 1972. Dr. Richason is a regional editor of *Remote Sensing of the Electromagnetic Spectrum.* He has served as a staff member in AAG and NCGE remote sensing workshops, and presently is chairman of the Remote Sensing Committee of the NCGE.

Benjamin F. Richason III is Assistant Professor of Geography at James Madison University, Harrisonburg, Virginia. He obtained his undergraduate degree in geography from Carroll College, his M.A.T. from Oregon College of Education, and his Ph.D. from Michigan State University. Dr. Richason has been involved in aerial photograph interpretation and remote sensing for ten years. He was associated with the NASA funded Remote Sensing Project at Michigan State University as a research assistant. He has written and presented several papers in remote sensing dealing with high altitude color infrared photography and land cover/use mapping.

Merrill K. Ridd is Professor of Geography and Chairman of the Department of Geography, as well as the director of the Center for Remote Sensing and Cartography at the University of Utah, Salt Lake City. He received the B.S. and M.A. degrees in geography from the University of Utah, and the Ph.D. in geography from Northwestern University. He served five years as editor of publications for the National Council for Geographic Education. He organized or co-edited four books on geographic education. He served on the Committee for geographic education of the AAG, and currently serves on the remote sensing committee of the NCGE. Dr. Ridd was associate editor for Earth Science Curriculum Project of the American Geological Institute, and chief editor of the textbook, *Investigating the Earth,* sponsored by the National Science Foundation. He is the principal investigator of three NASA sponsored projects, including a research grant, "Identifying Environmental Features for Land Management Decisions," and an educational study, "Studying the Earth from Space."

Noel Ring is the instructor in geography at Norwich University, Vermont, where cultural landscape studies highlight her research. Her remote sensing interests began with use of Gemini and Apollo photographs while a high school teacher in California. Recently she focused on land use studies under a U.S. Office of Education environmental education project, "Landscapes of Vermont." She serves as a regional editor of *Remote Sensing of the Electromagnetic Spectrum,* is a member of the NCGE Executive Board and of its Remote Sensing Committee, and is chairperson of the Remote Sensing Group of Northern New England. Her current comparative aerial archeological work in New England and Europe involves materials ranging from Skylab imagery to low-altitude multispectral aerial surveys.

Paul M. Seevers is Research Agronomist and Associate Professor in the Remote Sensing Center, Conservation and Survey Division, University of Nebraska-Lincoln. He holds a Ph.D. in agronomy from the University of Nebraska and has extensive experience in application of remote sensing to land use, range management, and agriculture.

Jack R. Villmow is Professor of Geography and University Director of Honors, Northern Illinois University, DeKalb. He was the first meteorology major at the University of Wisconsin-Madison (1948), and received his Ph.D. in climatology, from the same institution. He has taught either on the regular staff or as a visiting instructor at Wellesley College, Clark University, Boston University, University of Massachusetts, George Peabody College for Teachers, University of Wisconsin-Parkside, University of Wisconsin-Madison, and Ohio State University. In May 1977 Dr. Villmow received the Award for Excellence in Teaching, one of three such awards made each year at Northern Illinois University.

Robert S. Weiner received his first cartographic training under the late Dr. Frances M. Hanson, University of Pittsburgh. With an NSF Grant in Modern Cartographic Techniques he studied under Dr. John C. Sherman, University of Washington. He developed and taught the geography/cartography major at Briarcliff College, New York, for twenty-six years. He is presently teaching cartography at the University of Connecticut at Storrs. He has participated in AAG workshops in remote sensing at Bendix Corporation, Ann Arbor, Michigan; Milwaukee, Wisconsin; and Briarcliff College. He received a Certificate of Achievement in Remote Sensing from the State University of New York, College of Environmental Science and Forest, Syracuse.

Gary Whiteford is a Canadian, born in Toronto and is currently an associate professor at the University of New Brunswick teaching geography in the Faculty of Education. He holds an undergraduate degree in geography from York University (Toronto), an M.A. in geography from Clark University, and a Ph.D. in geography from the University of Oklahoma. He has taught at West Texas State University and Champlain College in Quebec and held summer school positions at St. Mary's University in Halifax and Concordia University of Montreal. For the past two summers he has conducted overseas geography courses in Jamaica and Europe. His research interests include topics in political geography and in the development of the spatial framework in children. At present he is working on an atlas of Maritime Canada. He serves as the NCGE coordinator for Maritime Canada and is actively involved with the Canada Studies Foundation.

Ruth I. Whitman was trained in physics at William and Mary College. She was appointed to the staff of NASA when it opened in 1958. She worked for NASA when she was appointed to the Langley Research Center. She spent one year at NASA headquarters. Since then she has served as Program Scientist for Landsat D, and Program Manager for Data Interpretation Techniques at NASA headquarters.

Index